Foodinformatics

Karina Martinez-Mayorga
José Luis Medina-Franco
Editors

Foodinformatics

Applications of Chemical Information to Food Chemistry

Springer

Editors
Karina Martinez-Mayorga
Instituto de Química
Universidad Nacional Autónoma de México
Mexico City
Mexico

Torrey Pines Institute for Molecular Studies
Port St. Lucie, FL
USA

José Luis Medina-Franco
Instituto de Química
Universidad Nacional Autónoma de México
Mexico City
Mexico

Torrey Pines Institute for Molecular Studies
Port St. Lucie, FL
USA

ISBN 978-3-319-10225-2 ISBN 978-3-319-10226-9 (eBook)
DOI 10.1007/978-3-319-10226-9
Springer Cham Heidelberg New York Dordrecht London

Library of Congress Control Number: 2014951057

Printed on acid-free paper

Springer is part of Springer Science+Business Media (www.springer.com)

In loving memory of

Orel Martínez Anzúres
and
Josefina Teresa de Jesús Franco y Moreno

Preface

The use of computers to collect, store, and manipulate chemical information is at the heart of chemoinformatics. The "tools of the trade" in this emerging area, whose main target thus far has been the pharmaceutical field, are general and can be applied to other types of chemical datasets, such as those containing food chemicals. *Foodinformatics: Applications of Chemical Information to Food Chemistry* collects together a number of studies where chemoinformatics tools have been applied in answering questions about food-related compounds. Chapter 1 presents a didactic introduction to the concepts of molecular similarity and chemical spaces, which are cornerstones of chemoinformatics. Chapters 2 and 3 discuss practical applications of chemical space and molecular similarity studies, respectively. Chapters 4 and 5 describe two concepts of current interest, namely, reverse pharmacognosy and epigenetics. While Chap. 4 concerns the discovery of new health-related applications for existing food ingredients, Chap. 5 focuses on the exploration of molecular determinants and the pharmacological role of food and food-derived compounds as modulators of epigenetics and metabolism. Chapters 6, 7, 8 and 9 exemplify the use of molecular and/or statistical models to analyze food-related compound collections for biological activities or organoleptic properties. Finally, Chap. 9 provides a compilation of software resources and databases that have been used or can be used in the food chemistry field; it also presents a perspective of *Foodinformatics*.

While the use of chemical information methodologies to address food-related challenges is still in its infancy, interest is growing and will continue to do so as the methods prove useful, particularly for providing practical solutions to food industry challenges. This book attempts to give an overview of basic concepts, applications, tools, and perspectives.

This book was made possible with the enthusiastic participation and efforts of all of the chapters' authors, and valuable support and discussions with Dr. Terry Peppard and Mr. John Sciré, who sadly passed away in November 2013.

K. Martinez-Mayorga
J. L. Medina-Franco

Contents

Contributors

Philippe Bernard Green Pharma, S.A.S, Orléans, France

Adrià Cereto-Massagué Research Group in Chemoinformatics & Nutrition, Departament de Bioquímica i Biotecnologia, Universitat Rovira i Virgili, Campus de Sescelades, Tarragona, Catalonia, Spain

Fernando B. Da Costa School of Pharmaceutical Sciences of Ribeirão Preto, University of São Paulo, Ribeirão Preto, SP, Brazil

Quoc Tuan Do Green Pharma, S.A.S, Orléans, France

Maureen Driscoll Department of Biomedical and Pharmaceutical Sciences, College of Pharmacy, University of Rhode Island, Kingston, RI, USA

Susan E. Ebeler Department of Viticulture and Enology, University of California, Davis, CA, USA

Santiago Garcia-Vallve Cheminformatics and Nutrition Group, Departament de Bioquímica i Biotecnologia, Universitat Rovira i Virgili (URV), Tarragona, Catalonia, Spain

Centre Tecnològic de Nutrició i Salut (CTNS), TECNIO, Reus, Catalonia, Spain

Laura Guasch Computer-Aided Drug Design Group, Chemical Biology Laboratory, Center for Cancer Research, National Cancer Institute, National Institutes of Health, Frederick, MD, USA

Hildegarde Heymann Department of Viticulture and Enology, University of California, Davis, CA, USA

Helene Hopfer Department of Viticulture and Enology, University of California, Davis, CA, USA

María José Ojeda Research Group in Chemoinformatics & Nutrition, Departament de Bioquímica i Biotecnologia, Universitat Rovira i Virgili, Campus de Sescelades, Tarragona, Catalonia, Spain

Gerald M. Maggiora University of Arizona BIO5 Institute, Tucson, AZ, USA

Translational Genomics Research Institute, Phoenix, AZ, USA

Karina Martinez-Mayorga Departamento de Fisicoquímica, Instituto de Química, Universidad Nacional Autónoma de México, Mexico City, Mexico, USA

Torrey Pines Institute for Molecular Studies, Port St. Lucie, FL, USA

José L. Medina-Franco Departamento de Fisicoquímica, Instituto de Química, Universidad Nacional Autónoma de México, Mexico City, Mexico

Torrey Pines Institute for Molecular Studies, Port St. Lucie, FL, USA

Miquel Mulero Cheminformatics and Nutrition Group, Departament de Bioquímica i Biotecnologia, Universitat Rovira i Virgili (URV), Tarragona, Catalonia, Spain

Terry L. Peppard Robertet Flavors Inc., Piscataway, NJ, USA

Gerard Pujadas Research Group in Chemoinformatics & Nutrition, Departament de Bioquímica i Biotecnologia, Universitat Rovira i Virgili, Campus de Sescelades, Tarragona, Catalonia, Spain

Ariadna I. Ramírez-Hernández Departamento de Fisicoquímica, Instituto de Química, Universidad Nacional Autónoma de México, Mexico City, Mexico, USA

Jean-Louis Reymond Department of Chemistry and Biochemistry, University of Bern, Bern, Switzerland

Alberto Del Rio Institute of Organic Synthesis and Photoreactivity (ISOF), National Research Council (CNR), Bologna, Italy

Department of Experimental, Diagnostic and Specialty Medicine (DIMES), Alma Mater Studiorum, University of Bologna, Bologna, Italy

Lars Ruddigkeit Department of Chemistry and Biochemistry, University of Bern, Bern, Switzerland

Navindra Seeram Department of Biomedical and Pharmaceutical Sciences, College of Pharmacy, University of Rhode Island, Kingston, RI, USA

Angela Slitt Department of Biomedical and Pharmaceutical Sciences, College of Pharmacy, University of Rhode Island, Kingston, RI, USA

Diana E. Terrazas-Álvarez Departamento de Fisicoquímica, Instituto de Química, Universidad Nacional Autónoma de México, Mexico City, Mexico, USA

Cristina Valls Research Group in Chemoinformatics & Nutrition, Departament de Bioquímica i Biotecnologia, Universitat Rovira i Virgili, Campus de Sescelades, Tarragona, Catalonia, Spain

Chapter 1
Introduction to Molecular Similarity and Chemical Space

Gerald M. Maggiora

List of Abbreviations

2-D	Two-dimensional
3-D	Three-dimensional
APFs	Atom pair fingerprints
CS	Chemical space
CSN	Chemical space network
DB	Database
ECFPs	Extended connectivity fingerprints
FPs	Fingerprints
HOMO	Highest-occupied molecular orbital
HTS	High-throughput screening
LBVS	Ligand-based virtual screening
LUMO	Lowest-unoccupied molecular orbital
MaxD	Maximum dissimilarity selection ('Dfragall') algorithm
MaxST	Maximum spanning tree
MinST	Minimum spanning tree
MDS	Multidimensional scaling
NLM	Nonlinear mapping
PC	Principal component
PCA	Principal component analysis
PCoA	Principal coordinate analysis
PSA	Post-search aggregation
P-S	Pearlman–Smith

G. M. Maggiora (✉)
University of Arizona BIO5 Institute, 1657 East Helen Street, Tucson, AZ 85721, USA
e-mail: gerry.maggiora@gmail.com

Translational Genomics Research Institute, 445 North Fifth Street, Phoenix, AZ 85004, USA

K. Martinez-Mayorga, J. L. Medina-Franco (eds.), *Foodinformatics*,
DOI 10.1007/978-3-319-10226-9_1

1.1 Introduction

It is estimated that the chemical universe associated with small organic molecules is nearly 200 billion [1]. An older estimate, which includes larger organic molecules up to a molecular weight of 500 Da, suggests that this number may be around 10^{60} [2] and constitutes what could be called the "small molecule universe." Enumerating and searching this set of compounds would be a daunting task. Recently, a new approach has been published that is based on the construction of what the authors claim is a "representative universal library" of drug-like compounds [3]. In any case, regardless of how the size of the chemical universe is assessed, there is no question that its size is immense. Because of the size of even "representative" subsets of that universe, computer-based methods are required to capture, manage, and search the massive amount of information, activities that fall under the rubric of chemical informatics.

While the chemical universe of molecules potentially relevant in food science is considerably smaller, it nonetheless is large enough to benefit from many of the chemical informatic concepts that have proved useful in medicinal chemistry and related fields of chemistry. Two of these concepts, molecular similarity and chemical space (CS), are dealt with in this chapter. Of the two, molecular similarity is more fundamental since it plays a crucial role in the definition of CS itself. Though important, activity or property landscapes, which provide the third leg of a triad of activities that play important roles in much of chemical informatics, will not be discussed here. Numerous recent publications describing the visual and statistical aspects of activity landscapes as well as the basic features of these landscapes should be consulted for details [4–8].

Similarity is a ubiquitous concept that touches nearly every aspect of our conscious lives and, no doubt, influences our subconscious thoughts as well. Although its earliest influence on scientific thinking can be traced to the Greek philosophers [9, 10], its impact in chemistry began in the nineteenth century, the most notable example being the development of the periodic table of elements by Mendeleev [11] and Meyer [12]. As noted by Rouvray (see Table 1.4 in [9]), the twentieth century saw a significant expansion in the number and variety of chemical applications of molecular similarity. However, it was not until late in that century that application of similarity flourished due in large measure to the greater availability of digital computers. This led to the development of a plethora of methods for computing molecular similarity, enabling medicinal chemists to address a growing need to search compound collections[1] of rapidly increasing size for molecules with similar properties or biological activities.

Underlying this effort was the *similarity-property principle* (SPP) [13–15], which simply states that "Similar molecules tend to have similar properties." Although perhaps intuitively obvious, it nonetheless provides an important rationale that has proved quite helpful as a basis for similarity searches of CSs.

[1] The term database (DB) will generally be used to describe large collection of compounds whether or not material exists for screening the compounds.

However, because similarity is a subjective concept ("Similarity like pornography is difficult to define, but you know it when you see it" [10]), an absolute standard to judge the effectiveness of similarity methods does not exist. As will be discussed in the sequel, this raises some significant issues that can seriously impact the effectiveness and reliability of similarity methods; chief among them is the fact that the similarity values depend on the method used to encode the relevant chemical or molecular information. Nevertheless, a large number of successful applications have shown that similarity methods, with all of their inherent flaws, can provide an effective means for carrying out a number of chemical informatic activities that facilitate the practice of medicinal chemistry and drug discovery (*vide infra*). There are two main approaches to similarity in chemistry, what is typically called *molecular* or *structural similarity*, which is the focus of this work, and *chemical similarity*. The chemical similarity typically, but not exclusively, utilizes representations associated with macroscopic chemical properties such as solubility, heat of vaporization, molar refractivity, and log*P*, although occasionally properties of individual molecules such as pi-electron densities, highest-occupied and lowest-unoccupied molecular orbital (HOMO and LUMO) energies, and dipole moments are also used.

Representations associated with molecular similarity are in general classified as one-dimensional (1-D), two-dimensional (2-D), or three-dimensional (3-D). 1-D representations generally refer to macroscopic (e.g., solubility, log*P*, sublimation energy, heat of formation, etc.) or microscopic (e.g., molecular orbital energy, atomic charges, spectra, etc.) scalar quantities (*vide supra*). 2-D features are derived from the 2-D structures typically used by chemists to represent molecules. Although such structures can encode stereochemical and conformational information, this is not generally the case in molecular similarity studies, which typically use what are called hydrogen-suppressed chemical graphs [16], where hydrogen atoms, except those on specific nitrogen and oxygen atoms, are not explicitly represented. Thus, chemical graphs primarily encode information on the types of atoms and the bonds between them—the latter is sometimes referred to as the *bond topology* of the molecule.

By contrast, 3-D features are generally derived from the overall 3-D geometric, and sometimes the electronic structure of molecules, which would seem to provide a more faithful representation of molecular information. Nevertheless, a number of substantive issues remain. This is especially true of molecules with multiple conformational states, since determining what conformational state or states have to be included in a given similarity analysis is not entirely straightforward. For example, in similarity studies aimed at identifying molecules with comparable biological activities to known active molecules, does one use the minimum energy conformation or the biologically active one, which in many cases is not known. What about the case, when there are multiple conformations of comparable energy? All of these issues can significantly complicate 3-D similarity studies.

Because of the greater simplicity of 2-D compared to 3-D representations, and because the corresponding functions used to evaluate similarities are generally easier to carry out as well, 2-D similarities tend to be much faster to compute than the

3-D similarities (see Sect. 1.2 for details). This raises the question of whether 2-D similarities perform equally or better than 3-D methods in tasks commonly carried out in chemical informatics. Conclusive results have not been achieved to date. Nevertheless, it appears that 2-D methods can in many cases perform equally well and in some cases outperform 3-D methods [17, 18] in a variety of tasks. These tasks include similarity-based searches designed to identify new, potentially active molecules based on previously determined actives and to identify molecules with potentially similar values for properties of interest in drug research such as log*P*—both are examples of the SPP. In addition, these workers showed that of the 2-D methods considered "molecular ACCess system" (MACCS) structural-key-based fingerprints (FPs) (*vide infra*) consistently exhibited the best performance.

Because of this, most applications of molecular similarity over large sets of compounds generally employ 2-D similarity methods. It should be emphasized, however, that procedures for comparing 2-D versus 3-D similarity methods are imperfect by their very nature since, as noted earlier, similarity is a subjective concept that does not admit to absolute comparisons of any type.

In simplistic terms, the concept of CS can be considered to be a multidimensional extension of the concept of a congeneric series. However, an important distinction between the two is that CS involves a *pairwise relation* that specifies the relationship of the molecules to each other, generally in terms of a molecular similarity or CS-distance function. A set of objects and a pairwise relation among them are the basic ingredients of a mathematical space. In the present case, the objects are molecules and the pairwise relation characterizes the similarity or distance of separation of each pair of molecules in the CS. Similarity and distance are inversely related; the more similar a pair of molecules, the closer they are in CS, and vice versa.

Because CSs are generally of high dimension, faithfully depicting them in 2-D or 3-D is not possible, and some type of approximation is required. This, however, is not generally a problem because their visual depiction is only used *qualitatively*. More *quantitative* results can be obtained simply by carrying out the computations with respect to the full dimension of the CS in question.

Importantly, CS provides a conceptual framework for organizing the structural and property relationships of vast numbers of molecules within a common framework. With the burgeoning amount of structural, chemical, and biological data currently being created and stored in publically accessible databases (DBs) such as ChEMBL [19], PubChem [20], ChemDB [21], and DrugBank [22], or in subscription-based DBs such as WOMBAT [23] and MDDR [24], a conceptual framework, such as that provided by CS, is essential if we are to gain insights from information stored in these DBs. A summary of many public and private compound DBs is given in [25].

The remainder of this chapter covers set- and vector-based representations of structural and molecular data and how this information is converted into the various similarity, dissimilarity, and distance measures that have found wide application in chemical informatics. Examples of some of the types of structural and molecular descriptors are also presented, along with a discussion of their essential features. Significant emphasis is given to the concept of CS, a concept that plays an absolutely essential role in almost all aspects of chemical informatics. Finally, examples of how similarity can be used to carry out many activities associated with CSs, such as comparing compound collections, acquiring new compounds to augment current collections, assessing the diversity of a collection, generating diverse subsets of compounds for high-throughput screening (HTS) campaigns, and ligand-based virtual screening (LBVS). The latter activity has risen in importance over the past decade as an important strategy in drug discovery. The words "molecule" and "compound," which are very similar and are quite prevalent throughout this work, are used essentially interchangeably.

Over the past decade, a number of books have provided a good overview of many aspects of the field of chemical informatics [26–30], and a number of reviews and papers on molecular similarity [31–34] and CS [35–40] have also been published. These sources should be consulted for additional details on any of the subjects discussed in this work.

This chapter is not meant as a comprehensive review of molecular similarity and CSs. Rather it is intended to be somewhat pedagogical and to present, in some detail, a number of their key features and the interrelationships among them. In this way, it is hoped that readers will have a basic feel for the nature of the concepts and will be able to move on from there to tackle more complex aspects of these concepts and to apply them in a practical setting.

1.2 Structural Similarity Measures

Structural similarity is a pairwise relation between molecules. Similarity values are determined by a *similarity measure* that has three key components: (1) a representation of the relevant chemical or structural features of the molecules being compared, (2) an appropriate weighting of these features, and (3) a function that maps the feature information for pairs of molecules to a value that lies on the unit interval of the real line [0,1]. As noted in the previous section, representations can utilize macroscopic chemical features, electronic structural features of individual molecules, and/or geometric features associated with the structure or substructures of molecules

A number of procedures for computing 2-D and 3-D molecular similarities have been described in great detail [10, 41]. In the current work, the focus is on the class of 2-D similarity measures based on molecular FPs that encode the substructural information in molecules and on measures derived from vectors whose components represent macroscopic and microscopic physicochemical properties or indices derived from the topological properties of their chemical graphs. These approaches are the most prevalent ones and have been applied in a wide range of applications cited in Sects. 1.1, 1.2.3, 1.3.1, 1.3.2.1, 1.3.5.2, and 1.3.5.3. Moreover, they provide clear examples of the general workings of the types of molecular similarity measures in wide use today.

1.2.1 Set-Based Similarity Measures

1.2.1.1 Set-Based Representations: Binary Structural FPs

Consider the set of n molecules

$$M = \{M_1, M_2, \ldots, M_i, \ldots, M_n\}. \tag{1.1}$$

A *binary molecular FP* for molecule M_i can be specified by a set of p substructural features

$$\mathbf{m}_i = \{m_i(1), m_i(2), \ldots, m_i(k), \ldots, m_i(p)\} \tag{1.2}$$

where the binary values of the indicator (characteristic) functions $m_i(k), k = 1, 2, \ldots, p$ in Eq. (1.2) determine whether a specific substructural feature is present or absent in the molecule, i.e.,

$$m_i(k) = \begin{cases} 1, \text{ if the } k\text{th structural feature is present;} \\ 0, \text{ if the } k\text{th structural feature is absent.} \end{cases} \tag{1.3}$$

Binary molecular FPs are sometimes called *bit vectors* or *bit strings* since their elements are “1s” and “0s”. In this work, the nomenclature *binary molecular FP* may also be given by *structural FP, molecular FP, binary FP*, or just *FP*. Multiple occurrences of structural features are not accounted for in binary FPs, although they can be as described later in this section.

Equation (1.4) depicts a hypothetical FP

$$\mathbf{m}_i = \overbrace{(1, 0, 0, 1, 1, \ldots, 1, 0)}^{p} \tag{1.4}$$

characterized by a *binary p*-tuple. This is a reasonably standard notation for FPs. However, because of their sparseness (i.e., relatively few 1-bits), it is not how they are generally handled in computers, where *index-based* and *run-length* encoding schemes are typically used [42]. The former scheme basically indexes all of the 1-bits of a given FP. By contrast, run-length encoding indexes the lengths of runs of 0-bits followed by a 1-bit. As an illustration, consider the following simple example of a binary structural FP (0,0,1,1,0,0,0,0,1,0,0,1,0,0,0). Hence, its index-based encoding is given by (2,0,4,2), while its corresponding run-length encoding is (3,4,9,12), an example that clearly shows that the encodings are not of fixed length. Except in a few instances where stereochemical information is represented, most molecular FPs are based on the 2-D structural features of the molecules they symbolize and, hence, the representations described in this work correspond to 2-D molecular FPs. The number of components or elements, p, in molecular FPs can be quite large and can be either fixed or variable. The former usually corresponds to *molecule-independent* FPs and the latter to *molecule-dependent* FPs.

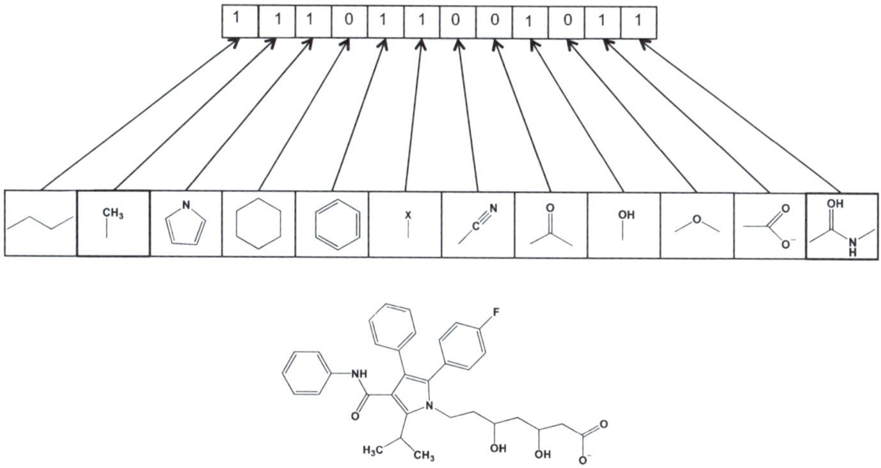

Fig. 1.1 An example based on the drug Lipitor of a simplified molecule-independent directory-based binary structural FP with its corresponding set of descriptors. The symbol 'X' corresponds to any of the halogen atoms (F, Cl, Br, I)

Molecule-Independent/Directory-Based FPs The number of structural features in molecule-independent FPs is fixed for all molecules, as exemplified by MACCS key FPs, which contain 166 structural features [43] and Barnard Chemical Information (BCI) FPs that contain more than 1000 features [44]. Figure 1.1 provides a simple example, based on the anticholesterol drug Lipitor, of a molecule-independent FP. Note that multiple occurrences of methyl groups, hydroxyl groups, and phenyl rings are not explicitly accounted for, nor is the elongated hydrocarbon chain that connects the nitrogen atom of the pyrrole ring with the terminal carboxylate fully accounted for, although a structural descriptor that represents a shorter hydrocarbon chain provides at least a partial account of the elongated chain.

Hence, structural information can be lost leading to similarity values of unity for pairs of molecules that are not structurally identical. Nevertheless, there is at least a partial correspondence between the descriptors in the directory and the binary molecular FP of a molecule, so that it may be possible in many instances to associate particular substructural features with molecular properties and/or biological activities, a characteristic that is not generally shared by molecule-dependent FP representations (*vide infra*). This can be partially ameliorated through the use of *weighted molecular FPs* that take account of the number of times a structural feature occurs in a molecule. However, since not all structural features that may be associated with a specific structure–property relationship (SPR) or structure–activity relationship (SAR) are necessarily accounted for in given FP, it may not be possible to infer SPR or SAR even when weighted FPs are employed.

Molecule-dependent FPs have variable numbers of elements that typically depend on the number of non-hydrogen atoms and functional complexity of molecules. Because of the rapid growth in the size and molecular complexity of modern compound DBs, molecule-dependent FPs have been growing in popularity since they can potentially handle a wider range of molecules than molecule-independent FPs. Two structural FPs that exemplify the types of molecule-dependent FPs in use today are the atom pair FPs (APFs) first developed by Carhart, Smith, and Venkataraghaven nearly 30 years ago [45] and the more recent extended connectivity FPs (ECFPs) developed by Rogers and Hahn [46] that are in widespread use today. Simple examples of APFs and ECFPs are depicted in Figs. 1.2 and 1.3, respectively.

Both of these FPs are referred to as "2-D FPs," since neither of them utilizes 3-D structural information. Although a number of FPs including AFPs and ECFPs can encode stereochemical information, they rarely do in common usage.

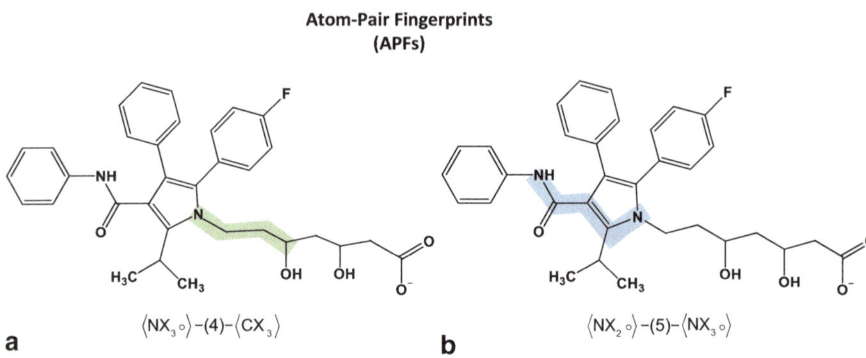

Fig. 1.2 Examples of molecule-dependent atom pair fingerprints (APF) descriptors depicted with respect to the drug Lipitor. Regions highlighted in *light green* and *light blue* correspond to substructures associated with two APFs; the labels below each figure correspond to respective designations given in reference [46] for these APFs

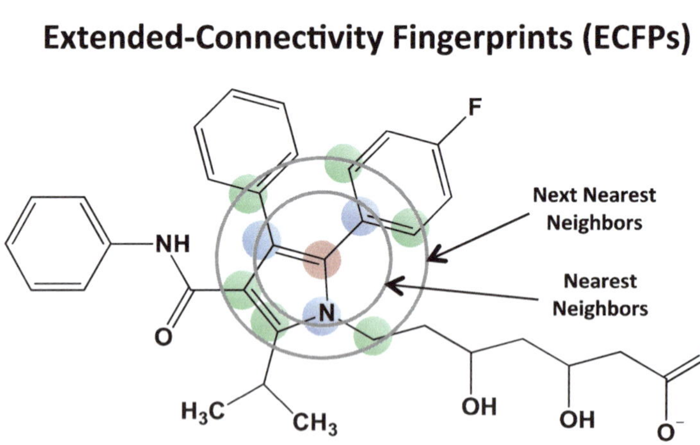

Fig. 1.3 Examples of molecule-dependent extended connectivity ECFP descriptors depicted with respect to the drug Lipitor. Atoms lying within the rings depicted in the figure correspond to nearest (*colored in light blue*) and next-nearest neighbors (*colored in light green*) to the central atom (*colored in light red*) of a given ECFP4 descriptor.

Atom Pair FPs Pairs of atoms and the minimum number of bonds linking them constitute the substructural components of APFs. Generally, only APs separated by seven or fewer bonds are considered. As described by Carhardt et al. [45], the general form of the substructure of an APF is given by Eq. (1.5):

$$\langle\text{“atom-}i\text{”}\rangle - (\text{atom-separation}) - \langle\text{“atom-}j\text{”}\rangle, \tag{1.5}$$

where “atom-*i*” and “atom-*j*” are descriptions that contain information on the atom type (e.g., C, N, O,…), the number of non-hydrogen atoms bound to it, and whether it possesses a bonding pi-electron. The “separation” between atoms is based on a count of all the atoms, including atom-*i* and atom-*j*, on the shortest through-bond path connecting the two terminal atoms of the chain. Consider, for example, the APF designation $\langle NX_3\bullet\rangle - (4) - \langle CX_3\rangle$ depicted in Fig. 1.2a. In the $NX_3\bullet$ term contained within the leftmost brackets, “ N ” designates the leftmost atom in the chain highlighted in light green, “ X_3” indicates that three atoms are bonded to it, and the “•” indicates the presence of bonding pi-electron on the nitrogen atom. Next, the “ 4 ” in the parentheses indicates the number of atoms in the chain including the terminal atoms. Last, in the CX_3 term contained within the rightmost brackets, “ C ” designates the rightmost atom in the chain and “ X_3” indicates that three atoms are bonded to it. A similar interpretation can be made for the designation corresponding to the APF highlighted in light blue in Fig. 1.2b.

Because of the way in which APFs are handled in a computer, it is not possible to associate substructural features with specific bits in an APF. An excellent discussion based on the closely related Daylight FPs [47] discusses this issue and many other of the technical details that must be addressed in order to effectively implement APFs.

Extended Connectivity FPs By contrast, ECFPs sample the molecular environment surrounding each non-hydrogen atom. Thus, the local “circular” environments surrounding each non-hydrogen atom constitute the substructural features of a given molecule as depicted in Fig. 1.3. Although not always employed, ECFPs can also encode stereochemical information, which can be important in many aspects of drug discovery research since all stereoisomers of a given compound may not be equally active.

For example, consider the pyrrolic carbon atom in Fig. 1.3 highlighted in light red. As seen in the figure, two layers of atoms surround it, the first, whose atoms are highlighted in light blue, corresponds to nearest neighbors and the second, whose atoms are highlighted in light green, corresponds to next nearest neighbors. Each non-hydrogen atom and its layers of surrounding atoms constitute substructural features. The maximum number of layers considered is given by the diameter of the largest circular environment surrounding the central atom. This is based on the number of bonds needed to connect two diametrically opposed atoms in that layer. In the case shown here, four bonds are required. Such FPs are designated by ECFP4.

From the above, it is easy to see that the number of possible FP descriptors that can be obtained for compound collections is quite large. For example, Rogers and

Hahn [46] have shown that sets of ~50,000 compounds can give rise to ECFP descriptors that number in the hundreds of thousands. For larger sets of compounds, the number of ECFP descriptors can potentially exceed 1 million. Hence, handling this amount of information efficiently presents some technical problems, the details of which are beyond the scope of this work. Interestingly, unlike AFPs whose substructural information cannot be retrieved, this is not the case for ECFPs, although the procedure for doing so requires several steps. The paper by Rogers and Hahn [46] provides a detailed discussion of many of these issues. They also note that ECFPs were designed primarily to characterize the activities of compounds. Hence, ECFPs contain information on features that are present as well as those that are not present. ChemAxon provides a very clear description of many of the technical details associated with application of ECFPs [48]. In addition, they offer a useful, albeit brief, comparative discussion of AFPs and ECFPs, pointing out that the former performs best for substructure searches while the latter appears to be more suitable for similarity searches. Several other papers also provide useful assessments of ECFPs [49, 50].

Weighted Structural FPs Weighting the features of structural FPs is not common practice in chemical informatics. Nevertheless, it has been shown in a number of studies to provide improved results in virtual screening experiments [51–53].

Although numerous schemes exist [54], weighting nowadays is typically accomplished by accounting in some fashion for the number of occurrences of each of the features in a molecule, as for example, the methyl, phenyl, hydroxyl groups depicted in Fig. 1.1 for the hypercholesterol drug Lipitor™.

Clearly, not accounting for multiple occurrences of features can lead to significant *degeneracies* that arise when different compounds have identical FPs. Sometimes the degeneracies can be quite large as shown by the following analysis based on Lipitor™. Consider each of the multiple structural FP descriptors in Lipitor™: three phenyl, two methyl, and two hydroxyl groups. There are seven possible descriptor patterns containing at least one phenyl group and three possible patterns containing at least one methyl group and three containing at least one hydroxyl group. Assuming that each of the three descriptor patterns is independent of each other, a quite reasonable assumption is that the total number of possible patterns is $7 \times 3 \times 3 = 67$. Hence, there are 67 different, albeit related, compounds that would all have exactly the same structural FP as Lipitor™. While this may be a somewhat extreme example, there are nonetheless numerous examples of compounds with multiple occurrences of specific substructural patterns. Surprisingly, the results obtained with unweighted FPs are quite good. And although both APFs and ECFPs can take account of multiple occurrences of substructural patterns, they are rarely if ever considered in actual applications.

In fact, most cheminformatic studies continue to use binary structural FPs.

1.2.1.2 FP-Based Similarity Coefficients

The third component of a similarity measure is the function that maps the structural information contained in the molecular FPs of each pair of compounds

Table 1.1 Set-theoretic expressions useful in molecular similarity analysis

Symbol	Set-theoretic expression[a]	Definition
N_i	$\text{Card}(\mathbf{m}_i)$	Number of features in molecule M_i
$N_{i,j}$	$\text{Card}(\mathbf{m}_i \cap \mathbf{m}_j)$	Number of features common to molecules M_i and M_j
$N_i - N_{i,j}$	$\text{Card}(\mathbf{m}_i) - \text{Card}(\mathbf{m}_i \cap \mathbf{m}_j)$	Number of features unique to molecule M_i

[a] "Card" refers to the cardinality (i.e., number of elements) of the set in question

being compared to the unit interval of the real line [0,1]. Such functions are called by a number of names—*similarity functions, similarity indices*, or *similarity coefficients*—the latter nomenclature will be adhered to in this chapter [10]. Although there are many types of similarity coefficients, only a limited number will be considered here. A summary of all types of similarity coefficients is given in a comprehensive review [31].

Based on his work in mathematical psychology, Tversky developed the most general form of similarity coefficient applicable to structural FPs [55]:

$$S_{\text{Tve}}(i,j|\alpha,\beta) = \frac{N_{i,j}}{\alpha(N_i - N_{i,j}) + \beta(N_j - N_{i,j}) + N_{i,j}}, \tag{1.6}$$

where the weighting parameters satisfy $\alpha, \beta \geq 0$, which ensures that the similarity values lie on the unit interval of the real line [0,1]. The various terms in Eq. (1.6) are described in Table 1.1.

As described in Table 1.1, the terms in parentheses in the denominator, $N_i - N_{i,j}$ and, $N_j - N_{i,j}$, can be interpreted as the number of features unique to molecules M_i and M_j, respectively, weighted by the corresponding values of α and β.

It is clear from the form of Eq. (1.6) that the Tversky similarity coefficient is generally *asymmetric* with respect to the interchange of its arguments, i.e., $M_i \rightarrow M_j$ and $M_i \leftarrow M_j$. This corresponds to interchanging the associated variables N_i and N_j in Eq. (1.6) so that $(N_i \rightarrow N_j \text{ and } N_i \leftarrow N_j)$, i.e.,

$$S_{\text{Tve}}(j,i|\alpha,\beta) = \frac{N_{i,j}}{\alpha(N_j - N_{i,j}) + \beta(N_i - N_{i,j}) + N_{i,j}} \tag{1.7}$$

which is equal to the expression in Eq. (1.6) and is symmetric *only* in cases where $\alpha = \beta$: Note that the variable $N_{i,j}$ is invariant to these interchanges. Such cases correspond to well-known similarity coefficients, three of which are described below.

For example, the currently most popular similarity coefficient, $S_{\text{Tan}}(i,j)$, is that due to Tanimoto and is obtained by setting $\alpha = \beta = 1$,

$$S_{\text{Tve}}(i,j|\alpha=1,\beta=1) = \frac{N_{i,j}}{(N_i - N_{i,j}) + (N_j - N_{i,j}) + N_{i,j}} = S_{\text{Tan}}(i,j). \tag{1.8}$$

The sum of the terms in the denominator is equal to the total number of features in common plus the number of unique features associated with molecules M_i and M_j, although the form of the expression differs from that usually used, namely, $N_i + N_j - N_{i,j}$, where the "$-N_{i,j}$" term corrects for double counting the features in both molecules. Thus, the Tanimoto similarity coefficient is the ratio of the number of features in common to both molecules over the total number of features (not the sum) in M_i and M_j.

Setting $\alpha = \beta = \frac{1}{2}$ leads to the *Dice similarity coefficient*:

$$S_{\mathrm{Tve}}(i,j|\alpha = \frac{1}{2}, \beta = \frac{1}{2}) = \frac{N_{i,j}}{\frac{1}{2}(N_i + N_j)} = S_{\mathrm{Dice}}(i,j) \tag{1.9}$$

where the term in the denominator is the *arithmetic mean* of the number of features in M_i and M_j. Thus, the Dice similarity coefficient is the ratio of the number of features in common to M_i and M_j over the arithmetic mean of the number of their features.

Although it cannot be obtained from $S_{\mathrm{Tve}}(i,j|\alpha,\beta)$ simply by choosing appropriate values for α and β, the well-known *cosine similarity coefficient* given by

$$S_{\cos}(i,j) = \frac{N_{i,j}}{\sqrt{N_i \cdot N_j}} = \frac{N_{i,j}}{N_i^{\frac{1}{2}} \cdot N_j^{\frac{1}{2}}} \tag{1.10}$$

can be obtained from a related but more general similarity function [56]. Interestingly, the denominator is the *geometric mean* of the number of elements in M_i and M_j, so that the cosine similarity coefficient is the ratio of the number of features in common to M_i and M_j over the geometric mean of the features.

Although not as general as the expression given in Eq. (1.6), a useful expression is obtained by setting $\beta = 1 - \alpha$, which gives

$$S_{\mathrm{Tve}}(i,j|\alpha) = \frac{N_{i,j}}{\alpha(N_i - N_{i,j}) + (1-\alpha)\cdot(N_j - N_{i,j}) + N_{i,j}} \tag{1.11}$$

so that $\alpha + \beta = 1$. Under such a constraint, it is not possible to transform Eq. (1.6) into the expression for Tanimoto similarity, Eq. (1.8), although the Dice coefficient given in Eq. (1.9) can still be obtained by setting $\alpha = 1/2$. Any value of $\alpha \neq 1/2$ leads to asymmetric similarity coefficients. This asymmetry has been applied to enhance the effectiveness of similarity searches of large compound DBs [57, 58].

An interesting pair of asymmetric similarity coefficients is obtained at the limits when $\alpha = 1$ or $\alpha = 0$:

$$S_{\mathrm{Tve}}(i,j|\alpha = 1) = \frac{N_{i,j}}{(N_i - N_{i,j}) + N_{i,j}} = \frac{N_{i,j}}{N_i} \tag{1.12}$$

and

$$S_{\text{Tve}}(i,j|\alpha=0)=\frac{N_{i,j}}{(N_j-N_{i,j})+N_{i,j}}=\frac{N_{i,j}}{N_j}. \tag{1.13}$$

Equation (1.12) can be interpreted as the fraction of M_i similar to M_j, while Eq. (1.13) can be interpreted as the fraction of M_j similar to M_i. By applying the "interchange rules" to Eq. (1.12), it is clear that the similarity coefficients are *asymmetric*, i.e.,

$$\begin{aligned} S_{\text{Tve}}(i,j|\alpha=1)&=\frac{N_{i,j}}{(N_j-N_{i,j})+N_{i,j}}=\frac{N_{i,j}}{N_j}\neq\frac{N_{i,j}}{N_i}. \\ &=S_{\text{Tve}}(j,i|\alpha=1) \end{aligned} \tag{1.14}$$

A similar argument can be applied to Eq. (1.13).

Symmetric similarity coefficients corresponding to the asymmetric coefficients are given in Eqs. (1.15) and (1.16) and can be obtained simply by changing the denominators using the "min" and "max" functions, which are symmetric to interchanges of variables N_i and N_j:

$$S_{\text{Min}}(i,j)=\frac{N_{i,j}}{\min(N_i,N_j)} \tag{1.15}$$

and

$$S_{\text{Max}}(i,j)=\frac{N_{i,j}}{\max(N_i,N_j)}. \tag{1.16}$$

As was the case for the other similarity coefficients, S_{Max} and S_{Min} are again ratios equal to the number of features common to M_i and M_j over the larger and smaller number of features of M_i and M_j, respectively.

It can be shown that all of the similarity coefficients described above lie on the unit interval [0,1]. Because the terms in the denominators satisfy the following inequalities:

$$0<\min(N_i,N_j)\le N_i^{\frac{1}{2}}\cdot N_j^{\frac{1}{2}}\le\tfrac{1}{2}(N_i+N_j)\le\max(N_i,N_j)\le N_i+N_j-N_{i,j} \tag{1.17}$$

and because their numerators are all identical and equal to $N_{i,j}$, the five symmetric similarity coefficients are ordered as:

$$0<S_{\text{Tan}}\le S_{\text{Max}}\le S_{\text{Dice}}\le S_{\text{Cos}}\le S_{\text{Min}}\le 1. \tag{1.18}$$

1.2.1.3 FP-Based Molecular Dissimilarity Coefficients

For FP-based representations, dissimilarity is the 1's *complement* of similarity, i.e.,

$$\text{Dissimilarity} = 1 - \text{similarity}. \tag{1.19}$$

Thus, dissimilarity values also lie on the unit interval $[0,1]$. For example, in the case of the Tanimoto similarity coefficient the corresponding dissimilarity coefficient is given by

$$D_{\text{Tan}}(i, j) = 1 - S_{\text{Tan}}(i, j) \tag{1.20}$$

which is symmetric because $S_{\text{Tan}}(i, j)$ is symmetric. Substituting Eq. (1.8) into Eq. (1.20) and simplifying terms yields

$$D_{\text{Tan}}(i, j) = \frac{(N_i - N_{i,j}) + (N_j - N_{i,j})}{(N_i - N_{i,j}) + (N_j - N_{i,j}) + N_{i,j}}. \tag{1.21}$$

Since the denominators, which normalize the similarity and dissimilarity values, in Eqs. (1.8) and (1.21), respectively, are the same for both coefficients, it is their numerators that provide the interpretation for these coefficients. In the case of Tanimoto similarity, the numerator, $N_{i,j}$, gives the number of features in common to both molecules, while the numerator for Tanimoto dissimilarity gives the number of features unique to M_i, $N_i - N_{i,j}$, and the number of features unique to M_j, $N_j - N_{i,j}$. This interpretation accords well with our qualitative notions of similarity and dissimilarity. *Features that do not appear in either molecule are not accounted for in any of these coefficients.*

It can also be shown that Tanimoto dissimilarity formally satisfies the three properties of an abstract distance [59]. In fact, the numerator is identical to the Hamming distance between two finite, classical sets [60] and the denominator ensures that the dissimilarity values satisfy $0 \le D_{\text{Tan}} \le 1$, as required by Eq. (1.20).

Based on Eq. (1.19), dissimilarity coefficients corresponding to the similarity coefficients given in Eqs. (1.9), (1.10), (1.15), and (1.16) can also be constructed. Interestingly, the terms in their denominators are unchanged from their corresponding similarity coefficients. However, the terms in their numerators are the same as those in their denominators with the important difference that $N_i \to N_i - N_{i,j}$ and $N_j \to N_j - N_{i,j}$. Thus, for example, the Dice dissimilarity coefficient becomes

$$D_{\text{Dice}}(i, j) = \frac{\frac{1}{2}\left[(N_i - N_{i,j}) + (N_j - N_{i,j})\right]}{\frac{1}{2}(N_i + N_j)} \tag{1.22}$$

which is the ratio of the arithmetic mean of the number of unique features in M_i and M_j to the arithmetic mean of the total number of features in M_i and M_j. Recall that the term in square brackets is the Hamming distance so, as was the case for Tanimoto dissimilarity, Dice dissimilarity also satisfies the distance postulates.

Analogous expressions for dissimilarity can be derived for the remaining similarity coefficients.

1.2.1.4 Size Dependence of FP-Based Similarity and Dissimilarity Coefficients

It is both intuitive and well known that the number of 1-bits in a binary molecular FP depends on the size and complexity of the molecule it is representing. More than 25 years ago, Flower noted a bias towards low similarity values in Tanimoto similarity-based searches when the *bit densities*, that is the ratio of 1-bits to the total number of bits in a binary FP, of the molecules being compared differed significantly [61]. Subsequently, a number of laboratories observed a bias in diversity analyses towards smaller compounds [31, 62–65]. A publication also in that period by Godden et al. [66] further elaborated the issue by showing that mean Tanimoto similarity values obtained from sets of compounds are inherently biased by statistically preferred similarity values.[2]

It is not difficult to see how molecular size may have a biasing effect on the Tanimoto coefficient given in Eqs. (1.8). Consider two molecules, a *query molecule*, M_Q, and a *retrieved molecule*, M_R, obtained from a similarity search. Now suppose that the query molecule is a small molecule such that the number of substructural features (1-bits) in the FPs of both molecules satisfies $N_Q < N_R$. Since the number of substructural features common to both molecules, $N_{Q,R}$, cannot be more than the number in the smaller of the two molecules,[3] i.e.,

$$N_{Q,R} \leq \max(N_{Q,R}) = N_Q. \tag{1.23}$$

In which case,

$$S_{Tan}(Q,R) = \frac{N_{Q,R}}{(N_Q - N_{Q,R}) + N_R} \leq \frac{\max(N_{Q,R})}{\left[N_Q - \max(N_{Q,R})\right] + N_R} = \frac{N_Q}{N_R} \tag{1.24}$$

The inequality obtains from Eq. (1.23) and the fact that the denominator of Eq. (1.24) satisfies

$$(N_Q - N_{Q,R}) + N_R \geq N_R \tag{1.25}$$

[2] Interestingly, since FP-based similarity coefficients are ratios of two integers, they represent a limited subset of rational numbers. Hence, they can by their very nature only yield restricted set of values on the unit interval of the real line.

[3] In that case, the set of features in M_Q are a subset of those in M_R.

Thus, for fixed values of N_Q and N_R, $S_{Tan}(Q,R)$ reaches its maximum when the features of the query molecule are a subset of those of the retrieved molecule, that is, when $N_{Q,R} \rightarrow \max(N_{Q,R}) = N_Q$. In this case, the smaller (or the closer in size) the retrieved molecule is to the query molecule, the larger the Tanimoto similarity value, and hence, the bias for small molecules in Tanimoto similarity searches when the query molecule is itself a small molecule. This type of bias should be called *algebraic bias* since it arises out of the form of the Tanimoto similarity coefficient and has no statistical component (cf. [67]).

If, on the other hand, the query is now a large molecule such that $N_Q > N_R$, then Eq. (1.26) can be obtained from Eq. (1.24) simply by interchanging the subscripts Q and R, i.e.,

$$S_{Tan}(Q,R) = \frac{N_{Q,R}}{(N_R - N_{Q,R}) + N_Q} \leq \frac{\max(N_{Q,R})}{\left[N_R - \max(N_{Q,R})\right] + N_Q} = \frac{N_R}{N_Q} \quad (1.26)$$

It is clear from the equation that since the query molecule is large and fixed, the only way to increase $S_{Tan}(Q,R)$ is to increase the size of the retrieved molecule. Hence, in Tanimoto similarity searches where the query molecule is large, something that rarely occurs in practice, the algebraic bias will be towards larger retrieved molecules. Holliday et al. [67] have significantly extended this analysis, providing an extensive and detailed treatment of a large number of similarity coefficients that are documented in Table 1.1 of their paper.

The algebraic bias in similarity searches has led some researchers to consider other possible similarity functions that might overcome this problem. An interesting work in this regard is that of Chen and Brown [57], which was based on asymmetric similarity searching. A detailed discussion of asymmetric similarity searching and how it might overcome, to some extent at least, the algebraic size bias described above was recently presented [10, 41].

Although the algebraic size bias discussed above is relatively straight forward, this is not case when dealing with dissimilarity-based searching as it is applied, for example, in diversity analysis. In each step of a typical iterative dissimilarity-based selection algorithm, the most dissimilar compound with respect to *all* of the previously selected compounds is chosen, a situation that differs significantly from that of similarity searching in a number of ways (see discussion in Sect. 1.3.5.3 for additional details). Moreover, the arguments presented above do not touch on some of the crucial issues that are statistical in nature. These were clearly described in a paper by Fligner et al. [65] and involved a statistical analysis of the discrete, hypercubical space in which binary structural FPs reside. Based on this analysis, they developed a modified version of the Tanimoto similarity coefficient that in addition to accounting for substructural features present in both molecules, also considered features that were absent. Basically, it is a weighted combination of Tanimoto similarity coefficients, one corresponding to the usual form of the Tanimoto coefficient associated with 1-bits and the other of essentially similar form but associated in this case with the 0-bits.

It was shown by both Fligner et al. [65] and Holliday et al. [67] that the modified Tanimoto coefficient did to a large extent ameliorate size bias associated with the Tanimoto similarity coefficient. More recently, Bajorath and his collaborators [58, 66] successfully introduced a related type of modified similarity measure that weights contributions associated with the presence and absence of substructural features. In their case, however, a Tverksy-type similarity coefficient was used rather than the Tanimoto expression employed by Fligner et al. [65].

1.2.2 Vector-Based Similarity Measures

Analogous expressions to the FP-based Tanimoto, Dice, and Cosine similarity coefficients (see Eqs. (1.8), (1.9), and (1.10), respectively) also exist for vectors with continuous, real valued components as described in the following section.[4] Since each of the vector components may be associated with properties that have different units, i.e., are not comparable, they can be *standardized* according to Eqs. (1.30) and (1.31), so that their values are mean centered and of unit variance. *Also, subscripts designating the similarity coefficient are given in bold face upper case type to distinguish them from the corresponding FP-based similarity coefficients.* Terms typically found in vector-based similarity and dissimilarity coefficients are described in Table 1.2.

Table 1.2 Vector-based expressions useful in similarity analysis

Operation	Vector expression	Corresponding set theoretic entities[a]
Scalar product of $\mathbf{x}_{\text{row}}(i)$ and $\mathbf{x}_{\text{row}}(j)$	$\langle \mathbf{x}_{\text{row}}(i), \mathbf{x}_{\text{row}}(j) \rangle = \sum_{k=1}^{p} x_{ik} x_{jk}$	N_i
Squared magnitude of $\mathbf{x}_{\text{row}}(i)$	$\lVert \mathbf{x}_{\text{row}}(i) \rVert^2 = \langle \mathbf{x}_{\text{row}}(i), \mathbf{x}_{\text{row}}(i) \rangle = \sum_{k=1}^{p} x_{ik}^2$	$N_{i,j}$
Squared Euclidean distance between $\mathbf{x}_{\text{row}}(i)$ and $\mathbf{x}_{\text{row}}(j)$	$\lVert \mathbf{x}_{\text{row}}(i) - \mathbf{x}_{\text{row}}(j) \rVert^2$ $= \langle \mathbf{x}_{\text{row}}(i) - \mathbf{x}_{\text{row}}(j), \mathbf{x}_{\text{row}}(i) - \mathbf{x}_{\text{row}}(j) \rangle$ $= \sum_{k=1}^{p} (x_{\text{row}}(i) - x_{\text{row}}(j))^2$	$(N_i - N_{i,j}) + (N_j - N_{i,j})$

[a] See Table 1.1

[4] Strictly speaking, these vectors should be called geometric vectors since they do not, in all cases, satisfy the properties of algebraic vectors (e.g., algebraic vectors satisfy the axioms of a linear vector space, namely, the addition of two vectors or the multiplication of a vector by a scalar should result in another vector that also lies in the space). Nevertheless, the terminology "vector," which is common in chemical informatics, will be used here to include both classes of vectors.

1.2.2.1 Vector-Based Representations

Vector-based representations provide another means for encoding the molecular and chemical information associated with molecule M_i and are of the general form of p-dimensional row vectors also called p-tuples:

$$\mathbf{x}_{\text{row}}(i) = (x_{i,1}, x_{i,2}, \ldots, x_{i,k}, \ldots, x_{i,p}), \quad i = 1, 2, \ldots n \tag{1.27}$$

Such vectors are in many instances given as column vectors. However, since the rows of data matrices generally correspond to points in a data space, the practice is continued here for consistency.

Each component of the vector represents the value of some macroscopic chemical property such as solubility, heat capacity, polarizability, pK_a [68], some molecular property such as molecular weight, ionization potential, pi-electron distribution, number of hydrogen bonding donors or acceptors, and HOMO or LUMO energies [69], or some properties that characterize topological aspects of molecules, such as branching and shape indices [70]. Martin [71] has discussed the computation of many physicochemical property descriptors in the context of computational drug design. Todeschini and Consonni have compiled an extensive compendium of them [72]; Guha and Willighagen have recently surveyed a wide variety of quantitative descriptors useful for the calculation of chemical and biological properties [73]. Labute has also developed an internally consistent set of 32 descriptors based on the surface properties of molecules such as $\log P$, molar refractivity, partial charges, and pK_as [74, 75]. They were shown to be weakly correlated with each other, able to represent much of the information in many "traditional" molecular descriptors, and capable of providing an effective means for carrying out a range of quantitative structure–activity relationship (QSAR) and structure–property relationship (QSPR) calculations.

BCUT Descriptors A particularly interesting set of descriptors is that developed by Pearlman and Smith [76–78]. Called BCUTS, they provide an internally consistent, balanced set of molecular descriptors that encode information on the electrostatic, hydrophobic, and hydrogen bonding features of molecules and are generated in a way that exploits information on through-bond or through-space interatomic distances and atomic properties related to intermolecular ligand–protein interactions. BCUT values are determined from matrices whose diagonal elements are associated with atomic properties and whose off-diagonal elements are associated with connectivity-related properties and a scale factor that balances both types of information. Different definitions of the off-diagonal elements differentiate the different classes of BCUTS from each other. For example, 3-D BCUTS use through space interatomic distances to determine off-diagonal elements, while 2-D BCUTS use Burden numbers [79], and 2-DT BCUTS use topological interatomic distances. The largest and smallest eigenvalue obtained from each matrix are retained as potential descriptors.

Since there are many ways to compute the diagonal and off-diagonal elements of BCUT matrices, the number of potential descriptors is quite large for any of the three BCUT classes. In order to deal with this issue, Pearlman and Smith developed an "auto-choose" algorithm based on a χ-squared statistic that selects an optimum subset of BCUT descriptors for a given set of compounds such that their distribution is as close to a uniform distribution as possible. Thus, intercompound correlations are reduced so that the compounds are maximally dispersed throughout CS in the minimum number of dimensions. Importantly, this shows that BCUT descriptors and their associated CSs depend on the set of compounds used to determine them. Thus, there are many possible CSs, most typically of dimension five and six. BCUT descriptor values are not standardized to zero mean and unit variance (see Eqs. 1.30 and 1.31) since their value ranges are all comparable.

BCUT descriptors have been shown to perform well in diversity-analysis-related tasks [80–82]. And although not originally intended for this purpose BCUT descriptors have, nonetheless, shown surprisingly good performance in QSAR and QSPR studies [83–85] and in selecting compounds for follow-on screening in drug discovery [86].

In general, the vectors associated with a set of n molecules can be combined into an $n \times p$–dimensional data matrix

$$\mathbf{X}_{nxp} = \begin{bmatrix} x_{1,1} & x_{1,2} & \cdots & x_{1,j} & \cdots & x_{1,p} \\ x_{2,1} & x_{2,2} & \cdots & x_{2,j} & \cdots & x_{2,p} \\ \vdots & \vdots & \ddots & \vdots & \ddots & \vdots \\ x_{i,1} & x_{i,2} & \cdots & x_{i,j} & \cdots & x_{i,p} \\ \vdots & \vdots & \ddots & \vdots & \ddots & \vdots \\ x_{n,1} & x_{n,2} & \cdots & x_{n,j} & \cdots & x_{n,p} \end{bmatrix} = \begin{bmatrix} \mathbf{x}_{\text{row}}(1) \\ \mathbf{x}_{\text{row}}(2) \\ \vdots \\ \mathbf{x}_{\text{row}}(i) \\ \vdots \\ \mathbf{x}_{\text{row}}(n) \end{bmatrix} \tag{1.28}$$

where ith row is the same as that given by Eq. (1.27) and jth column is given by

$$\mathbf{x}_{\text{col}}(j) = \begin{pmatrix} x_{1,j} \\ x_{2,j} \\ \vdots \\ x_{k,j} \\ \vdots \\ x_{n,j} \end{pmatrix}, \quad j = 1, 2, \ldots, p \tag{1.29}$$

Because the units associated with each of the descriptors are, in general, likely to differ, they should be normalized so that they all have equivalent units. This can be accomplished by standardizing the set of values for each descriptor to zero mean and unit variance using the well-known "z-transformation," i.e.

$$z_{i,j} = \frac{x_{i,j} - \overline{x}_{\text{col}}(j)}{\sqrt{s(j)}} \tag{1.30}$$

where the sample mean and variance of the jth variable are given by, respectively,

$$\bar{x}_{\text{col}}(j) = \tfrac{1}{n}\sum_{i=1}^{n} x_{i,j}$$
$$s(j) = \tfrac{1}{n}\sum_{i=1}^{n}\left[x_{i,j} - \bar{x}_{\text{col}}(j)\right]^2 \tag{1.31}$$

All of the variables are now unitless and, thus, on equal footing. Row vectors and data matrices corresponding to the new z-transformed variables are now given, respectively, by (cf. Eqs. 1.30 and 1.31)

$$\mathbf{z}_{\text{row}}(i) = (z_{i,1}, z_{i,2}, \ldots, z_{i,k}, \ldots, z_{i,p}),\ i = 1, 2, \ldots n \tag{1.32}$$

and

$$\mathbf{Z}_{nxp} = \begin{bmatrix} z_{1,1} & z_{1,2} & \cdots & z_{1,j} & \cdots & z_{1,p} \\ z_{2,1} & z_{2,2} & \cdots & z_{2,j} & \cdots & z_{2,p} \\ \vdots & \vdots & \ddots & \vdots & \ddots & \vdots \\ z_{i,1} & z_{i,2} & \cdots & z_{i,j} & \cdots & z_{i,p} \\ \vdots & \vdots & \ddots & \vdots & \ddots & \vdots \\ z_{n,1} & z_{n,2} & \cdots & z_{n,j} & \cdots & z_{n,p} \end{bmatrix} = \begin{bmatrix} \mathbf{z}_{\text{row}}(1) \\ \mathbf{z}_{\text{row}}(2) \\ \vdots \\ \mathbf{z}_{\text{row}}(i) \\ \vdots \\ \mathbf{z}_{\text{row}}(n) \end{bmatrix} \tag{1.33}$$

1.2.2.2 Vector-Based Similarity Coefficients

The vector-based Tanimoto similarity coefficient corresponding to the FP-based coefficient in Eq. (1.8) is given by

$$S_{\mathbf{TAN}}(i,j) = \frac{\left\langle \mathbf{x}_{\text{row}}(i), \mathbf{x}_{\text{row}}(j) \right\rangle}{\left\| \mathbf{x}_{\text{row}}(i) \right\|^2 + \left\| \mathbf{x}_{\text{row}}(j) \right\|^2 - \left\langle \mathbf{x}_{\text{row}}(i), \mathbf{x}_{\text{row}}(j) \right\rangle}$$
$$= \frac{\sum_{k=1}^{p} x_{ik} \cdot x_{jk}}{\sum_{k=1}^{p} x_{ik}^2 + \sum_{k=1}^{p} x_{jk}^2 - \sum_{k=1}^{p} x_{ik} \cdot x_{jk}}, \tag{1.34}$$

where the form of the continuous, real valued vectors is given in Eq. (1.27) and the nature of their components are described in the previous section. The vector-based similarity coefficient due to Hodgkin and Richards [87] is an analog of the FP-based Dice similarity coefficient given in Eq. (1.9):

$$S_{\mathbf{HR}}(i,j) = \frac{\left\langle \mathbf{x}_{\text{row}}(i), \mathbf{x}_{\text{row}}(j) \right\rangle}{\tfrac{1}{2}\left(\left\| \mathbf{x}_{\text{row}}(i) \right\|^2 + \left\| \mathbf{x}_{\text{row}}(j) \right\|^2 \right)}$$
$$= \frac{\sum_{k=1}^{p} x_{ik} \cdot x_{jk}}{\frac{1}{2}\left(\sum_{k=1}^{p} x_{ik}^2 + \sum_{k=1}^{p} x_{jk}^2 \right)} \tag{1.35}$$

The well-known *cosine similarity coefficient*, also called the *Carbo similarity index* [10], provides a measure of the cosine of the angle between two vectors

$$S_{\text{COS}}(i,j) = \frac{\langle \mathbf{x}_{\text{row}}(i), \mathbf{x}_{\text{row}}(j)\rangle}{\sqrt{\|\mathbf{x}_{\text{row}}(i)\|^2 \cdot \|\mathbf{x}_{\text{row}}(j)\|^2}} = \frac{\sum_{k=1}^{p} x_{i,k} \cdot x_{j,k}}{\sqrt{\sum_{k=1}^{p} x_{i,k}^2} \cdot \sqrt{\sum_{k=1}^{p} x_{j,k}^2}} \tag{1.36}$$

A variety of function and vector-based similarity coefficients have also been described [10], and a detailed analysis of their interrelationships has been presented [56].[5]

1.2.2.3 Vector-Based Dissimilarity Coefficients and Distances

Vector-based dissimilarity coefficients can also be defined in analogy to those given in general for FP-based dissimilarities in Eq. (1.19). Tanimoto dissimilarities are given by

$$\begin{aligned} D_{\text{TAN}}(i,j) &= 1 - S_{\text{TAN}}(i,j) \\ &= \frac{\langle \mathbf{x}_{\text{row}}(i) - \mathbf{x}_{\text{row}}(j), \mathbf{x}_{\text{row}}(i) - \mathbf{x}_{\text{row}}(j)\rangle}{\|\mathbf{x}_{\text{row}}(i)\|^2 + \|\mathbf{x}_{\text{row}}(j)\|^2 - \langle \mathbf{x}_{\text{row}}(i), \mathbf{x}_{\text{row}}(j)\rangle} \\ &= \frac{\|\mathbf{x}_{\text{row}}(i) - \mathbf{x}_{\text{row}}(j)\|^2}{\|\mathbf{x}_{\text{row}}(i)\|^2 + \|\mathbf{x}_{\text{row}}(j)\|^2 - \langle \mathbf{x}_{\text{row}}(i), \mathbf{x}_{\text{row}}(j)\rangle} \\ &= \frac{\sum_{k=1}^{p} (x_{ik} - x_{jk})^2}{\sum_{k=1}^{p} x_{ik}^2 + \sum_{k=1}^{p} x_{jk}^2 - \sum_{k=1}^{p} x_{ik} \cdot x_{jk}} \end{aligned} \tag{1.37}$$

Again, the terms are analogous to those for the FP-based dissimilarity given in Eq. (1.21) and summarized in Tables 1.1 and 1.2. As was the case for FP-based dissimilarities, the value of the vector-based dissimilarity is complementary (see Eq. 1.20) to the corresponding similarity value and, hence, lies on the unit interval

[5] An interesting relationship between the FP- and vector-based similarity coefficients occurs when both have binary component values, e.g. $\mathbf{m}_l = (1,0,0,0,1,1,0,1,0,1)$ and $\mathbf{x}_{\text{row}}(l) = (1,0,0,0,1,1,0,1,0,1)$. In such cases, but only in such cases, the similarity coefficients based on binary FPs or binary vectors yield exactly the same similarity value for all of the similarity coefficients described above. However, this limitation has not been consistently adhered to and similarity values computed using continuous vectors or weighted FPs based on Eqs. (1.27)–(1.29) yield values that may differ significantly from their corresponding FP-based similarity coefficients.

[0,1]. Importantly, the numerator is just the square of the *Euclidean distance* (see also Table 1.2):

$$\begin{aligned} d_{\text{Euc}}(\mathbf{x}_{i\cdot},\mathbf{x}_{j\cdot})^2 &= \langle(\mathbf{x}_{i\cdot}-\mathbf{x}_{j\cdot}),(\mathbf{x}_{i\cdot}-\mathbf{x}_{j\cdot})\rangle \\ &= \left\|\mathbf{x}_{i\cdot}-\mathbf{x}_{j\cdot}\right\|^2 \\ &= \sum_{k=1}^{p}(x_{ik}-x_{jk})^2 \end{aligned} \tag{1.38}$$

Since the denominator is just a constant factor that scales the distance so that distance lies on the unit interval [0,1], it again follows that D_{TAN} satisfies the distance axioms as was true in the corresponding FP-based case for D_{tan}.

Similarly, it can be shown that Hodgkin–Richards dissimilarity,

$$D_{\text{HR}}(i,j) = \frac{\sum_{k=1}^{p}(x_{ik}-x_{jk})^2}{\frac{1}{2}\left(\sum_{k=1}^{p}x_{ik}^2+\sum_{k=1}^{p}x_{jk}^2\right)} \tag{1.39}$$

accords well with the FP-based case for $D_{\text{Dice}}(i,j)$. Note that the numerator is the squared Euclidean distance of the two molecular feature vectors, so dissimilarity again satisfies the distance axioms and is a normalized distance whose values lie on [0,1].

Thus, it is clear from the above discussion that there is an underlying consistency to the FP- and vector-based similarity coefficients. Moreover, for the case of binary FPs and binary feature vectors, the two approaches yield identical results (*vide supra*). However, for *integer-weighted FPs* (see Sect. 1.2.1.1) such as arise in cases where the number of occurrences of substructural features is considered, methods for treating vectors with continuous, real-valued components are no longer appropriate and multiset procedures provide a better, more consistent approach for dealing with such FPs [10, 41].

1.2.3 Fusing ("Aggregating") Similarity Measures

Although molecular similarity studies have been carried out for more than two decades, it is generally recognized that no one similarity measure is capable of providing high-quality results for all classes of compounds. This has raised the possibility that aggregating or fusing multiple similarity measures may in some fashion lead to improved results [88]. Based on the pioneering works of Sheridan and his colleagues at Merck [89, 90] and Peter Willett and his colleagues in Sheffield, a number of procedures have been developed for combining similarity measures based on data fusion methods [91–93]. A recent review by Willett provides a comprehensive overall summary and analysis of similarity-based data fusion methods [94].

Table 1.3 Examples of fusion rules

Fusion rule	Applicable fusion method[a]	Mathematical expression
MAX	Group fusion	$\max\left\{S_i^{\text{Ref}_1}, S_i^{\text{Ref}_2}, \ldots, S_i^{\text{Ref}_q}\right\}$
MIN	Similarity fusion	$\min\left\{R_i^{\text{Sim}_1}, R_i^{\text{Sim}_2}, \ldots, R_i^{\text{Sim}_p}\right\}$
MEAN	Similarity fusion	$(1/p)\sum_{k=1}^{p} S_i^{\text{Sim}_k}$
RRF	Similarity and group fusion	$\sum_{k=1}^{p}(1/R_i^{\text{Sim}_k})$ or $\sum_{l=1}^{q}(1/R_i^{\text{Ref}_l})$

Only the most effective fusion rules are included in the table, where "S_i" corresponds to a similarity value and "R_i" to a specific rank. See text for details

[a] See [94] for a detailed discussion of the performance of the different fusion rules

Data fusion methods [95, 96] fall under the more general rubric of *data aggregation* methods that are widespread in many applications of multiparameter decision making [97]. The basic idea behind data fusion is that combining data from multiple sources will lead to improved results over data obtained from a single source. Data fusion can be implemented as an *unsupervised* or *supervised* procedure, the former being the most well studied of the two approaches, since the latter requires experimental activity data in addition to computed similarities [94]. The focus in this work is on unsupervised procedures, and the previous reference should be consulted for details of supervised procedures. The description of similarity searching given in Sect. 1.3.3.3 is complementary to that presented here, where the emphasis is on issues associated with data fusion procedures.

Although there are many possible unsupervised ways to combine multisource data, those typically applied in chemical informatics are relatively limited (see Fig. 1.2 of [94]). Table 1.3 provides a summary of the most effective data fusion rules associated with the different fusion procedures typically employed in chemical informatics applications (*vide infra*). In certain applications, as seen in the table, data are best treated as similarity values or as rankings—specifics are described below. Mathematical expressions corresponding to the different fusion rules given in Table 1.3 are relatively straightforward except for the reciprocal rank fusion (RRF) rule, which is directly related to the mean of the harmonic mean of the rank values [98].[6] Because the RRF rule treats rank values reciprocally, compounds near the top of a ranked list will have lower values, and thus will be given more influence in the RRF rule than those further down the list. Recent studies in information retrieval [99] and chemical informatics [100] suggest that the RRF rule may be more generally applicable than heretofore had been suspected. Thus, it may be suitable as a replacement for the other fusion rules considered in Table 1.3 (i.e., MAX, MIN, and MEAN), which have enjoyed widespread use in the past [94]. Finally, it should be noted that fusions can also be effected using similarity values computed with any of the similarity measures although most studies have been confined to FP-based measures.

[6] The RRF rule works best with rank values since similarities can in certain cases have zero values leading to undefined values for the reciprocals, a situation that can be overcome by the addition of a small positive constant to the denominator of each term.

1.2.3.1 Similarity Fusion

The initial approach to data fusion, called *similarity fusion*, combines the results of searches using multiple similarity measures with respect to a single reference molecule. The data generated in this procedure can be envisioned in the form of a data table such as that depicted in Fig. 1.4a, where the columns correspond to the p different similarity measures, and the rows correspond to the n molecules in a DB—at this point the ordering of the molecules is arbitrary. Each of the similarity elements in the table, $S_i^{\mathrm{Sim}_k}$, is designated by the DB molecule with which it is associated, as indicated by the set of subscripts $\{1,2,\ldots,n\}$. The corresponding similarity measures used to calculate its value are indicated by the set of superscripts $\{\mathrm{Sim}_1,\mathrm{Sim}_2,\ldots,\mathrm{Sim}_p\}$. *All of the similarity values are computed with respect to the same reference molecule.*

The similarity values in each row can be aggregated in various ways to yield a fused similarity value S_i^{SF}. For example, as shown in Table 1.3, the arithmetic mean values of the similarity values in each row can be computed and placed in the corresponding column "fused sim" of Fig. 1.4a. Once this process is complete the rows can be reordered, as depicted in Fig. 1.4b, such that the first row contains the most similar molecule to the reference molecule based on its fused similarity value, the second row contains the next most similar molecule, and the process continues until all of the molecules are reordered with respect to their fused similarity values. This procedure effectively *permutes* the order of the molecules given in the first column of Fig. 1.4a, which as noted earlier is arbitrary, to that shown in the first column of

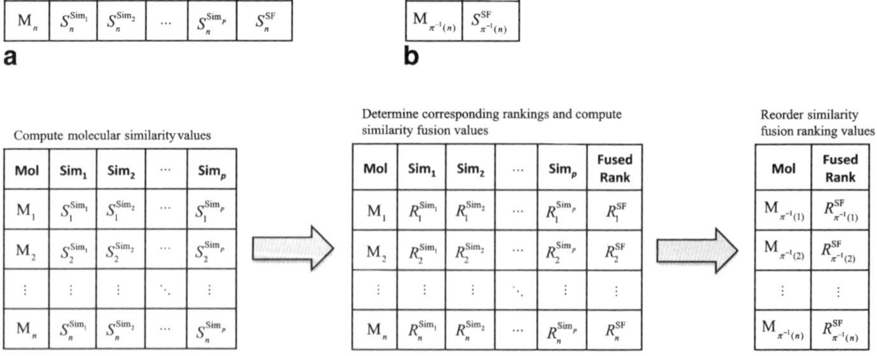

Fig. 1.4 Data tables illustrating *similarity fusion* of similarity and rank values: **a** and **b** depict the procedure for fusing similarity values. **c**, **d**, and **e** depict the corresponding procedure for fusing rank values

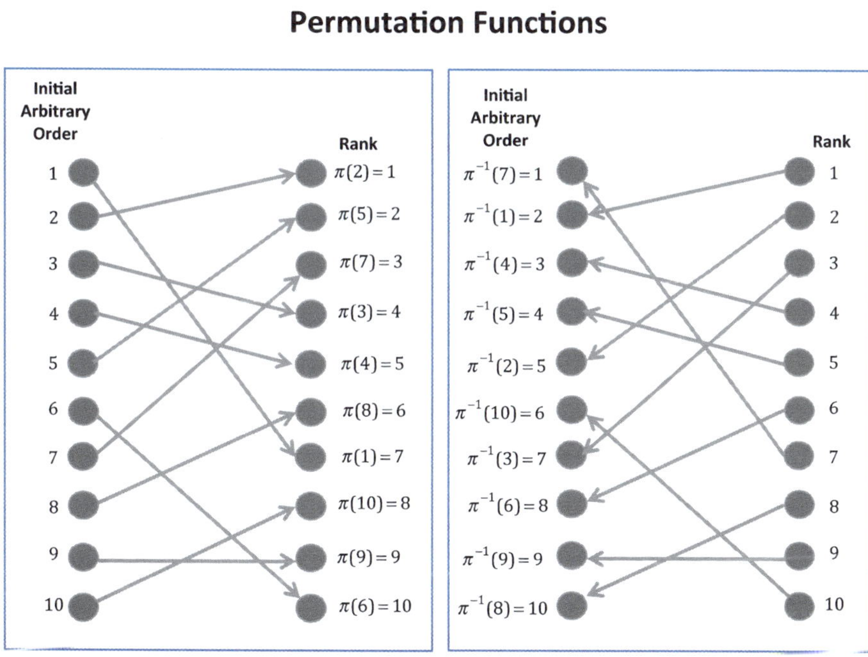

Fig. 1.5 Graphical example of mappings produced by the permutation functions π and their inverses π^{-1} (see text for additional details)

Fig. 1.4b, which is based on the decreasing fused similarity values in the second column of Fig. 1.4b, i.e.,

$$S^{SF}_{\pi^{-1}(1)} \geq S^{SF}_{\pi^{-1}(2)} \geq \cdots \geq S^{SF}_{\pi^{-1}(n)} \tag{1.40}$$

The subscript notation $\pi^{-1}(i)$ in the mathematical expression given in Eq. (1.40) is based on the mathematical theory of permutations [101], where the permutation function value $\pi(i)$ gives the rank of the ith molecule and the unique inverse $\pi^{-1}(j)$ designates the jth molecule in the overall ranking. A graphic example of how these functions operate is provided in Fig. 1.5. It is important to note that while the permutations determine the rank order of the compounds, it is the similarity values themselves that are combined using the MEAN fusion rule in similarity fusion.

Alternatively, data fusion procedures can also be directly applied to rankings themselves as seen in Fig. 1.4c. In this case, the computation of similarities is followed by a determination of the rank of each of the compounds with respect to each of the similarity measures as illustrated in Fig. 1.4d, and an appropriate data fusion procedure, in this case the MIN rule given in Table 1.3 is applied. Lastly, the resulting MIN fused rankings are permuted, i.e., $R^{SF}_{i} \rightarrow R^{SF}_{\pi^{-1}(i)} = i$, in increasing order

$$R^{SF}_{\pi^{-1}(1)} < R^{SF}_{\pi^{-1}(2)} < \cdots < R^{SF}_{\pi^{-1}(n)} \tag{1.41}$$

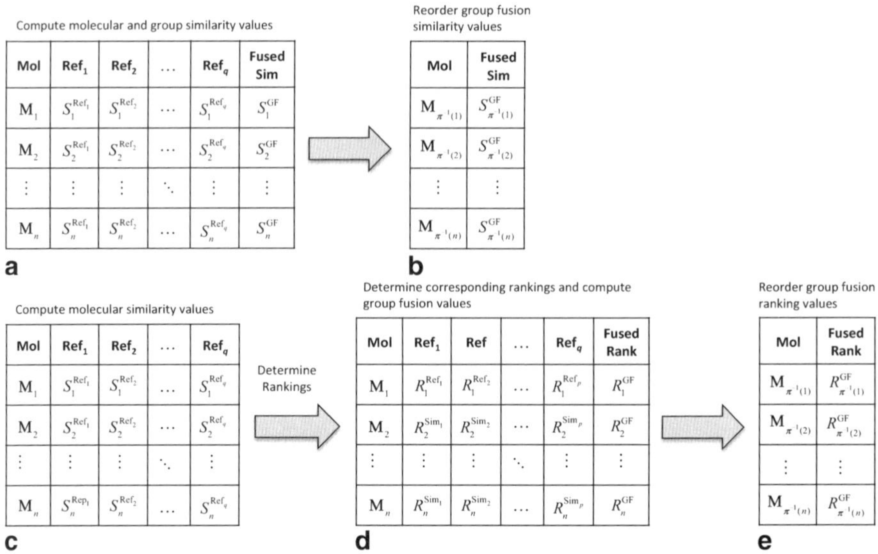

Fig. 1.6 Data tables illustrating *group fusion* of similarity and rank values: **a** and **b** depict the procedure for group fusion of similarity values. **c**, **d**, and **e** depict the corresponding procedure for fusing rank values

1.2.3.2 Group Fusion

The development of *group fusion* [91, 92, 102] quickly followed that of similarity fusion. In contrast to latter, a *single similarity measure but multiple reference molecules* are used. This is illustrated in Fig. 1.6a, b, which are quite similar to the previous figure except that the similarity measures in the top row of Fig. 1.4a are replaced by a set of q reference molecules $\{\text{Ref}_1, \text{Ref}_2, \ldots, \text{Ref}_q\}$ in Fig. 1.6a. Similarity values are computed for each DB compound with respect to each of the reference compounds using a single similarity measure, and the values in each row of the table are fused, yielding the similarity values in the last column of Fig. 1.6a. As was the case for similarity fusion, the next step is to reorder the fused similarity values from largest to smallest as indicated in Fig. 1.6b and Eq. (1.42):

$$S^{\text{GF}}_{\pi^{-1}(1)} \geq S^{\text{GF}}_{\pi^{-1}(2)} \geq \cdots \geq S^{\text{GF}}_{\pi^{-1}(n)} \tag{1.42}$$

Numerous studies have shown that applying the MAX rule to similarity values provides excellent overall performance in similarity searches that are designed to assess the efficacy of group similarity for retrieving known actives from compound DBs [91, 92, 94, 100]. Although, in general, the MAX rule works well, the RRF rule for combining rank values (see Table 1.3) appears to perform even better [100]. Figure 1.6c–e describes the rank-based group fusion process, which is similar to that given in Fig. 1.4c–e for the corresponding similarity fusion procedure. The fused values obtained by the RRF rule are given in the far right column of Fig. 1.6d, and the combined values are then permuted, i.e., $R_j^{\mathrm{GF}} \rightarrow R_{\pi^{-1}(i)}^{\mathrm{GF}} = i$, in increasing order

$$R_{\pi^{-1}(1)}^{\mathrm{GF}} < R_{\pi^{-1}(2)}^{\mathrm{GF}} < \cdots < R_{\pi^{-1}(n)}^{\mathrm{GF}} \tag{1.43}$$

to yield the final fusion-based ranking. Whichever rule is used, the superior performance of group fusion makes it the preferred method for carrying out similarity searches [103].

Either the reordered similarity values or compound rankings can be used as a basis for subset selection. Furthermore, although group fusion provides improved results over both single similarity and similarity fusion approaches, it requires multiple reference compounds, which may not always be readily available. Even when such data are available, they usually are the result of early-phase HTS experiments and hence may, to a degree, be suspect. However, as discussed in the following section, a modification of group fusion called turbo similarity suggests that even somewhat erroneous data may not unduly affect the results obtained using group fusion.

1.2.3.3 Turbo Similarity

As noted in the previous section, a variant of group fusion called *turbo similarity* has also shown promise [104–106]. Turbo similarity provides a procedure for applying group fusion when only a single active is known and is based on the following procedures: (1) compute the similarity of the known (reference) active with respect to all of the molecules in a DB of unscreened compounds; (2) order the list with respect to decreasing similarity or increasing rank values; (3) choose a subset of the highest scoring or ranked compounds that, based on the SPP [13–15] (see Sect. 1.3.1 for details), are *assumed* to be active; and (4) use these putative active compounds as the set of reference compounds in a group-fusion-based similarity search as described in the previous section (see also Table 1.3 and Fig. 1.6). Note that either the MAX rule with respect to similarity or the RRF rule with respect to rank values can be applied with nearly comparable effectiveness (*vide supra*). A recent study [52] has shown that frequency weighting the components of structural FPs leads to improved results obtained with turbo similarity searching.

Interestingly, turbo similarity is reminiscent of library search procedures, where a given query yields a set of hits, each of which is used in a subsequent query to broaden the search [107].

1.2.4 Validating Similarity-Based Approaches

Although model validation is an important requirement in the development of computational methods, there are cases where it can become problematic. One such case is molecular similarity. Due to its subjective nature, well-defined values of molecular similarity do not exist. Hence, directly assessing the results of similarity calculations is not possible, and indirect methods must be used. These methods are typically based on the SPP noted in Sect. 1.3.1 (see also [13–15]) and assess the recovery rates (or some related measure) obtained from similarity searches of large compound DBs containing known actives [108, 109]. Two such measures are the *recall* and *precision* of compound retrievals given, respectively, by

$$\begin{aligned} \text{Recall} \quad &= \frac{\text{Number of actives retrieved}}{\text{Total number of actives}} = \frac{n^*_{\text{Act}}}{n_{\text{Act}}} \\ \text{Precision} &= \frac{\text{Number of actives retrieved}}{\text{Total number of compounds}} = \frac{n^*_{\text{Act}}}{n} \end{aligned} \tag{1.44}$$

These measures, although relatively widespread, have a number of deficiencies, one of which is that they do not sufficiently account for "early enrichments" in sets of retrieved compounds. This issue can be dealt with using cumulative recall curves, which plot the fraction of actives against the number of compounds retrieved [108, 109]. These curves are similar to receiver operating characteristic (ROC) curves. Truchon and Bayly [110] have provided a detailed analysis of their application to virtual screening methods.

Significant issues remain that can confound attempts to assess the validity of similarity measures: (1) Untested DB compounds are *assumed* to be inactive, an assumption that is problematic at best. (2) The presence of *activity cliffs* [111–113], which arise when small changes in structure are associated with large changes in biological activity, although rare, represent violations of the SPP giving rise to what Stahura and Bajorath call the "similarity paradox" [114]. (3) The surprising prevalence of *similarity cliffs* [7, 8], which in contrast to activity cliffs occur when small changes in activity are accompanied by large differences in similarity, suggests that active compounds tend to be scattered throughout CSs, although they are likely to be found in multiple clusters of actives, not as singletons, dispersed throughout those spaces.[7] (4) As noted earlier, similarity measures are not invariant to the representation and similarity coefficient used. This lack of invariance leads, either directly or indirectly, to the notion that combining the results obtained from multiple similarity measures, as discussed in Sect. 1.2.3, can yield improved results in molecular similarity analyses.

The prevalence of similarity cliffs noted above also provides a rationale, albeit a tentative one, as to why group fusion (Sect. 1.2.3.2) performs as well does. Numerous analyses by Willett and his colleagues show that it appears to work best with diverse rather than highly similar reference sets [92, 94, 106]. Their conclusion

[7] It should be noted that similarity cliffs are more general than scaffold hops since all scaffold hops do not result in compounds that are highly dissimilar, as may be the case when the scaffolds associated with scaffold hops are approximate bioisosteres or compounds with dissimilar scaffold nonetheless have similar overall structures.

is consistent with the unexpectedly high occurrence of similarity cliffs in pairs of active compounds. In fact, in more than 50% of the cases where both compounds in a compound pair are active (i.e., $pK_i \sim 7$), the compounds are also dissimilar [7] (cf. [115][8]).

The significant presence of similarity cliffs suggests that similarity search methods that rely on single active reference compounds, regardless of whether single or multiple similarity measures—as in similarity fusion—are used, will by their very nature miss a significant portion of potentially active compounds because only the top scoring or highest-ranked compounds obtained in similarity searches are typically chosen—compounds located further down the ordered list are routinely ignored.

Group fusion, on the other hand, employs multiple reference actives and, as noted above, performs best when the reference compounds are as diverse as possible. Hence, the dispersion of active compounds is explicitly accounted for by the method, although the available reference set may not, in many cases, provide sufficient coverage of all of the regions of CS that contain active compounds with respect to the given assay, and some actives will undoubtedly be missed.

Because group fusion uses either the MAX rule for similarities or the RRF rule for rankings, compounds located close to the reference compounds are given preference over more distant, less similar compounds, a situation that accords well with the SPP since compounds located close to known actives are more likely to also be active than are less similar compounds. Thus, the performance of group fusion can be rationalized by the significant presence of similarity cliffs in activity landscapes.

1.2.5 Computational Versus Perceptual Aspects Molecular Similarity Measures

The computational methods described above provide algorithms for computing molecular similarities, albeit imperfect ones, due to the inherently subjective nature of similarity. This, however, begs the question as to how these similarity measures accord with the perceptions of chemists, an issue that has been discussed in more detail in several recent publications [10, 34]. An important question in this regard is whether similarity scales used intuitively by chemists agree with those obtained computationally. The answer, as we shall see, is that they do not.

Essentially, all computed similarity values lie on the unit interval [0, 1] of the real line (more correctly the unit interval of the rational line). Highly similar molecules have values at the high end of this scale, while dissimilar molecules tend to lie at the lower end. Humans can, in general, assess the similarity of very similar objects, as chemists can assess the molecular similarity of molecules with similar structures. But what happens when molecules become less similar (more dissimilar)? There

[8] Even though the overall percentage of active compounds in large DBs is usually quite small, since most compounds are inactive in a given assay, the fraction of those actives where both compounds of a compound pair are approximately of equal activity can be significant.

is basically no issue with computational similarity measures, but humans, on the other hand, find it increasingly difficult to assess the degree of similarity of highly dissimilar objects. Beyond some point, all that can be said is that the objects are "not very similar," but the degree of similarity becomes moot. This is also true for chemists' assessments of highly dissimilar molecules.

Is this difference between computational and perceptual measures of similarity important? Since chemists are unable to perceive low degrees of similarity among molecules, low values of computed similarity do not have any explicit "structural meaning," at least to chemists. Because of this, it is difficult for chemists to make meaningful structural inferences as would be required when, for example, assessing the diversity of or clustering a compound library, or evaluating compounds for acquisition [116–118]. Computers, on the other hand, are not saddled with this perceptual limitation, and thus can handle similar and dissimilar molecules with equal ease.

Another matter bears on the issue of computation versus perception of structural similarity. In the former case, as described in previous sections, the similarity value obtained depends on the molecular representation used, the weighting of its components, and the similarity coefficient. Changing any or all of these can result in significant changes to the computed similarity values. By contrast, perception of molecular similarity depends on a chemist's training, experience, and the field of chemistry in which they work. For example, a synthetic organic chemist might focus on likely sites of substitution, a medicinal chemist on the placement and nature of pharmacophoric groups, and a physical chemist on the electron distribution or the energy of a molecule's highest-occupied and lowest-unoccupied orbitals.

1.3 Chemical Spaces

The amount of chemical information is growing exponentially. Thus, a framework is needed for dealing effectively with the flood of information. The concept of CS provides such a framework. In analogy to mathematical spaces, CSs are specified by a set of molecules and a binary relation that characterizes the relationship of one molecule to another and is typically based on some type of similarity, dissimilarity, or distance measure. Importantly, the notion of CS provides a basis for the well-known SPP that explicitly or implicitly underlies many applications of similarity in chemical informatics (*vide infra*) and is discussed in the following section.

CSs come in three flavors: (1) coordinate based, (2) cell based, and (3) graph or network based. Multidimensional vectors with continuous, real-valued components define the positions of molecules in coordinate-based CSs. The value associated with each of the coordinates is obtained from one of a wide variety of property descriptors discussed in Sect. 1.2.2.1. A simple 3-D example is given in Fig. 1.7a, but since these spaces are generally greater than dimension three, their graphic portrayal requires some type of reduction in the dimensionality of the space. Details of how this can be accomplished will be described in Sect. 1.3.2.

By contrast, compounds in cell-based CSs reside in p-dimensional hypercubes called cells that *partition* the original p-dimensional coordinate-based CS. Cell-based

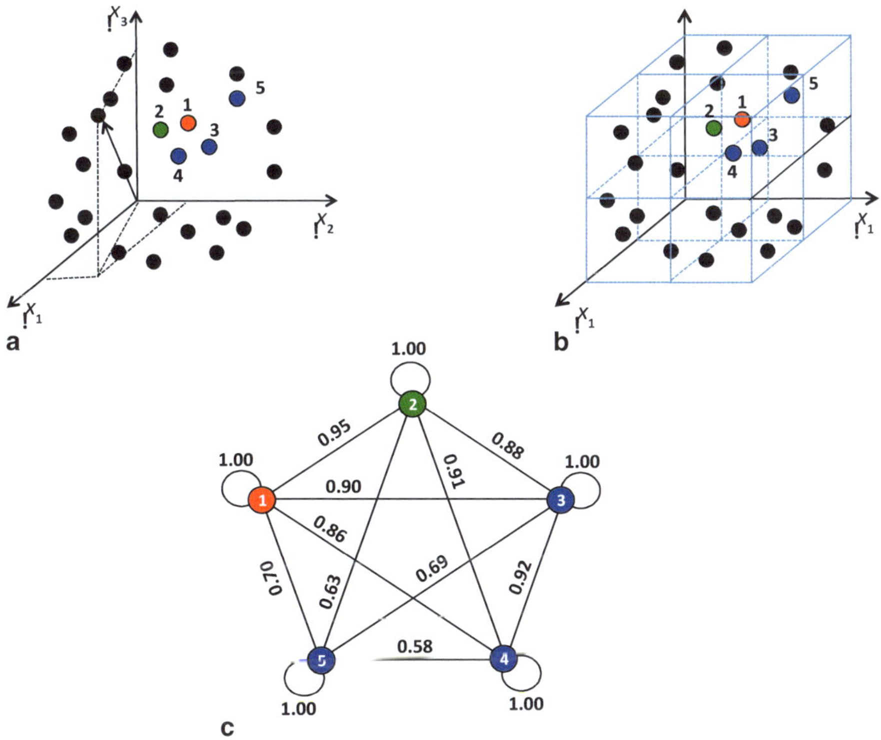

Fig. 1.7 Examples of different CS representations: **a** coordinate-based CS. **b** cell-based CS, and **c** depiction of a complete, self-similar CSN representation of the chemical subspace of the five numbered compounds. The *red filled circle* corresponds to an active compound, the *green filled circle* to its nearest neighbor, and the three *blue filled circles* to next-nearest neighbors

partitioning is a *coarse-grained* approach that lowers the resolution of the space but does not necessarily reduce its dimensionality. Nevertheless, it offers potential advantages for handling a number of procedures commonly carried out in chemical informatics, some of which will be discussed in more detail in Sect. 1.3.3. Figure 1.7b depicts a cell-based CS associated with the coordinate-based space illustrated in Fig. 1.7a. Other types of partitioning methods such as recursive partitioning have also been applied to molecular systems [119–121]. However, it should be noted that recursive partitioning and other tree-based decision methods generally fall in the class of supervised machine learning methods, while cell-based and clustering methods generally, but not always, fall into the class of unsupervised methods.[9]

The third type of CS representation is illustrated by the mathematical graph depicted in Fig. 1.7c called a *reflexive, labeled* or *simple, labeled graph*. The term

[9] Supervised machine learning methods typically try to model the relationship of a set of predictor (independent) variables to a set of known values (e.g., biological activities and/or solubilities) associated with one or more dependent variables. Unsupervised methods only require information associated with predictor (independent) variables (e.g., physicochemical descriptors).

"reflexive" indicates that that each vertex possesses a graph loop, while the term "labeled" indicates that each vertex and edge may be labeled by a set of alphanumeric characters or numbers that describe the properties of these graph entities. In the present application, each vertex corresponds to a molecule and each pair of molecules may or may not be connected by an edge labeled by the value of a pairwise property associated with the two molecules. In contrast to the previous two CS representations, this is a *relational model* that provides a faithful, discrete representation of CSs. More specifically, the edges represent binary relations associated with similarities, dissimilarities, or distances among compound pairs and the nodes are associated with individual compounds. Because CSs typically contain many compounds, the graphs representing them are quite large and generally fall under the rubric "networks." Network research has experienced extremely rapid growth over the past decade in a number of fields from social science [122] to biology and medicine [123–126] as well as in the popular literature [127, 128]. In this regard, the book by Newman not only provides an excellent overview of many aspects of networks but also addresses a number of algorithmic issues associated with them that are critical to their effective application [129].

Although the network model of CS is not in extensive use today, it corresponds closely to the data model of a new graph-based DB technology [130], and thus may provide an additional incentive for adopting this model for future work in chemical informatics. Details of how networks can be applied to the study of CSs are provided in Sect. 1.3.4.

1.3.1 Similarity-Property Principle

The SPP plays a major role in chemical informatics since it provides a crucial link between the similarity of molecules and their corresponding bioactivities or properties. Wilkins and Randic formally described this principle, which now seems intuitively obvious, in a seminal paper published more than three decades ago [13]. Although "similarity" arguments had been advanced in chemistry before this time (cf. [9]), none directly addressed the structural similarity between molecules in a computationally amenable form. In the late 1980s and the early 1990s, the SPP was reiterated [14, 15] and since that time has played a substantive role, explicitly or implicitly, in numerous studies associated with similarity searching and virtual screening.

While the SPP obtains in most cases, there are some notable exceptions such as the presence of activity cliffs [111–113], which arise when pairs of similar compounds exhibit significantly different activities leading to *quasi-discontinuities*[10] in their corresponding CSs [131, 132]. Although statistically rare [7, 8], activity cliffs provide significant SAR information because they afford a means for identifying

[10] Since CSs are inherently discrete, the concept of discontinuity, which applies to continuous systems, is only approximate. Thus, "discontinuities" in these spaces, such as those arising from the presence of activity cliffs, are denoted as quasi-discontinuities.

small structural changes, for example, the presence or absence of a functional group, that are associated with correspondingly large changes in activity.

Another quasi-discontinuous feature occurs in the case of *similarity cliffs* that, in contrast to activity cliffs (*vide supra*), represent compound pairs where small changes in biological activity are associated with large changes in similarity (*vide supra* Sect. 1.2.4). Thus, these cliffs are related to the notion of *target promiscuity* that stands in sharp contrast to the better-known notion of *compound promiscuity* associated with polypharmacologies [124, 133]. The fact that similarity cliffs are the most prevalent feature observed in activity landscapes for active compounds [7, 8] implies that target promiscuity is also more prevalent than heretofore had been assumed. Taken together, both concepts reinforce the idea that compound specificity may be a difficult goal to attain in many instances.

As noted earlier, since similarity measures are not invariant to the representation or similarity coefficient employed, small differences with respect to one measure may not be comparably small with respect to another measure. In such cases, activity cliffs themselves will not be invariant to similarity measure [10, 34, 41], an uneasy state of affairs that raises the question of whether activity cliffs actually exist [134]. Alternative representations based on *matched molecular pairs* (MMPs) have sought to address this question using the 2-D structural representation favored by chemists, but entirely quantitative results have yet to be obtained [135, 136]. Because of its inherent subjectivity, it is unlikely that invariant values (absolute values) of molecular similarity can ever be obtained. Nevertheless, while it may be difficult to quantitate the magnitudes of activity cliffs, there is no doubt that they exist since many examples of "small" structural changes, as perceived by medicinal chemists, have resulted in relatively large activity differences [134].

Based on earlier work by Brown and Martin [17, 18], Martin et al. [137] have provided an updated assessment of the SPP in medicinal chemistry. They examined a large dataset containing the results from more than 100 different HTS assays and concluded that there is only about a 30 % chance that a compound with a Tanimoto similarity value ≥ 0.85 (based on daylight FPs [138]) to a known active is also active, significantly revising an earlier estimation of 80 % [139] (cf. [140]). However, a recent publication [34] has shown that such thresholds may not, in any case, be statistically significant

Steffen et al. [141] have described a novel approach to the SPP that differs significantly from typical FP methods. In their work, these authors employed a vector representation, where the vector components are categorical variables [41] and are based on the activities of compounds with respect to each one of a fixed set of assays. Hence, the vectors live in "biological activity space" not, as is usually the case, in some form of structure space. This enables the potential identification of compounds with similar biological activity profiles that are structurally dissimilar—compound pairs that fall into this class are related to *similarity cliffs* (*vide supra*) [7, 8]. These authors also showed that representations that included physicochemical or pharmacophoric features were generally better able to retrieve dissimilar compound pairs with similar biological activity profiles. Importantly, this work opens up new possibilities in the study and application of the SPP.

Given the caveats described above, it is important to remember that the SPP is applicable to any type of CS regardless of how it is represented (*vide infra*). Today, the SPP is applied explicitly in many areas of chemistry, but particularly in medicinal chemistry. It might be said that the SPP, whether it is used explicitly or implicitly, is one of the foundations of medicinal chemistry.

1.3.2 Coordinate-Based CSs

The most common representation of coordinate-based CSs is as a set of points, each representing an individual molecule, embedded in a multidimensional Euclidean space much like the stars and planets in our galaxy. In general, p-dimensional Euclidean spaces have p orthogonal coordinate axes, and each of the n points occupying the space is described by a p-dimensional vector as that given in Eq. (1.27). The set of row vectors can then be combined into the $n \times p-$dimensional data matrix given in Eq. (1.28), which contains the molecular and/or chemical information associated with the entire set of compounds. The relationship between any compound pair can be assessed in several ways: (1) by any of the vector-based similarity coefficients described in Eqs. (1.34)–(1.36), (2) by any of the corresponding dissimilarity coefficients described in Eqs. (1.37) and (1.39), or (3) by the Euclidean distance in CS between two molecular feature vectors as described in Table 1.2 and Eq. (1.38).

Figure 1.7a provides an illustration of a simple model 3-D CS. The five color-coded compounds are, respectively: Cpd-1, an active colored in red; Cpd-2, its nearest neighbor colored in green; and Cpd-3, Cpd-4, and Cpd-5, the three next nearest neighbors, colored in blue, are ordered with respect to decreasing similarity (or increasing dissimilarity or distance) with respect to Cpd-1. Thus, Cpd-3 is nearer to Cpd-1 than Cpd-4, which is closer than Cpd-5.

Figure 1.8a portrays a 3-D projection of a real, six-dimensional (6-D) 3-D BCUT CS; additional details on its construction are supplied in Sect. 1.3.3.2. The projection is with respect to the three most significant BCUT descriptors that are derived from the electronic ("Elec"), hydrophobic ("HPhob"), and hydrogen-bonding ("HBond") features of atoms (see Sect. 1.2.2.1 for a more detailed description of BCUT descriptors). A diverse set ("Diverse") containing approximately 175,000 compounds is depicted in yellow; a combinatorially generated set ("Combi") containing approximately 150,000 compounds constructed from a set of 40 different scaffolds, is depicted in red. It is clear from the figure that Combi, which is of nearly comparable size to Diverse, covers only a small fraction of the CS covered by the latter. Figure 1.8b shows a magnified version of the CS shared by both collections.

The fact that many data spaces including CSs possess more than three dimensions has, over the years, generated a significant amount of effort in the development of dimensionality reduction techniques. There are three main reasons for reducing dimensionality. The first and most obvious is that graphical depiction of the space is restricted to three or fewer dimensions. The second and more important reason is due to the "curse of dimensionality" [142] that occurs because the data

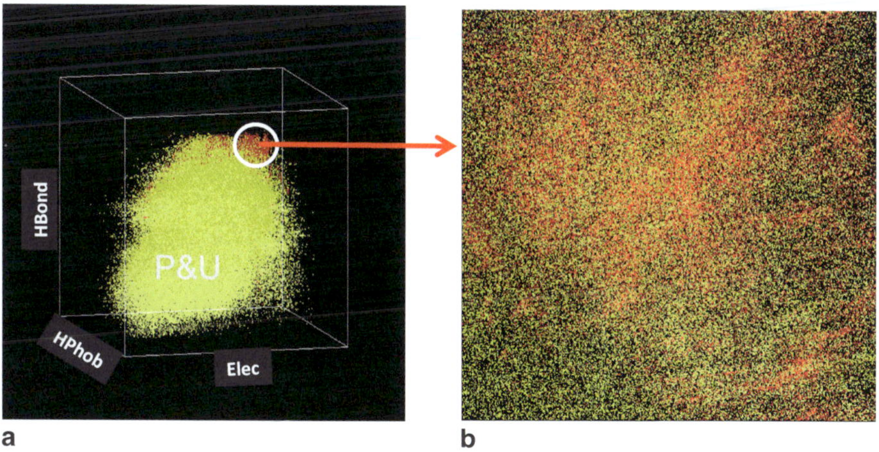

Fig. 1.8 **a** Example of a three-dimensional projection of a six-dimensional 3-D BCUT CS containing *ca.* 175,000 molecules depicted in *yellow*, and a combinatorial library of *ca.* 150,000 molecules depicted in *red*. **b** Magnified version of the region of CS shared by both sets of molecules (See Section *2.2.1* for description of BCUT descriptors). (Figure kindly provided by Veer Shanmugasundaram)

distribution becomes more sparse as the dimension of the space increases. Thus, in order to ensure balanced or comparable coverage of the resulting higher-dimensional space requires an increase in the amount of data, which becomes more difficult to achieve as the dimension increases. Higher-dimensional spaces can also exhibit idiosyncratic behaviors that are difficult to comprehend [143]. The third reason is that the intrinsic dimension of the data may be considerably lower than its apparent dimension and may in some cases be confined to a non-Euclidean subspace, which could also be nonlinear. As discussed in Sect. 1.3.2.3, distances between points in non-Euclidean subspaces are generally different than they are in the Euclidean space in which they are embedded.

1.3.2.1 Coordinate-Based CSs Derived from Structural FPs

Constructing coordinate-based CSs from low-dimensional vector representations, which is relatively straightforward, is exemplified by BCUT descriptors described in Sect. 1.2.2.1. Figure 1.8 depicts an example of a 3-D BCUT chemical subspace projected from the original 6-D BCUT CS. Today, a common means for representing molecules is by their structural FPs. However, their direct use in the construction of coordinate-based CSs is beset by a number of problems that include: (1) they are generally of very high dimension, usually in the range of ~ 150–2000, and hence are plagued by the curse of dimensionality [142] and (2) their coordinates are generally binary or integer valued and thus are not compatible with the types of continuous, real-valued CS representations described above. Nevertheless, structural FPs can be transformed into continuous, real-space coordinates in a number of ways usually through the computation of some pairwise measure that characterizes the relationships among the molecules of the set. These relationships are typically associated with the similarity or dissimilarity coefficients described in Sect. 1.2 or with some type of CS distance such as the Hamming distance [60].

A distinct advantage of this approach is that any type of representation can be used that affords a means for computing a similarity, dissimilarity, or distance measure. For example, chemical graphs [16], which cannot be treated using a purely coordinate-based approach, can be handled in a straightforward, albeit somewhat computationally demanding, manner [41]. Recent work on graph-based Kernel methods provides a novel means for extending and generalizing methods for computing similarity coefficients [144].

Given that a matrix of similarity, dissimilarity, or distance values can be computed for each unique pair of molecules, the question now becomes, "How can this array of values be transformed into a set of coordinates that define the positions of molecules in a coordinate-based CS?" In this regard, most efforts in chemical informatics have generally focused on five main techniques: (1) principal component analysis (PCA) [145], (2) principal coordinate analysis (PCoA) [145], (3) multidimensional scaling (MDS) [146], (4) nonlinear mapping (NLM) [147], and (5) factor analysis [148]. All five methods provide the means for constructing low-dimensional representations of CSs. A recent review by Shanmugasundaram and Maggiora [41] provides additional details and references to these methods.

Although any of the five methods would suffice, PCA, a method used in many chemical informatics applications, will be employed here as an example of how CSs can be constructed from several varieties of structural FPs. Consider the similarity coefficient values of a set of n molecules computed with respect to some type of structural FP that generates an $n \times n$ – dimensional symmetric matrix of similarity coefficients[11]

$$\mathbf{S}_{n\times n} = \begin{bmatrix} s_{1,1} & s_{1,2} & \cdots & s_{1,j} & \cdots & s_{1,n} \\ s_{2,1} & s_{2,2} & \cdots & s_{2,j} & \cdots & s_{2,n} \\ \vdots & \vdots & \ddots & \vdots & \ddots & \vdots \\ s_{i,1} & s_{i,2} & \cdots & s_{i,j} & \cdots & s_{i,n} \\ \vdots & \vdots & \ddots & \vdots & \ddots & \vdots \\ s_{n,1} & s_{n,2} & \cdots & s_{n,j} & \cdots & s_{n,n} \end{bmatrix} \tag{1.45}$$

where $s_{i,j}$ corresponds to any of the similarity coefficients described in Sect. 1.2. There is no need to scale these values since they are all on the same scale and lie on the unit interval [0,1] of the real line.

Although the matrix does not have the form of a typical data matrix, the similarity values can, nevertheless, be thought of as descriptor values. Consider, for

[11] In mathematics these are generally called Gram matrices and in statistics are usually called association matrices.

example, the *i, j*th element of $\mathbf{S}_{n\times n}$, which can be interpreted as the similarity of the *i*th molecule in the set of *n* molecules with respect to the *j*th "descriptor molecule". In this case, the *n* "descriptor molecules" are taken from the same set of *n* molecules under study—a generalization of this approach was recently described [149]. As was suggested by Kruscal [150], square symmetric matrices such as $\mathbf{S}_{n\times n}$ can be handled in exactly the same manner that general data matrices are treated using PCA:

$$\mathbf{S}_{n\times n} \Rightarrow \underset{\text{Mean center}}{\overline{\mathbf{S}}_{n\times n}} \Rightarrow \underset{\text{Compute covariance matrix}}{\mathbf{C} = \tfrac{1}{n-1}\overline{\mathbf{S}}^{\mathrm{T}}_{n\times n}\overline{\mathbf{S}}_{n\times n}} \Rightarrow \underset{\text{Diagonalize } \mathbf{C}}{\mathbf{V}^{\mathrm{T}}\mathbf{C}\mathbf{V} = \Lambda} \tag{1.46}$$

the eigenvalues

$$\Lambda = \begin{pmatrix} \lambda_1 & & 0 \\ & \ddots & \\ 0 & & \lambda_n \end{pmatrix} \tag{1.47}$$

which are ordered from largest to smallest, are related to the variances in the new, transformed coordinate system[12]

$$\mathbf{Z}_{n\times n} = \overline{\mathbf{S}}_{n\times n}\mathbf{V}_{n\times n} \tag{1.48}$$

such that the percent of the total variance corresponding to the *i*th eigenvalue is given by

$$\text{Percent-Variance}(\lambda_i) = \frac{\lambda_i}{\sum_{j=1}^{n}\lambda_j} \times 100 \tag{1.49}$$

Thus, to graphically depict a CS in three dimensions, the transformed coordinates associated with the first three eigenvalues will suffice, i.e.,

$$\mathbf{Z}_{n\times 3} = \overline{\mathbf{S}}_{n\times n}\mathbf{V}_{n\times 3} \tag{1.50}$$

Note, however, that the entire mean-centered similarity matrix $\overline{\mathbf{S}}_{n\times n}$ is required.

Although this procedure provides a reasonably straightforward approach to the construction of low-dimensional CSs, the number of compounds that can be handled is somewhat limited because determining the transformed coordinates requires diagonalization of the $n\times n$ covariance matrix, which becomes difficult for $n > 2500$, although there are ways that this limitation can be overcome, for example, by using real time PCA [151].

Figure 1.9 shows examples of CSs constructed with respect to four different binary FP representations using the similarity-based PCA procedure described in the previous section. The first two examples are based on atom pair and MACCS key FPs that were discussed in some detail in Sect. 1.2.1.1. Of the latter two, both

[12] Note that the coefficient $(n-1)^{-1}$ would, if ignored, merely scale the eigenvalues by $n-1$; the eigenvectors are unaffected.

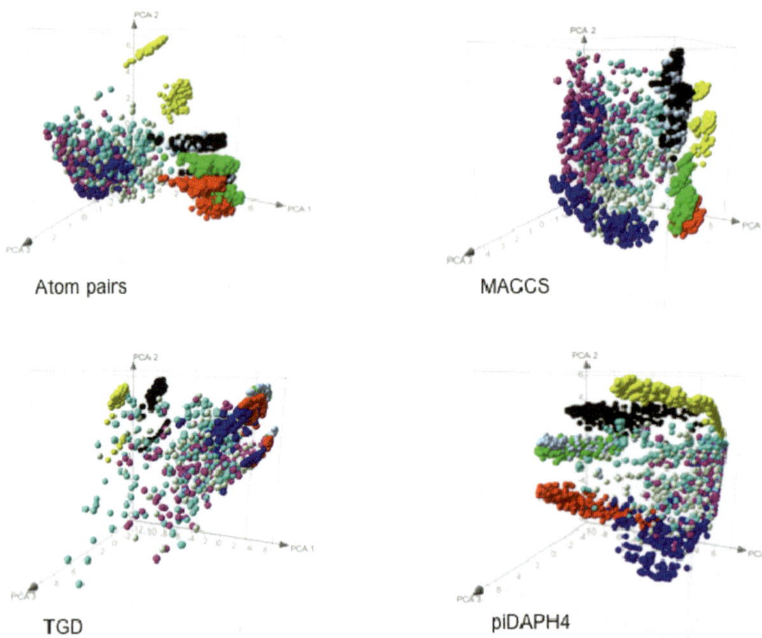

Fig. 1.9 Depictions of CSs generated from Tanimoto similarity coefficients computed with respect to binary FPs associated with four different types of descriptors—APF, MACCS key, TGD, and piDAPH4. (Adapted from Medina-Franco & Maggiora, Molecular Similarity Analysis [10])

of which are available in molecular operating environment (MOE) [152], TGD FPs are similar to those in atom pairs, while the piDAPH4 are related to FPs whose components are 3-D pharmacophores [153]. Hence, in contrast to the first three, piDAPH4 FPs contain some 3-D structural and stereochemical information. The Tanimoto similarity coefficient given in Eq. (1.8) was used to compute the similarity value in all four cases.

A total of 2250 molecules comprising nine classes of 250 molecules each were considered. The molecules in each class are color coded as follows: approved drugs (cyan), natural products (light green), a general screening collection from two vendors (magenta), compounds targeted to adenosine receptors (blue), and five in-house combinatorial libraries from the Torrey Pines Institute for Molecular Studies (depicted red, yellow, green, black, and light blue). The first three PCs account for 80.8, 85.9, 90.3, and 73.0% of the total variance in the data associated with the atom pair, MACCS key, TGD, and piDAPH4 FPs, respectively.

Although some of the variance in the data is not accounted for in the 3-D plots, a significant portion of it is. Hence, it is possible to draw some conclusions, albeit qualitative ones, from the distribution of compounds associated with the four different FP representations. It is quite obvious from the figure that the four different FP representations lead to dramatically different graphical portrayals of the CS distributions of the same set of compounds, a not unexpected but visually dramatic example of the non-invariance of similarity measures and its consequences. Interestingly, in some cases, substantial differences arise even within individual compound classes, as shown, for example, by the class of approved drugs colored in cyan. Of even greater interest is the graphical depiction in Fig. 1.10 of the distribution of the same set of compounds with respect to similarity fusion based on the mean of the similarity values (see Table 1.3). The results depicted in Fig. 1.10 differ significantly from any of those depicted in Fig. 1.9, which are based on the values of individual, "unfused" similarity measures obtained with respect to four different binary FPs.

It is important to note that graphical depictions described in this section are meant primarily as a means for enhancing intuition about the relationships among molecules in CSs. If quantitative analyses are required, detailed computations can be carried out using the full, multidimensional representation of the molecules in a dataset, as noted earlier.

1.3.2.2 Non-Euclidean Coordinate-Based CSs

The fact that CSs must have fewer than four dimensions for their graphical depiction is obvious. A less-well-known and much more subtle point is that high-dimensional data, in general, and CSs, in particular, may lie on lower-dimensional curved (i.e.,

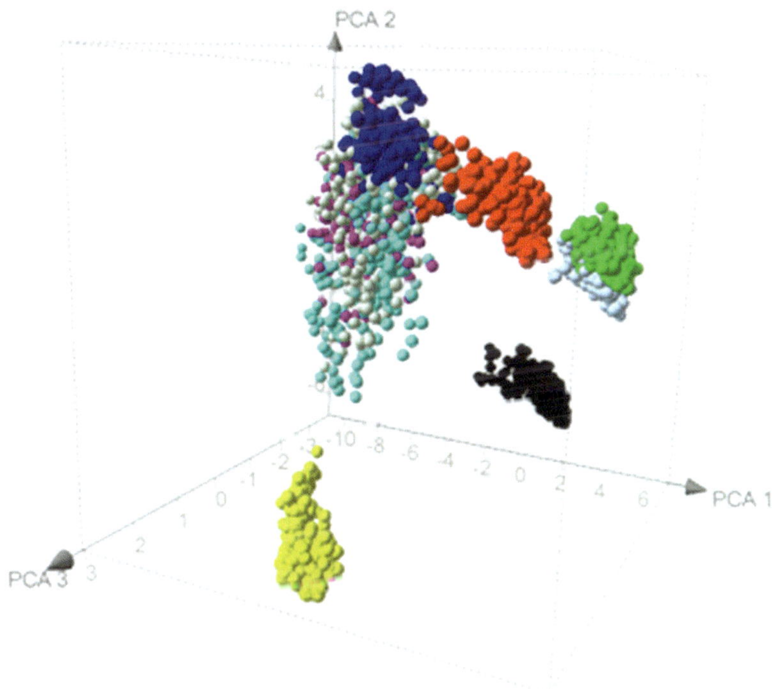

Fig. 1.10 Depiction of a CS generated from mean fusion of similarity values obtained from Tanimoto similarity coefficients computed with respect to binary FPs associated with APF, MACCS key, TGD, piDAPH4 descriptors. (Adapted from Medina-Franco & Maggiora, Molecular Similarity Analysis [10])

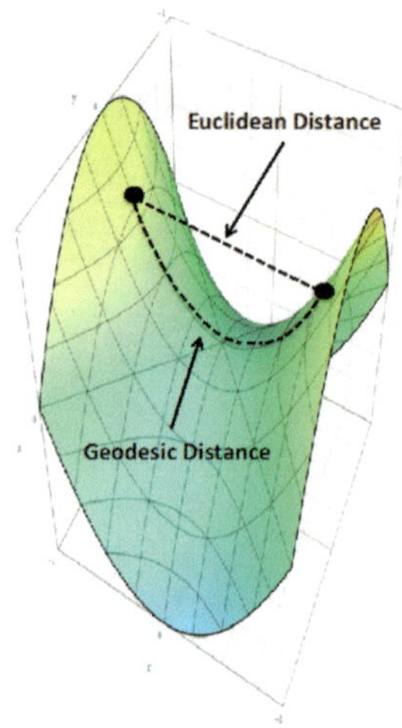

Fig. 1.11 Example of a Euclidean distance and the corresponding geodesic distance of two compounds in a model CS. The surface rendition is by Sam Derbyshire, http://creativecommons.org/licenses/by-sa/3.0

non-Euclidean) manifolds that are embedded in higher-dimension Euclidean spaces (*vide supra*). A simple example is given in Fig. 1.11, which depicts a 2-D hyperbolic manifold embedded in a 3-D Euclidean space. The important point is that the distance between points A and B depends on the space in which the distance is being evaluated. In the example, the Euclidean ("straight line") distance is clearly less than the geodesic distance measured along curved surface of the 2-D manifold. Thus, molecules A and B are judged more similar if considered in Euclidean CS than if their similarity was assessed on the 2-D manifold defined by the hyperbolic surface depicted in Fig. 1.11 that more accurately represents the data (in this toy example).

Figure 1.12 provides a more "down to Earth" example that clearly illustrates the difference between the two distance measures. In this case, the Euclidean distance is given approximately by the air miles between the American cities of Seattle, Washington and Miami, Florida which is about 2730 miles. By contrast, the geodesic distance between these two cities, measured along the US highway system is about 3300 miles, which represents about a 20 % increase in miles by car.

The paper by Agrafiotis and Xu provides a number of examples illustrating geodesic distances [154]. Although these authors published two more papers on this subject [155, 156] very little else has been published in the chemical information literature. This is obviously an important area of future research since it is one of several factors that can significantly influence the computed values of CS distances and, hence, the inferences that can be made about the compounds in a CS.

Lastly, it is well to point out that similarity values that lie on the unit interval [0,1] of the real line can be obtained by transforming Euclidean distances, d, or non-Euclidean geodesic distances, $\hat{d}$, using any one of a number of different mathematical expressions, one possibility being

$$s_{i,j} = 1/[1 + \eta \widehat{d_{i,j}}], \tag{1.51}$$

Fig. 1.12 Car (*blue-grey*) vs. air (*red*) routes from Seattle, Washington to Miami, Florida. (Adapted from Google Maps)

where the parameter $\eta > 0$ controls the rate at which the similarity value changes as a function of distance.

1.3.3 Cell-Based CSs

Cell-based partitionings of CSs [76, 157] are identical to partitions of mathematical spaces into families of nonintersecting subsets that cover the spaces. Thus, the set of N_{cells} cells that constitutes a cell-based CS is given by:

$$\mathrm{C}_{\text{cells}} = \left\{\mathrm{C}_1, \mathrm{C}_2, \ldots, \mathrm{C}_i, \ldots, \mathrm{C}_{N_{\text{cells}}}\right\} \tag{1.52}$$

and satisfies

$$\begin{aligned} &\mathrm{C}_i \cap \mathrm{C}_j = \varnothing, 1 \le i, j(< i) \le N_{\text{cells}} \quad \text{(Non-Intersecting cells)} \\ &\bigcup_{i=1}^{N_{\text{cells}}} \mathrm{C}_i = \mathrm{C}_{\text{cells}} \text{ (Set Cover)} \end{aligned} \tag{1.53}$$

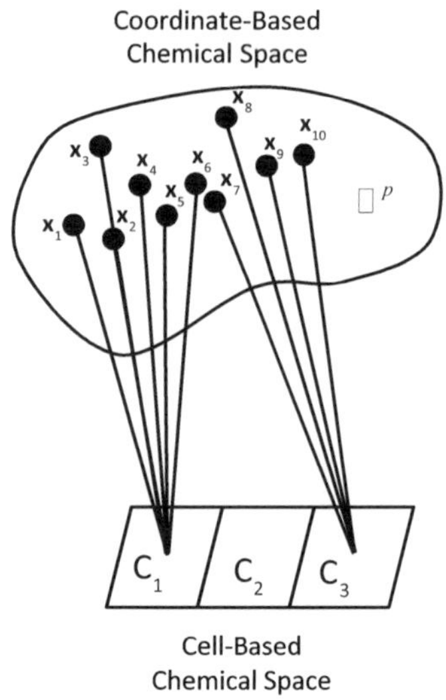

Fig. 1.13 Schematic depiction of a many-to-one set-valued mapping. Note that most cells are not generally occupied in cell-based CSs (see text for additional details)

Each cell corresponds to an *equivalence class*, and the molecules within it are hence, in some fashion at least, equivalent. The *many-to-one set-valued mapping*[13] depicted in Fig. 1.13 takes molecules in a p-dimension coordinate-based CS to one of the cells of the corresponding cell-based space, i.e.,

$$\Phi : \mathbb{R}^p \rightarrow C_{\text{cells}} \tag{1.54}$$

Thus, the location of compounds in cell-based CSs is given in two ways, namely, by their coordinates in the underlying coordinate-based CS, and by the address of the cell in which they reside. Figure 1.13 also shows that some cells in cell-based spaces are empty since only 15–20% of the cells in cell-based CSs are typically occupied. It is also interesting to note that cell-based CSs are very similar to the multi-way contingency tables used in many statistical applications [158], except for the fact that contingency tables rarely have cells with zero values.[14]

The procedure for constructing virtually all cell-based CSs is basically a two-step process:

- Generation of an appropriate low-dimensional coordinate-based CS
- Binning each of the axes of that space in such a way that the occupancy of the bins optimally covers the CS

The first and perhaps most important step in the process is the selection of suitable sets of reference compounds and descriptors, since they both play major roles in

[13] In function notation, the mapping in Eq. (1.54) is given by $\Phi(\mathbf{x}_i) = C_k, i = 1, 2, \ldots, n; k = 1, 2, \ldots, N_{\text{cells}}$.

[14] Note that there are a number of "correction factors," such as the well-known Laplace correction, that can be applied to the cells of a contingency table to correct for empty cells.

determining the nature of the CSs ultimately generated. While it is well appreciated that descriptor selection is important, the role played by the reference set of compounds is perhaps less well appreciated but is nonetheless crucial to the final form of the CS generated. Potential compound sets include corporate compound collections, publically available collections [25] such as ChEMBL [19], PubChem [20], ChemDB [21], and DrugBank [22], or sets of compounds suited to some specific tasks. In the latter case, for example, if the goal is to compare two large sets of compounds, it is desirable to combine the sets since the resulting CS will be more "balanced" and, hence, will take better account of the influence that molecular features missing in one of the two collections may have on the overall representation of the resulting CS. Alternatively, if the goal is to generate diverse subsets for an HTS campaign, the corporate compound collection from which the sample will be drawn, may be the best choice. These are just two of the many possibilities that can be considered, some of which will be presented in the sequel.

The second step in the process involves binning each axis of the coordinate-based CS yielding a total number of cells given by

$$N_{\text{cells}} = N_{\text{bins}_1} \times N_{\text{bins}_2} \times \cdots \times N_{\text{bins}_p} \tag{1.55}$$

As an example, consider a typical 6-D coordinate-based CS with seven bins per axes, which will generate a cell-based CS containing 117,649 cells. Although bins generally are of equal size on each axis, this is not required as discussed by Bayley and Willett [159]. Choosing an appropriate number of bins per axis is also important: If the number is too large, numerous cells will be unoccupied—normally a number of "occupied" cells around 15–20% appears to be reasonable. In this regard, it is important to note that in many types of cell-based analyses, including the above, the specific number of compounds in a given cell is not enumerated, only if the cell is occupied by at least some number of compounds (usually one) called the *cell occupancy threshold value*.[15]

Lastly, while cell-based CSs used in cheminformatic studies are generally partitioned into hypercubes, other possibilities exist that may offer more effective ways to partition these spaces. Rush [160] has mathematically explored some of the possibilities, but practical applications in chemical informatics have not to my knowledge been carried out to date.

Figure 1.7b portrays a model cell-based CS for the same set of compounds depicted in Fig. 1.7a. Although this example is oversimplified, cell-based CSs, nevertheless, are typically around 3-D to 6-D. Cpd-1, the active compound indicated by the red dot, its nearest-neighbor Cpd 2 indicated by the green dot, and two of its next nearest neighbors, Cpd-4, and Cpd-5 indicated by the blue dots, all reside within the same cell. Hence, from a cell-based perspective, all four compounds are considered to be roughly equivalent. On the other hand, Cpd-3, which is nearer to Cpd-1 than either Cpd-4 or Cpd-5, resides in a neighboring cell, and thus, from a

[15] A similar situation exists in the case of threshold graphs obtained from labeled graphs when the edge values exceed some threshold value. Details of this are described in Sect. 1.3.7 on graph-based CSs.

cell-based perspective, is not considered to be equivalent to any of the compounds in the neighboring cell. This illustrates one of the limitations of the cell-based approach, which does not *explicitly* employ the concept of nearest neighbor cells, although the position of compounds in the underlying coordinate-based CS does afford the possibility for identifying nearest neighbors.

Clustering provides an additional way to partition CSs into a set of nonintersecting subsets that cover the space [161]. Although clustering methods have some advantages over cell-based partitioning, they are difficult to apply to datasets as large as those that can be handled relatively easily using a cell-based approach. For example, the addition of large numbers of new molecules can significantly alter clusterings. This is not a problem in the cell-based case since the CS partitioning scheme is effectively compound independent—adding new compounds does not change the partitioning scheme. Moreover, many methods such as k-means clustering require specification of the number of clusters and hierarchical methods produce similarity (or distance)-dependent clusterings [161]. Lastly, because the clustering methods are a vast subject, even when only considered with respect to cheminformatics applications, no further discussion on this topic is provided in this work.

1.3.3.1 Representations of Cell-Based CSs

The BCUT descriptors described in Sect. 1.2.2.1 have proved to be a popular choice for directly constructing low-dimensional CSs. There are, of course, many other types of suitable descriptors that, in many cases, cannot be used directly since they lead to spaces whose dimension are too high. This can be ameliorated, as discussed by Xue, Stahura, and Bajorath [157], using a dimensionality reduction technique such as PCA.

The power of the cell-based description lies in its ability to simplify the representation of CS, and thus to enhance the speed at which a number of the tasks, such as compound acquisition [162], diversity analysis [163], comparison of compound collections [77], and LBVS [164] can be performed. But the enhanced speed comes at a cost, which may or may not, significantly impact the results obtained. As discussed above, the cell-based partitioning leads to a coarse-grained representation of CS and, importantly, can introduce significant effects at cell boundaries. For example, molecules located near a common boundary in adjacent cells are generally more similar to each other than to many other molecules in their own cells (cf. Figs. 1.7 and 1.13). Obviously, this can lead to significant bias depending on the actual (not cell based) distribution of compounds in the CS, a problem that is also encountered in a number of clustering methods.

1.3.3.2 Example of Cell-Based CSs

The CS was constructed by combining the four compound collections given in Table 1.4 into a single, large collection. Determining the optimal set of 3-D BCUT descriptors for that augmented collection yielded a 6-D CS upon which all subsequent analysis is based. Each axis was then partitioned into seven bins, giving a total of 117,649 cells in the 6-D space.

The difference between the Diverse and Combi collections depicted graphically in Fig. 1.8 is verified. Several key features in the table supporting this conclusion are the comparative number of occupied cells (18,731 and 2434, respectively) and the average cell occupancies (9.4 and 61.5, respectively), all of which clearly point to the more restricted and dense distribution of compounds in Combi compared to that in Diverse. The MDDR collection exhibits similar behavior to that of Diverse, although the absolute values of the cell-based parameters are somewhat lower than those of Diverse, which is not surprising given that Diverse is nearly twice as large as MDDR. Micros is a small, diverse collection of known drugs and related substances. Given its size, it nonetheless is relatively diverse since only slightly more than one compound on an average occupies each of the 516 occupied cells. On the other hand, its 516 cells occupied cells are almost insignificant when compared to the 18,371 occupied cells in Diverse. Moreover, each occupied cell in Micros contains on an average only 1.3 compounds, which again pales in comparison to Diverse's average cell occupancy of 15.6.

These data illustrate two important points about diversity. First, small compound collections, which may be relatively diverse with respect to their own set of compounds, may not in an absolute sense contain anywhere near the diversity that can *potentially* be obtained from much larger compound collections. Second, while diversity may confer some advantage in identifying active compounds in HTS campaigns, if the diversity is sparsely distributed the chance of identifying actives is significantly diminished even if the diversity is widespread in a large compound collection. This follows from the fact that in a given assay the percentage of actives within "active regions" of CS is still surprisingly small, generally around 10–15% or less.

The cell-based CS data summarized in Table 1.4, while helpful, are not sufficiently detailed to address more specific questions regarding the similarity or difference between different compound collections. This is remedied in Sect. 1.3.5.1 where details for comparing compound collections are described.

Table 1.4 Summary of compound collections in six-dimensional 3-D BCUT chemical space with seven bins per axis (total cell count = 117,649)

Compound collection	Number of compounds	Number of occupied cells	Percent occupied cells	Average cell occupancy	Largest cell population
Diverse[a]	173,375	18,371	15.6	9.4	738
Combi[b]	154,474	2434	2.1	61.5	5694
MDDR[c]	97,409	10,203	8.7	8.5	349
Micros[d]	799	516	0.4	1.3	7

[a] Subset of diverse compound collection (see text)

[b] Combinatorial chemistry library (see text)

[c] Subset of MDDR collection—Molecular Drug Data Report (MDDR), Version 2005.2; Symyx Software. San Ramon, CA, 2005

[d] Small discovery oriented library—MicroSource Discovery Systems, Inc., Gaylordsville, CT 06755

1.3.4 Chemical Space Networks

In addition to the coordinate and cell-based representations just described, CSs can also be represented by *mathematical graphs*. Such graphs provide information that is comparable to that provided by similarity, dissimilarity, or distance matrices and, as will be seen in the sequel, afford an intuitive as well as solid conceptual basis for analyzing many relationships among the compounds populating CSs. Since compound collections can be quite large, their corresponding graphs are also quite large and generally fall under the rubric of "Networks." The development and application of network theory, which has burgeoned over the two decades, has been applied in numerous fields, including social science, physics, computing, biology, and medicine. A number of "chemically oriented" examples have been reported (see e.g., [123–126, 165–167]), and five papers describing the application of networks to the analysis of compound collections have been published [168–172]. An investigation that examines power laws in chemical systems, as do several of the just cited publications, has also been published. However, it does not directly address issues related to similarity-based networks that describe compound collections [173].

The present section provides a number of examples that elucidate the underlying features of networks such as their patterns of vertex connectivity. An understanding of these feature patterns is required in order to comprehend the nature of the large, complex networks such as those needed to represent CSs; because these networks are large, their feature patterns are usually analyzed in statistical terms. An important aspect of the network representation of CSs is that it facilitates navigation of those spaces since there are powerful graph-based network algorithms for determining paths between vertices [129] in contrast to the situation in more traditionally represented CSs [174].

In order to facilitate understanding of networks, a number of simple examples based on the graphs depicted in Figs. 1.7c and 1.14 are presented in the following sections. These examples, though simple, illustrate a number of the most important network features needed to interpret the statistical data and to understand the nature of the CSs being analyzed.

1.3.4.1 Simple Example of a CS Network

As an illustration of the basic features of graphs, consider the *reflexive, labeled graph* $\hat{\mathcal{G}}$ depicted in Fig. 1.7c that represents the similarity relations among "hypothetical" compounds 1–5 depicted in Fig. 1.7a, b. A compound identifier, which is a number in the present case, labels each vertex and a similarity value labels each edge of $\hat{\mathcal{G}}$. Since the vertices represent distinct molecules they are distinguishable, a feature that influences the statistical mechanical features of networks (*vide infra*) [175]. As noted earlier, the graph is reflexive because each vertex has an associated graph loop labeled by the value of the self-similarity[16] of the molecule that

[16] Self-similarity is the similarity of the molecule with itself, and thus, its value is always unity. Graphs without self-loops and multiple edges between vertices are also called simple graphs.

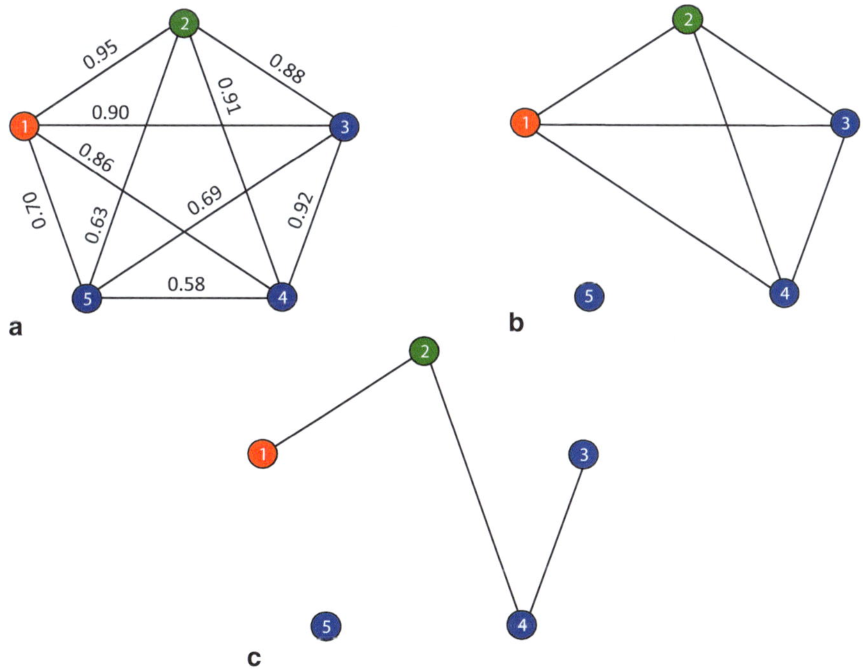

Fig. 1.14 Other CSNs related to that depicted in Fig. 1.7c: **a** simple, complete CSN, **b** threshold CSN ($S_t > 0.85$); the CSN linking compounds 1–4 is a complete *subgraph/network* called a *clique*, and **c** threshold CSN ($S_t > 0.90$); while compounds 1–4 are still linked they no longer form a clique

corresponds to that vertex. In most practical implementations, edges corresponding to self-similarities are omitted for clarity (*vide infra*). Since similarity coefficients are generally symmetric, i.e., $S(i, j) = S(j, i)$, the edges of the corresponding graph do not have directionality. Hence, the networks typically employed can be classified as *undirected, unlabeled*, and *simple networks*.

There are, however, cases when the use of directed graphs may be desirable as in the representation of activity cliffs [112] or where asymmetric similarity coefficients such as those given in Eqs. (1.6) and (1.11)–(1.13) are employed. Graphs where each vertex is connected to every other vertex connected are called *complete*. Thus, a complete graph with *n* vertices has $n(n-1)/2$ edges, and each vertex has $n-1$ edges called its vertex degree.

The similarity matrix given in Eq. (1.56) contains the same information as $\hat{\mathcal{G}}$ in Fig. 1.7c.

$$\mathbf{S} = \begin{bmatrix} 1.00 & 0.95 & 0.90 & 0.86 & 0.70 \\ 0.95 & 1.00 & 0.88 & 0.91 & 0.63 \\ 0.90 & 0.88 & 1.00 & 0.92 & 0.69 \\ 0.86 & 0.91 & 0.92 & 1.00 & 0.58 \\ 0.70 & 0.63 & 0.69 & 0.58 & 1.00 \end{bmatrix} \tag{1.56}$$

Hence, the similarity matrix provides a means for treating graphs algebraically [176]. For example, the eigenvalues associated with the matrix representations characterize a variety of graph invariants that have seen many useful applications in chemical graph theory [16], and although they have not yet been applied extensively in the study of CSs, they, nonetheless, have the potential to provide new and interesting insights in graph-based CSs.

The example in Fig. 1.7c is, of course, a great simplification of "real" CSs that may contain millions of vertices each corresponding to a specific molecule and billions of edges linking the pairs of vertices each labeled by an appropriate similarity, dissimilarity, or distance value. In this work, the networks are called "CS networks" (CSNs) to emphasize their relationship to CSs. Hence, the graph in Fig. 1.7c can be described as a *complete-reflexive-labeled* CSN. The reflexive character of the graph is captured by the values of diagonal elements of similarity matrix, $S(i,i)=1, i=1,2,\ldots,n$. Since the self-similarities do not add any new information since they are all the same and of value 1.00, graph loops are routinely omitted yielding the simple graph $\widehat{\mathcal{G}}$, as illustrated in Fig. 1.14a. Such networks will be called *complete* CSNs since each vertex is connected to every other vertex except itself as the graph loops have been removed.

Because CSs are so large, their graphical display as CSNs can become visually "noisy" and difficult to comprehend for all but the smallest sets of compounds. Nevertheless, as in the case of the coordinate-based portrayal of CSs, the graphical depictions are only meant to provide an intuitive feel for the underlying relationships associated with the CSN of a large compound collection. Alternative ways exist, however, for characterizing and handling the information contained in CSNs. Because matrices can provide faithful representations of graphs and networks, this affords the possibility that many powerful algebraic techniques can be applied to their analysis [177]. Algorithmic techniques, some but not all of which are based on the properties of graph matrices, have provided numerous other ways for analyzing the properties of graphs and networks. However, because of their size and complexity, information on the characteristic features of networks obtained using these methods is commonly reported in terms of the statistical properties of the features, as will be described in Sect. 1.3.5.1 [129, 178].

All of the existing publications that describe applications of networks to CS analysis [168–172] do not use labeled graphs or networks, but rather rely on simpler entities called *threshold graphs*, which are generated by keeping only those labeled edges whose values satisfy some threshold as illustrated in Fig. 1.14b, c. In the first case, shown in Fig. 1.14b, a similarity threshold value of $S_t > 0.85$ is used. Vertex 5 is now isolated from the vertices 1–4, which remain fully connected, and thus form a complete subgraph of the original graph called a *clique*. Figure 1.14c provides another example based on a higher threshold value of $S_t > 0.90$. Not surprisingly, fewer edges remain, and although vertices 1–4 are still connected, they no longer form a clique.

An important type of matrix that plays a role in many procedures designed to determine graph/network properties is the *adjacency matrix* of mathematical graphs and networks. The adjacency matrix corresponding to the CSN in Fig. 1.14b is given by

$$\mathbf{A}_{0.85} = \begin{pmatrix} \begin{bmatrix} 0 & 1 & 1 & 1 \\ 1 & 0 & 1 & 1 \\ 1 & 1 & 0 & 1 \\ 1 & 1 & 1 & 0 \end{bmatrix} & 0 \\ 0 & [0] \end{pmatrix}, \tag{1.57}$$

Where

$$a_{i,j} = \begin{cases} 1 \text{ if an edge exists between Cpd-}i\text{ and Cpd-}j \\ 0 \qquad\qquad \text{otherwise} \end{cases} \tag{1.58}$$

As noted above, the subset of compounds $\{\text{Cpd-1,Cpd-2,Cpd-3,Cpd-4}\}$ forms a complete subgraph of the threshold graph called a *clique,* i.e., $\mathcal{H}_{0.85} \subset \mathcal{G}_{0.85}$. Thus, the four compounds are all linked in the threshold CSN, while Cpd-5 is an isolated vertex as reflected by the block diagonal structure of the adjacency matrix in Eq. (1.57). Because of the block diagonal structure, each block can be treated independently of the others, a form of dimensionality reduction.

If the threshold is raised, to say $S_t > 0.90$, the subset of compounds remains linked, but the subgraph induced by the higher threshold $\mathcal{H}_{0.90}$ no longer forms a clique and $\mathcal{H}_{0.90} \subset \mathcal{H}_{0.85}$. Cpd-5, of course, remains an isolated node. In this case, the adjacency matrix simplifies to

$$\mathbf{A}_{0.90} = \begin{pmatrix} \begin{bmatrix} 0 & 1 & 0 & 0 \\ 1 & 0 & 0 & 1 \\ 0 & 0 & 0 & 1 \\ 0 & 1 & 1 & 0 \end{bmatrix} & 0 \\ 0 & [0] \end{pmatrix} \tag{1.59}$$

Although the block diagonal structure remains, the main 4×4 block is simpler (i.e., has fewer nonzero elements) than that in Eq. (1.57). In any case, whether a graph-based or matrix-based representation is used, threshold CSNs provide a comprehensive representation of the global "pathways" that connect compounds with respect to a given threshold similarity value. As an example, it is possible to determine the minimum number of edges that must be traversed to go from any given compound to another compound given that the similarities of compounds along the pathway exceeds the similarity threshold value, a feature that can be useful in large screening campaigns but is difficult to carry out in coordinate or cell-based CSs.

As will be seen in the sequel, statistical analyses also play a major role in assessing the characteristic features of networks [129, 177, 178]. In addition, algorithms for treating very large systems such as the Internet as networks has given rise to the development of many powerful methods for handling mega-networks [179]. Thus, representing CSs as CSNs has some distinct advantages as is seen below.

1.3.4.2 Statistical Aspects of CSNs

Vertex Degrees and Degree Distributions Because of their extremely large sizes and complexities, networks are typically characterized in terms of the statistical properties of their vertices and the relationships among subsets of them. One of the most important features of networks illustrated by the simple examples below is *vertex degree*—the number of edges incident on a vertex. [17] The distribution of vertex degrees for large random networks follows a *Poisson distribution* [129] that for networks with very large numbers of vertices becomes

$$\Pr(k) = e^{-\bar{k}}\left(\frac{\bar{k}^k}{k!}\right) \tag{1.60}$$

where k is the degree of a randomly chosen vertex and $\bar{k}$ is the mean vertex degree of a large random network. Although it remains finite, for large values of $\bar{k}$ $\Pr(k)$ approaches a normal distribution.

It will be seen in the sequel that such networks do not describe typical CSNs. As illustrated in Fig. 1.14a, b, the degree of each vertex in a complete graph is given by $k_i = n-1, \quad i = 1,2,\ldots,n$, where n is the number of vertices in the complete graph; $n = 5$ in the current example. In Fig. 1.14a, $k_i = 5-1 = 4$, while for the complete subgraph $\mathcal{H}_{0.85}$ in Fig. 1.14b, $k_i = 4-1 = 3, \quad i = 1,\ldots,4$, while the vertex degree of the isolated vertex is, of course, zero. In larger, more complex networks, vertex degrees are typically given by statistical distributions as illustrated by the simple example in Fig. 1.14c, where

$$\begin{aligned} k_5 &= 0 \\ k_1 &= k_3 = 1 \\ k_2 &= k_4 = 2 \end{aligned} \tag{1.61}$$

The degree distribution is the probability a given vertex has k incident edges, i.e.,

$$\Pr(k) = \frac{\sum_{i \in k_i = k} k_i}{\sum_{l=1}^{5} k_l}, k = 1,\ldots,5 \tag{1.62}$$

where the term in the numerator is a sum over all vertices of equal degree, and the values corresponding to the example in Fig. 1.14c are

$$\begin{aligned} \Pr(k=0) &= 1/6 \\ \Pr(k=1) &= 2/6 \\ \Pr(k=2) &= 4/6 \end{aligned} \tag{1.63}$$

[17] Although it is not addressed here, the vertex degree of directed graphs/networks can be handled by assessing the "in-degree" and "out-degree" of a vertex that corresponds, respectively, to the number of edges directed towards the vertex and the number directed away from the vertex.

Degree Correlations: Assortativity Coefficients Degree correlations, also called assortativity coefficients, provide a measure of the correlation of vertex degrees between pairs of directly connected vertices. It is obvious from Fig. 1.14a, b that degree correlations for vertices in complete graphs or subgraphs are unity since all vertices in these graphs have identical vertex degrees and hence are maximally correlated. However, the situation in Fig. 1.14c is more complex. The average vertex degree based on the values in Eq. (1.61) is $\bar{k} = \frac{1}{5}(0+1+1+2+2) = 1.2$ and the assortativity coefficients are given by a modified version of the Pearson correlation coefficient [180][18]:

$$\Delta(\mathcal{G}_{0.90}) \doteq \frac{\sum_{i=1}^{5}\sum_{j=i+1}^{5} \mathrm{A}_{0.90}(i,j)\cdot(k_i - \bar{k})\cdot(k_j - \bar{k})}{\sum_{i=1}^{n}(k_i - \bar{k})^2} \tag{1.64}$$

where $\mathcal{G}_{0.90}$ is the threshold graph of $\mathcal{G}$ with respect to a similarity threshold value of 0.90, and $\mathrm{A}_{0.90}(i,j)$ is the *i, j*th element of the adjacency matrix corresponding to that threshold graph. Because of the block structure of the adjacency matrix in Eq. (1.59) only, the vertices corresponding to Cpd-1 through Cpd-4 need be considered in Eq. (1.64).

Carrying out the computation yields a value for the degree correlation of

$$\Delta(\mathcal{G}_{0.90}) = 0.24.$$

Transitivity: Mean Clustering Coefficient Another coefficient of interest is the transitivity or mean clustering coefficient, $\bar{C}(k)$, of all vertices with *k* edges, which can be computed according to:

$$\bar{C}(k) = \frac{1}{N_k}\sum_{i=1}^{N_k} C_i(k) \tag{1.65}$$

where N_k is the number of vertices with *k* edges and $C_i(k)$ is the *local clustering coefficient*

$$C_i(k) = \frac{\varepsilon_i}{\frac{1}{2}k(k-1)} \tag{1.66}$$

with ε_i being the number of edges connecting the *k* neighbors of the *i*th vertex to each other and $\frac{1}{2}k(k-1) = \binom{k}{2}$ is the number of unique pairs of neighbors. Thus, the local clustering coefficient is the ratio of the number of edges connecting the *k* neighbors with each other divided by the total number of *possible* edges among the set of *k* neighbors.

It is clear from Fig. 1.14c that the transitivity in all cases is zero. By contrast, the transitivity of the complete graph in Fig. 1.14a is unity since each vertex has an

[18] Note that the summations are over all unique pairs of vertices (i.e., molecules) and that the coefficient cancels out of the numerator and denominator of Eq. (1.64).

identical number of edges and the vertices connected to that vertex are fully connected with each other, hence, $C_i(k) = 4 \cdot 3 / \left[\frac{1}{2}4(4-1)\right] = 1$, which when substituted into Eq. (1.65) gives $\bar{C}(4) = 1$.

Shortest (Geodesic) Path Lengths/Distances In general, a path between vertices can be quite complex as it can include vertices or edges that have been traversed previously. Here, a special kind of path called a *shortest path* is considered. Such paths, also called geodesic paths, are the shortest distance between two vertices based on a count of the number of unlabeled edges in the path. They are not necessarily unique since several paths of equal length may exist in the same graph or network. Shortest path values are entirely equivalent to *graph distances,* $d_{i,j}$, and hence satisfy the well-known distance axioms [177]. A number of algorithms that exist for determining shortest paths have been clearly described in Newman's book [129].

Mathematically, the mean geodesic distance between all unique pairs of vertices is given by

$$\bar{L} = \frac{1}{\frac{1}{2}n(n-1)} \sum_{i=1}^{n} \sum_{j=i}^{n} d_{i,j} \tag{1.67}$$

As can be seen in Fig. 1.14b, the shortest (geodesic) path between two vertices of a complete, unlabeled graph is unity in all cases. This is not the case for the threshold graph in Fig. 1.14c. Computing shortest path lengths in this case is simple since a single path connects the four vertices. Hence, for example, the shortest path between vertex-1 and vertex-4 is of length two and that between vertex-1 and vertex-3 is three. The corresponding mean shortest (geodesic) path length is, from Eq. (1.67),

$$\begin{aligned} \bar{L}(\mathcal{H}_{0.90}) &= \frac{1}{\frac{1}{2}n(n-1)}\left[d_{1,2} + d_{1,3} + d_{1,4} + d_{2,3} + d_{2,4} + d_{3,4}\right] \\ &= \frac{1}{\frac{1}{2}4(4-1)}[1+3+2+2+1+1] = \frac{1}{6}[10] \\ &= 1.67 \end{aligned} \tag{1.68}$$

Another feature of shortest (geodesic) paths is of note, namely, they are self-avoiding, as they do not cross themselves. If they did a loop would be formed that could be removed without interrupting the traversal of the path between the specified vertices. Determining shortest paths can be a challenge for large networks, but as noted above, robust path algorithms exist for mega-networks such as the Internet, so dealing with CSs while challenging is not out of the realm of possibility.

Small World Effect The small world effect, namely, that the mean geodesic distance between the vertices in networks defined by Eq. (1.69) is proportional to log n, and thus, is generally small for a number of real-world networks (see e.g., Table 8.1 in [129]). A common feature of many small-world and random networks is that their vertex degree distributions tend to be homogeneous with a peak at the mean value of the distribution and an exponential decay, $\Pr(k) \sim \exp(-k)$, in its tail, giving rise to what are called *exponential networks*. Interestingly, there are a number of types of small world networks including ones discussed below that also exhibit scale-free behavior (*vide infra*) [181].

One consequence of the small world effect is the famous "six degrees of separation" hypothesis, namely, that everyone on Earth is separated by no more than five individuals (vertices) and hence six links (edges). That this is not an entirely unreasonable hypothesis is based on the following overly simplistic argument. Suppose I have 100 friends each of which has 100 friends, each of which has 100 friends, etc. Thus, with only one degree of separation I can connect to 100 individuals, with two degrees I can connect to $100 \times 100 = 10,000$ individuals, and with only three degrees of separation I can connect to $100 \times 100 \times 100 = 1,000,000$ individuals. If all six degrees of separation are considered, I could potentially connect to one trillion individuals, 50 times more than required to connect to everyone on Earth. Although, as pointed out by Watts [127] this argument has significant practical flaws, it nonetheless captures some essential features of small-world networks.

Networks exhibiting small-world behavior, hence, can facilitate many processes such as communication, the spread of disease, and the speed of inter-server access on the Internet. Not surprisingly, as will be discussed in Sect. 1.3.5.2, CSNs tend to exhibit small world behavior as well. This is not surprising given the nature of molecular and chemical similarity, which in general does not exhibit transitive behavior. i.e., if A is similar to B and B is similar to C, it does not in all cases follow that A is similar to C. This same phenomenon exists in social networks as well, i.e., if A knows B and B knows C it does not mean that A and C also know each other, although the likelihood that they do is higher than random chance. As discussed by Newman [129], transitivity is related to various forms of clustering coefficients.

Scale-Free Networks The vertex degree distributions of scale-free networks differ from those of large random networks and many small world networks, which are Poisson distributed (*vide supra*). By contrast, scale-free networks described by Barabási and Albert [182] are nonhomogeneously distributed and follow power laws, such that the probability that a random vertex has degree *k* [19] is inversely related to a power of vertex degree, i.e.,

$$\Pr(k) = \kappa \cdot \left(\frac{1}{k}\right)^{\alpha} = \kappa \cdot k^{-\alpha} \tag{1.69}$$

where κ is a constant and the exponent $\alpha > 1$ is a *scaling coefficient,* which usually lies in the range $2 \le \alpha \le 3$ for many real-world networks (see e.g., Table 0.1 in [129]). Van Steen gives a clear description of why the power law given by Eq. (1.69) is scale free [180]. In addition, the mean shortest path length of scale-free networks is proportional to log log *n*, a value that is much less than the log *n* behavior noted above for many small world networks.

Two important properties of scale-free distributions are that they do not have peaks and they decay at much slower rates than the corresponding Poisson and

[19] Note that this can also be interpreted as the fraction of vertices of degree k.

normal distributions. The second property is especially important because it indicates a higher probability that more extreme events may occur than can occur in the latter distributions. In this regard, an important example in the case of scale-free networks is the presence of vertices with exceptionally high vertex degrees, a situation that gives rise to highly connected "hubs" interconnected by relatively small numbers of edges, a rather extreme form of small world behavior to say the least.

Because of its form, depicting Eq. (1.69) as a $\log \Pr(k)$versus$\log k$ plot should result in a straight line if the distribution does follow a power law, at least asymptotically. Proving that it does is not necessarily easy, since some values of k in the tail of the distribution may not satisfy the power law relationship. However, as pointed out by Newman among others [129], alternatives exist that provide a means for accomplishing this, although sometimes it requires removing some of vertex degrees that do not follow the power law.

1.3.4.3 Topologies of CSN

As noted earlier, five papers have been published that address various aspects of *similarity-based networks* of CSs [168–172], all of which differ from the related work on power laws in CSs by Benz et al. [173] that predates these papers. Both of the latter reports have presented evidence of the small world behavior of CSNs and in some cases scale-free behavior as well. Because the edges of the CSNs are unlabeled, threshold graphs were generated for different similarity threshold values. Not surprisingly, statistical features related to vertex degree tend to decrease as the similarity threshold is raised as is nicely illustrated in Table 1.2 of reference [169].

Although this behavior seems intuitive, it can be rationalized as follows. Due to the central limit theorem [183], the set of similarity values associated with large compound DBs is normally distributed with a mean around, say for example, 0.50. Now arrange the set of similarity values in *descending* order and determine the corresponding *cumulative probability distribution* depicted in Fig. 1.15, where the abscissa corresponds to the threshold similarity value for a given CSN, and the ordinate corresponds to the *fraction,* f_{edge}, of the $n(n-1)/2$ possible edges that can be drawn between the n compounds that constitute the vertices of the network. It is clear from the figure that for a threshold similarity value of 0.75 less than 10 % of the compounds will be connected directly. Even at a threshold similarity of 0.5 only about half the possible number of edges are present.[20] In order to gain a sense of the magnitude of the problem, consider a DB of only $n = 10{,}000$ compounds. In this case, the *complete* CSN would have ~50 million edges. However, even at a similarity threshold value of 0.75 about 8 % of the total possible edges (~4,000,000 edges) will be formed. As this is more than 400 times the minimal number of edges needed to connect all of the vertices with one another (~10,000), it is certainly sufficient to introduce significant and interesting structure in the CSN. Hence, it easy to see

[20] This argument is, of course, oversimplified since it depends on the width (standard deviation) of the probability distribution.

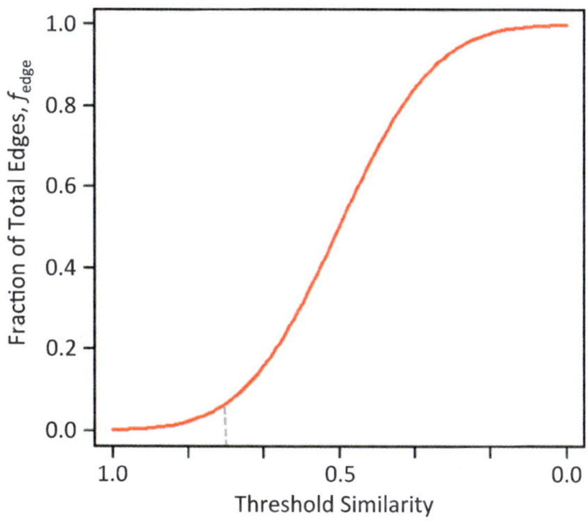

Fig. 1.15 Cumulative distribution curve showing the fraction of possible edges formed as a function of similarity threshold value. The light grey dashed line corresponds to a threshold similarity value of 0.75

that expanding to a DB of say 200,000 compounds can prove to be a challenging enterprise.

The paper by Tanaka et al. [168] investigates small world phenomena in several libraries obtained directly from the ZINC DB [184] and from virtual libraries constructed from structurally diverse fragments. By contrast, the paper of Krein and Sukumar [169] undertakes a much more comprehensive analysis based on a number of different sets of CS descriptors applied not only to CSs but also to their subspaces associated with activity cliffs. A recent paper from Bajorath's group [172] also addresses subnetworks associated with activity cliffs. Obviously, these analyses can be extended to other landscape features such as similarity cliffs (see Sect. 1.2.4).

The approximately scale-free nature of CSNs observed by Krein and Sukumar led them to infer the existence of hubs, highly interconnected regions of CSNs linked together by relatively sparse paths. Hubs represent regions of CS associated with different structural motifs. Hence, paths linking hubs may provide a means for addressing the problem of scaffold hopping, a process associated with the presence of similarity cliffs, which are more general since they include scaffold hops as a special case.

Another application of threshold CSNs is exemplified by the work of Bajorath's group on network-like similarity graphs (NSGs). NSGs are threshold graphs they developed as a means for analyzing the SARs of large, diverse sets of compounds Figure 1.16 provides an example of an NSG that characterizes the activities of a set of lipoxygenase inhibitors taken from the paper by Wawer et al. [170]. Compound potencies are color coded from red for the most active (1 nM) to green for the least active (100 μM). Links are drawn between compound pairs if their MACCS Tanimoto similarity exceeds 0.65. Additional annotation corresponds to SAR index scores (decimal values) associated with compound clusters. The index ranges from 0.00 to 1.00, the larger the value the more "discontinuous" a given compound cluster—activity cliffs correspond to high levels of discontinuity.

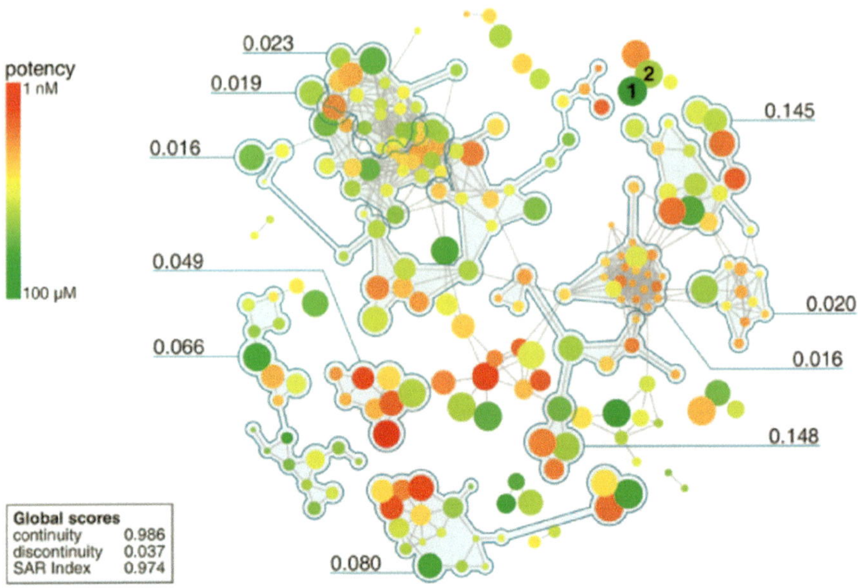

Fig. 1.16 Network-like similarity graph (NSG) depicting the CS and activity relationships of a set of lipoxygenase inhibitors taken from the work of Wawer et al. [170]. Compound potencies are color coded as shown by the colored bar on the upper left hand side of the figure, red being the most active and green being the least active. Compounds are connected by an edge if the MACCS Tanimoto similarity value of a given compound pair exceeds 0.65. The *decimal numbers* associated with clusters of compounds correspond to SAR Index scores (See text for additional details)

1.3.5 Exploring CSs

The concepts of structural similarity and CS, which are ubiquitous in medicinal chemistry, are finding a place in other chemically related sciences such as materials science and engineering [185]. A question that now arises is how can we develop procedures and algorithms that exploit these concepts to facilitate the discovery of new drugs and bioactive agents? Or, more appropriate to the book in which this chapter resides, how can these concepts be applied in food science and in aroma and flavor chemistry? Although the examples presented in this section do not represent a comprehensive set of the many possible methods that are available, they will at least provide a sample that should afford sufficient information to help answer this question.

1.3.5.1 Comparing Compound DBs

It is obvious from previous discussion in this chapter that compound DBs play an extremely important role in many aspects of chemical informatics. Thus, it is important that methods exist for assessing their similarities and differences. As has been noted by a number of investigators cell-based methods are particularly suited to this task.

For example, consider the compound DBs listed in Table 1.4 and discussed in Sect. 1.3.3.2. While the numerical values in the table provide a reasonable summary of the cell-based characteristics of each collection, they are not specific enough to afford a detailed *comparative* assessment, as they do not account for relationships between the cells in collections being compared. Pearlman and Smith [76] developed an approach that is able to address this deficiency, albeit only partially.

The procedure is as follows. First, a cell occupancy threshold is chosen; in the example discussed here, an occupancy value ≥ 1 is used, i.e., each occupied cell contains at least one compound. Obviously this is a potential source of error since an occupied cell in one collection could contain a single compound, while the corresponding cell in another collection could be occupied by, say, more than a 100 compounds. Hence, the Pearlman–Smith (P–S) procedure only compares patterns of occupancy, but this may be sufficient when very large compound collections of comparable size are being compared, or if only a coarse-grained estimate is required. Carrying out the analysis for a sequence of occupancy thresholds, e.g., $t_{occ} \geq 1, \geq 2, \geq 3, \ldots$, would provide a measure of the sensitivity of the results to the chosen occupancy threshold, but such an approach to my knowledge has not been carried out.

The P–S procedure can be viewed in a manner that is entirely equivalent to that described earlier for binary FPs since the set of cells in a cell-based CS can be thought of as one long FP. How the cell-based CS is unfolded into the linear array of cells is unimportant; what is important is that all equivalent cell-based CSs that are compared be unfolded in exactly the same way. Occupied cells are labeled with a "1" if they are occupied by at least one compound and by a "0" if they are unoccupied. Hence, any of the FP-based similarity coefficients can now be used to assess the similarity of any pair of compound collections or libraries described by the same cell-based CS. These "DB FPs" are on the order of 100,000 or more cells, and hence, many times larger than typical binary structural FPs that usually have less than 2000 elements. And, as seen in Table 1.4, only a small fraction of the cells are occupied so that these FPs are very sparse. The discussion in Sect. 1.2.1.1 shows that they can be handled using run-length encoding, or a similar procedure. Additional compression, such as is the case for some large molecular FPs, is not necessary in this case since the number of DBs being compared is many times smaller than the number of molecular FPs typically dealt with in similarity search-based activities.

The P–S procedure defines two measures for assessing the similarity of two compound DBs, nominally A and B, residing in the same CS:

$$\begin{aligned} &1.\text{Fraction of } A's \text{ cells occupied by } B \\ &2.\text{Fraction of } B's \text{ cells occupied by } A \end{aligned} \tag{1.70}$$

These definitions are completely equivalent to the *asymmetric* Tversky measures given in Eqs. (1.12) and (1.13), respectively, and can be interpreted in a like manner,

Table 1.5 Comparison of percent occupancies of compound collections in six-dimensional 3-D BCUT chemical space based on the P–S procedure

A\B	Diverse	Combi	MDDR	Micros
Diverse		11.7	43.8	2.8
Combi	88.5		85.2	6.0
MDDR	78.9	20.3		44.0
Micros	98.5	28.3	86.6	

See Table 1.4 for details of compound collections. Cell occupancies ≥ 1

but any of the similarity coefficients described in this work that are based on binary structural FPs can be used. Note that the two expressions given in Eq. (1.70) can also be interpreted probabilistically.

Since the set of cells in a cell-based CS are analogous to binary structural FPs, other similarity measures such as those based on the Tanimoto or Dice similarity coefficients given in Eqs. (1.8) and (1.9) can be used. Alternatively, the corresponding dissimilarity coefficients given in Eqs. (1.21) and (1.22) also can be used. As noted in Sect. 1.2.1.3, the numerator of the Tanimoto dissimilarity coefficient is just the Hamming distance, which is a measure of the number of differences between the two DB FPs.

Table 1.5 provides an example of how the similarity measures given in Eq. (1.70) can be applied to a more detailed assessment of the similarity of pairs compound collections. For example, 0.885 of the occupied cells in the Combi collection are also occupied in the Diverse collection. Conversely, only 0.117 of the occupied cells in the Diverse collection are also occupied in the Combi collection, a clear example of the much greater diversity inherent in the Diverse collection. In contrast, 0.985 of the occupied cells in the Micros collection are also occupied by the Diverse collection, while only 0.028 of the occupied cells in the Diverse collection are also occupied in the Micros collection—not a surprising result given that only 516 cells are occupied by the entire Micros collection. Thus, although in relative terms the Micros collection is diverse, in absolute terms it does not compare with that of the Diverse collection.

1.3.5.2 Subset Selection and Compound Acquisition

Subset Selection Subset selection is used primarily for assembling diverse subsets of compounds for HTS campaigns. Another form of subset selection called *similarity searching* or LBVS also requires activity data, albeit on a small subset of compounds, as will be discussed in Sect. 1.3.5.4. Hence, subset selection usually takes places in early screening while similarity searching or LBVS is typically used in subsequent follow-on screening activities. Because in the former case activity data are generally unavailable, constructing appropriate subsets of compounds for the initial phases of an HTS campaign can be challenging [186–189].

While there are many variations, the underlying strategy for generating initial screening sets almost always relies on maximizing their diversity by minimizing the similarity (or maximizing the dissimilarity) of the compounds in the putative screening set. It is important to note that unlike similarity or dissimilarity, which are pairwise measures, diversity is a population-based measure associated with the dissimilarity of the entire subset of compounds [10, 41]. In this regard, a number of authors have addressed the issue of how to estimate the diversity of a large collection of compounds [190–192]. Willett [193, 194] and Agrafiotis [191] have presented descriptions of many aspects of diversity-related methods and procedures. An interesting discussion of the early history of the concept of molecular diversity was published in 2001 [195].

Although the field of molecular diversity is vast, the focus in this work is on two approaches: on cell-based sampling of CS [76] and on a maximum dissimilarity/distance algorithm called "Dfragall" [63]. Here the terminology MaxD will be used in place of Dfragall to indicate the generality of the procedure. Both approaches generally use 2-D structural information, although the use of 3-D BCUTS does account, albeit in a somewhat limited fashion, for 3-D information. Matter has presented a more detailed comparison of the role of 2-D and 3-D descriptors in selecting diverse subsets of compounds [196]. As will be seen in the following subsection on compound acquisition, the cell-based approach is clearly superior in its ability to identify and fill so-called "diversity voids," which can be important in a number of instances.

A variety of cell-based sampling schemes can be employed in order to obtain a subset of the desired size and diversity [76, 78]. These schemes include *simple sampling*, where a single compound is obtained from each occupied cell, *threshold-based sampling*, where the number of compounds selected from each cell is less than (if the cell has fewer compounds than the threshold value) or equal to the threshold value, *proportional sampling*, where the size of the sample is proportional to the number of compounds in the cell, or *property-based sampling*, where compounds are selected based on a range of values for one or more properties such as molecular weight or log*P*. Property-based sampling can, of course, be applied simultaneously with any of the other sampling procedures. If the size of the desired sample is less than the number of compounds obtained by a given sampling procedure, either fewer cells can be sampled or the number of compounds per cell can be reduced. In the former case, since neighborhood relations among cells are not considered in cell-based CSs, a random selection of sampled cells could be considered. By contrast, the subset selection procedure based on MaxD is much more computationally demanding and does not explicitly fill diversity voids, although it may inadvertently do so to some degree. In the MaxD case, a typical selection procedure is shown in Table 1.6.

An example that illustrates, but of course does not generally prove, the superior performance of cell-based compared to dissimilarity-based subset selection is depicted in Fig. 1.17. The computations were carried out in 3-D BCUT CS based on the Diverse DB (see Table 1.4) described earlier. The cyan dots in the 2-D projection of the CS depicted in Fig. 1.17a, b represent the compounds in the DB, while the yellow dots represent the compounds obtained in each of the sampling procedures. In the MaxD subset selection depicted in Fig. 1.17a, only about 36 % of the

Table 1.6 MaxD subset selection procedure

Step	Procedure
1	Choose a compound, x_1, at random from the compound collection of interest
2	Determine x_2, the compound most dissimilar to or most distant from x_1
3	Determine x_3, most dissimilar to or distant from compounds x_1 and x_2
4	Repeat the process until the desired number of compounds is obtained or the chosen dissimilarity or distance value falls below the chosen threshold value or reaches a plateau

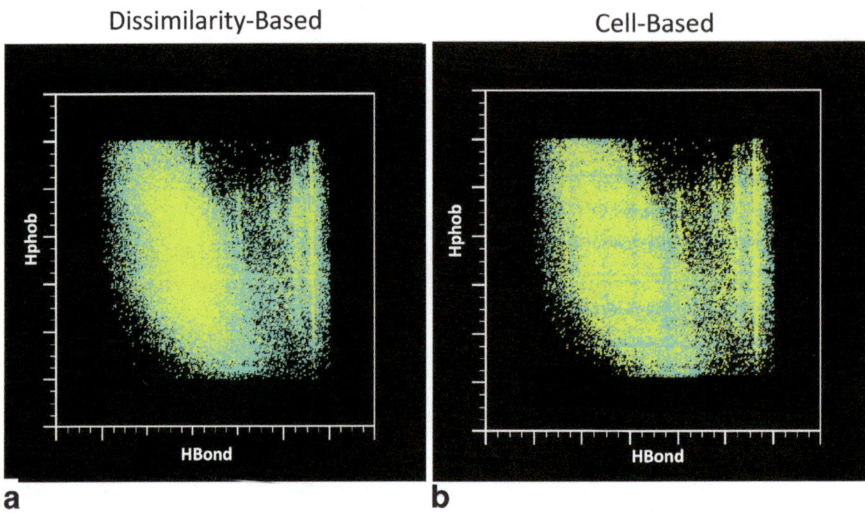

Fig. 1.17 Comparison of subset selection procedures based on compounds in the Diverse collection depicted in cyan (see Table 1.4 and Sect. 3.6.1 for details). Yellow dots represent compounds obtained by the subset selection procedures: **a** dissimilarity-based selection. **b** Cell-based subset selection. (Figure kindly provided by Veer Shanmugasundaram)

original 18,371 occupied cells in the associated cell-based CS are occupied by at least one sampled compound. By contrast, 100 % of the available cells are occupied in the cell-based procedure by a similar number of compounds to that obtained by the MaxD algorithm, which is not surprising since the cell-based procedure is based on sampling each cell of the CS. This affirms, but certainly does not prove, what is intuitively expected, namely, that the cell-based procedure results in broader sampling than the corresponding MaxD procedure.

Compound Acquisition There are two general goals associated with compound acquisition—enhancing the *diversity* of an existing collection and maintaining its *integrity*. While the focus is generally on the former, the latter is also important due to the rate at which compounds can be used up in assays and related activities or can decompose over time. Enhancing diversity usually involves filling unoccupied or partially occupied regions of CS. Maintaining DB integrity, on the other

Table 1.7 Compound acquisition procedure

Step	Procedure
1	Identify vendor collections from which to purchase compounds and preprocess them to remove "undesirable" compounds
2	Generate a cell-based chemical space containing the combined original compound DB and appropriate vendor DBs
3	Select the initial set of vendor compounds by filling diversity voids
4	Additional diversity assessment of the initially selected set of vendor compounds using a modified MaxD algorithm (see Table 1.8)
5	Apply compound filters that were developed based on the knowledge of experienced medicinal chemists
6	Direct review by medicinal chemists
7	Submit compounds for purchase

hand, involves replenishing DB compounds that have become depleted or, if exact replacements are unavailable providing compounds that are, at least to some degree, similar to the original ones. A number of papers addressing compound acquisition have been published over the years, a sampling of which is given by the following references [162, 197–199].

The following is a brief description of the acquisition process based on the work reported in [162]. It illustrates a number of the general issues that must be dealt with, but since there are many ways to do so, what is given here should only be considered a rough outline of an acquisition process. The papers just cited should be consulted for additional examples. Table 1.7 provides a summary of the compound acquisition procedure.

A number of issues arise in step-1, especially when the purchase of large sets of compounds is desired. Some of which include the presence of compounds with undesirable features (e.g., nitro groups) in a vendor's collection and whether the compounds are "Lipinski compliant," i.e., obey the rule of five [200]. Although the rule of five was intended primarily to address potential drug delivery and bioavailability issues, it has become a surrogate for drug likeness, and its application has far exceeded the developers' initial intentions as to its domain of applicability. A recent procedure suggests a modification of the rule of five that increases its robustness to small differences in the parameter values, although it does not extend its domain of applicability [201]. In a related study, Bickerton et al. [202] developed a similar, but more comprehensive procedure that takes account of additional features, namely, molecular polar surface area, number of rotatable bonds, number of aromatic rings, and number of structural alerts, typically associated with drug likeness. In addition, diversity and structural novelty of a collection, timely availability of compounds, and compound purity are other desirable characteristics of vendor compound collections.

In step-2, there are several choices of methods to carry out the initial selection of compounds. The cell-based approach is employed here because of its computational speed and ease of application. Figure 1.18 depicts a model of a cell-based sampling scheme similar, but not algorithmically identical, to that implemented in *Diverse Solutions*™ [78] (cf. [63]) and presented in a way that is designed to clarify the

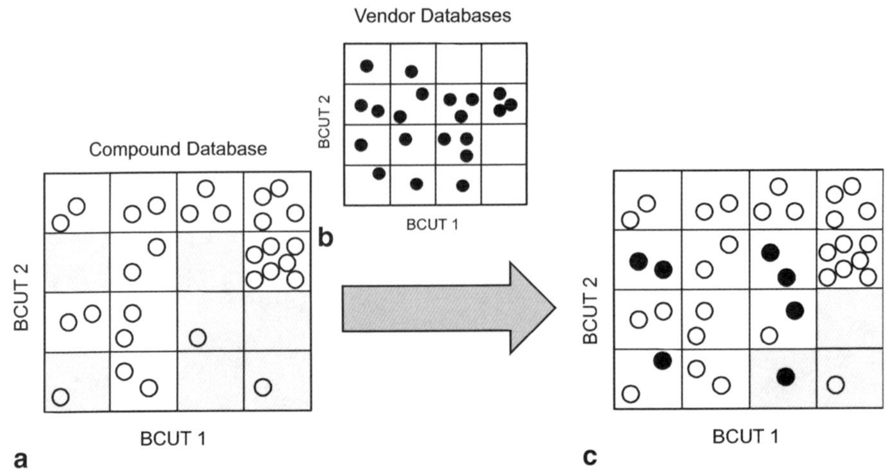

Fig. 1.18 Schematic depiction of a model 2-D cell-based selection process for compound acquisition (Cf. [162]). In **a** *unfilled circles* represent compounds in the original compound DB; in **b** *filled circles* represent compounds in the combined, pre-processed vendor DB; **c** depicts the augmented compound DB after the initial selection process has been completed. Cells shaded in *light grey* represent diversity voids for cells containing fewer than two compounds. (See text for addition details)

compound selection process. A two-dimensional BCUT CS is generated by *combining* (using set-theoretic union) the set of compounds in the original compound DB, O^{DB}, and the compounds in the set of vendor DBs $V^{DB} = \{V_1^{DB}, V_2^{DB}, V_3^{DB}, \dots\}$, where V_i^{DB} is the set of compounds in the *i*th preprocessed vendor DB:

$$\begin{aligned}\widehat{M} &= O^{DB} \cup V_1^{DB} \cup V_2^{DB} \cup V_3^{DB} \cup \cdots \\ &= O^{DB} \cup V^{DB}\end{aligned} \tag{1.71}$$

$\widehat{M}$ is then used as a basis for constructing a CS that includes all of the original and preprocessed vendor compounds, which can be written symbolically as $\widehat{M} \Rightarrow CS(\widehat{M})$.

Figure 1.18a shows the distribution of the original set of compounds in the newly constructed CS. Likewise, Fig. 1.18b shows the distribution of the vendor compounds in the same CS. In the cell-based approach, empty cells as well as those with very few compounds, say less than two or three, can be considered to be diversity voids. Such cells are suitable candidates for compound acquisition. In the example in Fig. 1.18a, there are four empty cells and three cells containing single compounds, all shaded in light grey, which can be classified as diversity voids in this model DB. Now compounds from the combined vendor DB depicted in Fig. 1.18b are used to fill the diversity voids in in Fig. 1.18a until the cell occupancy of all cells in the DB is at least two. This is illustrated in Fig. 1.18c, where the cells shaded in light gray indicate diversity voids that remain after compound acquisition. As seen in the figure, some of the empty cells are now populated with vendor's compounds

Table 1.8 Diversity assessment using a modified MaxD subset selection procedure

Step	Procedure
1	Determine vendor compound, x_1, that is most dissimilar to all of the compounds in the original compound database (C-DB) and add it C-DB giving C-DB + x_1
2	Determine the vendor compound, x_2, that is most dissimilar to C-DB + x_1 and add it yielding C-DB + x_1 + x_2
3	Repeat steps 1 and 2 until the desired number of compounds is obtained or until the dissimilarity value falls below a specified threshold

and some remain unoccupied, as no vendor compounds existed for those cells. The third cell from the left in the bottom row of Fig. 1.18c, which was unoccupied originally, is now occupied by a single vendor compound since only one such compound was available to fill that cell as seen in Fig. 1.18b.

The basic idea here is to populate unpopulated cells and those of low occupancy with commercially acquired compounds. As was the case in subset selection, there are a number of ways in which cells can be populated with new compounds, the simplest being to populate all unpopulated cells with at least one compound. While such an approach is straightforward, it is not, in general, a practical strategy. An examination of Table 1.4 clearly shows why this is the case. In that example, the 6-D CS contains 117,649 cells, 18,371 of which are occupied by at least one compound. This leaves 99,278 empty cells. Even if a set of sufficiently diverse compounds were available for purchase the cost would be significant—at an average price of $ 25 per sample, this would amount to nearly $ 2.5 million, an amount that would test the budget of all but the largest pharmaceutical companies. Thus, additional strategies need to be implemented to address compound acquisition in a way that ensures an optimal, albeit incomplete, selection is made [162].

Although the number of cells in cell-based CS is large, the hyper-dimensional volume of each of the cells is also large. Hence, compounds within a given cell may be quite dissimilar. In contrast, compounds located near a common boundary between two cells may be quite similar even though they reside in different cells (*vide supra*). Because of this type of "idiosyncratic" behavior associated with cell-based CSs, and additional level of similarity analysis may be warranted to ensure that the selected compounds are as dissimilar to each other as possible. This can be accomplished in step-4 using a modified form of the MaxD ("Dfragall") algorithm [63] based on Euclidean distance computed with respect to the BCUT coordinates or, as is traditionally done in the algorithm, using some form of similarity/dissimilarity measure, a procedure that further reduces the number of compounds.

An alternative approach to that described above has been described by Lajiness [63]. It is a variant of the MaxD ("Dfragall") algorithm presented earlier and is summarized in Table 1.8. One clear deficiency of this algorithm is that it is difficult to fill specific diversity voids.

In step-5 of Table 1.7, a set of compound filters based on the knowledge of experienced medicinal chemists is applied further reducing the size of the set of potential compounds for acquisition. Examples of these filters include a number of compound characteristics such as number of rings (2–4), molecular weight (200–400), number of rotatable bonds (0–5), log*P* (−1 to 2). Finally, in step-6, medicinal chemists directly evaluate the remaining molecules [116], and those that survive this final review are submitted for purchase.

1.3.5.3 Similarity Searching and LBVS

Basically, there are three in silico approaches used to the identify compounds with potential biological activity all of which fall under the rubric of virtual screening methods:

- Ligand–protein docking
- Similarity searching based on 2-D molecular descriptors (2-D LBVS)
- Similarity searching based on 3-D molecular descriptors (3-D LBVS)

A number of edited volumes [164, 203–205] and reviews [104, 206–215] have addressed many aspects of virtual screening; and Parker and Bajorath have discussed an important but rarely touched upon issue concerning the effect of errors on both HTS and LBVS [216].

Ligand–Protein Docking[21] Docking involves two basic steps, finding an optimal structure of the ligand–protein complex and scoring, in some fashion, the fitness of that complex. An advantage of this approach is that it does not require any prior knowledge of biological activity. On the other hand, it does require knowledge of the 3-D structure of the target protein, or of some closely related protein that can serve as a model of the desired target protein, to which the ligand can be docked. However, this is just the tip of the iceberg, as there are many complex issues that must be dealt with in ligand–protein docking including protein flexibility, ligand sampling, and effective scoring functions. In addition, if biological activity requires specific changes in protein structure induced by ligand binding and/or if the solution environment plays a crucial role in the functioning of the protein, then these added complications must also be addressed. And there are other factors some known and some unknown that can further complicate the docking process [217–219].

Similarity Searching There are two types of similarity searching procedures—also called LBVS—that are classified according to the dimensionality of their feature descriptors. 2-D methods employ structural FPs or vector-based descriptors as described in Sects. 1.2.1 and 1.2.2, while the corresponding 3-D methods involve matching pharmacophores [153, 220–223] or molecular shapes [224–226]. Since 3-D methods appear to contain more structural information such as stereochemistry, which in many cases is important for activity, it is surprising that 2-D methods tend to outperform or at least perform comparably to 3-D methods. There are

[21] There are, of course, other docking processes that are of importance in biology including protein–protein, ligand–nucleic acid, nucleic acid–nucleic acid docking to name a few. Ligand–protein docking is highlighted in this work because of its importance in drug discovery and its widespread application in that field.

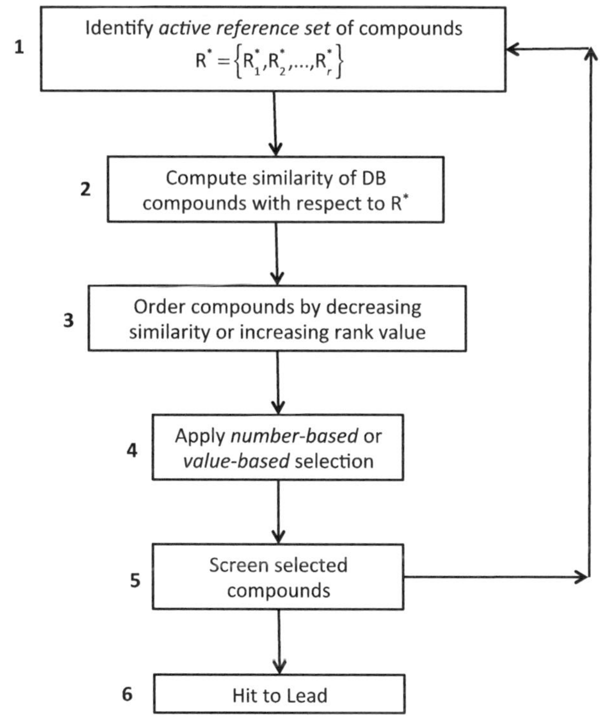

Fig. 1.19 Ligand-base virtual screening procedure

many possible reasons for this observation including the fact that the topological structure encoded in 2-D representations may more than compensate for missing 3-D information [10, 18, 88, 227, 228]. In addition, determining the ensemble of biological active conformations can be a difficult and uncertain task [229], and the many approximations made to increase computational efficiency and reduce computing time, also contribute to the somewhat problematic performance of 3-D-based approaches. Hence, in keeping with the discussion in the rest of this chapter, the focus here is on the simpler and faster 2-D LBVS methods.

2-D LBVS[22] Although Stanton et al. [230] were, perhaps, the first group to explore the application of similarity-based techniques in HTS, many examples of LBVS have been published since then, especially in the first decade of the twenty-first century as can be seen from the following references [32, 33, 86, 104, 231–233] and those cited at the beginning of Sect. 1.3.5.3.

As depicted in Fig. 1.19, LBVS is typically an *iterative process*. In step-1, an *active reference set* of compounds is identified in some manner, usually in an HTS campaign. In step-2, the similarity values with respect to each of the actives in R^* are computed. Several cases arise in this regard. First, consider the simplest case of a single active reference compound, which may obtain in many instances, at least

[22] See Sect. 1.2.3 for related discussion.

in the initial iteration of the LBVS process. The compounds are then arranged in decreasing order of their similarity values, or in ascending order by their ranks, one being the highest rank. If, on the other hand, a distance-based measure of similarity is used, the list of compounds will be ordered from smallest distance to the largest distance value. The rank ordering will remain the same, one again being the highest rank. A subset of the top-"scoring" compounds (i.e., compounds with the largest similarity or smallest rank values) is selected. This can be accomplished in two ways, *number based* or *value based*. In the former case, a number of compounds, say the top 100, are selected for follow-on screening regardless of their similarity values or rankings, whereas in the latter, a subset of compounds all of whose similarity values or rankings with respect to R^* are less than or greater than, their respective threshold similarity or ranking values. Regardless of how the compounds are selected, they are screened yielding a new set of actives, and the process is repeated.

This, however, raises a new issue, namely, how are multiple active reference compounds handled in the LBVS process? There are several approaches to this problem. One way is through the use of group fusion described in Sect. 1.2.3.2, which is ideally suited to deal with this problem since multiple active reference compounds are an inherent feature of the method. And, as discussed in Sects. 1.2.3.2 and 1.2.4, group fusion exhibits excellent performance as a means for identifying new actives. Interestingly, group fusion based on the fusion maximum similarity or minimum distance values is essentially identical to an approach called list-based searching [76, 78, 86].

This completes step-3 regardless of whether singleton or multiple active reference compounds were dealt with in that step. Obtaining a subset of the compounds from the resultant ordered list using either number- or value-based selection then completes step-4. In step-5, the resulting set of compounds is then screened. At this point, a choice must be made. If, after screening is completed, it is determined that a sufficient number active compounds of appropriate quality have been obtained, the process may then move to step-6 where the hit-to-lead phase of the drug discovery process can commence, otherwise the process moves back to step-1 and the process is repeated. It is well to note that identifying active reference sets may also include additional assays designed to more firmly establish the biological or pharmacological characteristics of the compounds, and thus to help in determining whether compounds active in HTS should be considered further.

Aggregating the Results of Individual Similarity Searches As discussed in Sect. 1.2.3, combining ("fusing") similarity values, which falls within the class of data aggregation methods [97], has been shown to yield improved results in similarity searches. Generally, fusion methods combine similarity (distance) values or rankings to yield new fused values prior to any similarity search. An alternative approach is to carry out multiple similarity searches on the same set of active reference compounds using different similarity or distance measures and then combine the sets of compounds obtained in this way [86], employing what can be called *post-search aggregation* (PSA). Although related, this differs from similarity fusion that, as discussed in Sect. 1.2.5.1, combines the similarity values and then carries out a similarity search using the fused values.

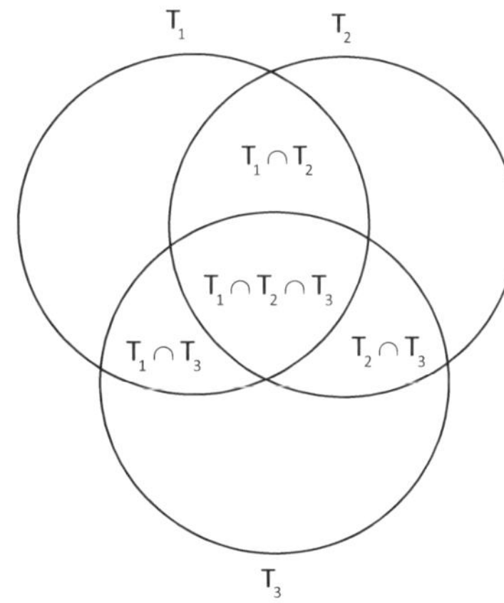

Fig. 1.20 Venn diagram representing the possible joint subsets obtained from three sets of compounds T_1, T_2, and T_3 retrieved by three different similarity or distance-based search methods of a compound DB

A difficulty with PSA methods is that the subset of compounds retrieved in each of the similarity- or distance-based searches may differ significantly. As an example, consider the family of three subsets of compounds retrieved by three corresponding similarity or distance-based searches of a compound DB, i.e.,

$$T = \{T_1, T_2, T_3\} \tag{1.72}$$

where the size of each of the subsets may be taken to be the same and can be determined by a number- or value-based procedure, or the sizes can, if desired, all be different. It is possible and, in fact, occurs frequently that some compounds may be found in more than one of the subsets. The Venn diagram depicted in Fig. 1.20 indicates this. As will be seen in Eq. (1.73), the smaller the "overlap" among the subsets, as measured by set intersection, the broader the sampling of the CS represented in a compound DB.

The basic assumption underlying this approach is that multiple searches using different similarity or distance measures will give rise to higher enrichment factors in a common assay than would be obtained using a single search method. To see this, consider the *background enrichment factor* for a given assay, $E_{\text{Background}}$, which is basically the *estimated* fraction of active compounds in a DB, an estimate usually arrived at by the assay of compounds randomly selected from the DB.

When considering all three subsets, the *breadth or diversity* of the search can be defined as

$$\Delta = \frac{\text{Card}(T_1 \cup T_2 \cup T_3}{\text{Card}(T_1) + \text{Card}(T_2) + \text{Card}(T_3)} \tag{1.73}$$

which satisfies $0 < \Delta \leq 1$, where "Card" refers to the cardinality (i.e. number of elements) in a given set (see also footnote *a* in Table 1.1). The union of the three subsets is the set of compounds unique to all three subsets. Similar expressions can be constructed for the pairwise case by removing the extraneous subset(s).

The singleton case is trivial since $\Delta = 1$. As can be seen from Eq. (1.73), as the breadth approaches unity, i.e., as $\Delta \to 1$, the sampling of CS increases reaching a maximum at unity. However, this procedure is of real value only if it leads to enhanced enrichment factors. The enrichment factor for the three sets of retrieved compounds can be obtained as follows:

The fraction of actives obtained from the three samples is given by

$$f_{\text{sample}} = \frac{\text{Card}\left(\text{T}_1^* \cup \text{T}_2^* \cup \text{T}_3^*\right)}{\text{Card}\left(\text{T}_1 \cup \text{T}_2 \cup \text{T}_3\right)} \tag{1.74}$$

where the asterisks in the numerator denote subsets of actives, such that $\text{T}_i^* \subseteq \text{T}_i$ for $i = 1,2,3$ and 'Card' refers to the cardinality, that is the number of elements in the sets. The *enrichment factor* is then given by

$$EF = \frac{f_{\text{sample}}}{f_{\text{background}}} \tag{1.75}$$

where $f_{\text{background}}$ is the fraction of actives obtained from a random sampling of the compound collection of interest.

Interestingly, the procedure appears to be a combination of group fusion (i.e., list-based searching) and similarity fusion. The reasons, the first two of which are associated with group and similarity fusion, are as follows: (1) multiple active reference compounds are used, (2) the most similar (closest) compounds to each active reference compound are retained, and (3) multiple similarity measures are applied.

This approach was described in Shanmugasundaram et al. [86], who investigated its application to a number of targets including those associated with anxiety, Alzheimer's disease, and pathogenic bacteria. The data provided below are based on a bacterial enzyme target and a set of 12 well-characterized active reference compounds. A distance measure based on three different sets of BCUT descriptors and a structural FP procedure based on the Tanimoto similarity coefficient were all employed in the analysis, yielding a breadth value of $\Delta = 132/159 = 0.83$. This shows that the approach covered a wider region of CS than could have been achieved using a single similarity (distance) measure. Moreover, the ratio of the fraction of actives in the three samples, $f_{\text{sample}} = 23/132 = 0.174$, to the fraction of actives obtained from a random sample of the database, $f_{\text{background}} \approx 0.04$ yields an enrichment of $EF \approx 0.174/0.04 = 4.4$. Thus, nearly four and a half times as many actives were obtained than would be expected by randomly sampling and screening compounds in the DB—more details can be obtained in the paper.

While this enhancement may not seem like a significant improvement over background, it is if a *Las Vegas model* of drug discovery is considered. As is true for many of the gambling activities in Las Vegas such as roulette and craps, the odds of winning are "shaved" slightly in the House's favor. Given that enough people place bets, statistically the House will almost certainly win over time. This has a close parallel to the HTS in drug discovery. If the odds of finding actives are even slightly better than those for random screening, and if enough compounds are screened, active compounds will almost certainly be found given that the compound DB is not highly biased, that is filled with biologically unsuitable compounds. Even an enhanced enrichment factor of two can still yield actives, but the smaller the factor the more compounds that need to be screened.

Target (Activity) Class-Specific Similarity Searching The basic idea behind target (activity) class-specific[23] similarity searching is that particular feature descriptors may exhibit some bias for specific classes of bioactivity such as, for example, HMG Co-A Reductase inhibitors, COX2 inhibitors, and 5HT (serotonin) receptor ligands. Since work in this area is based primarily on molecule-independent structural FPs, their bit positions can be unequivocally associated with specific structural features. The probability that a given feature is associated with a specific activity is estimated essentially by computing its relative frequency of occurrence in the set of molecules associated with that target class. Bits associated with features having high probabilities of occurrence, which may be called *characteristic bits*, are generally, but not always, weighted in some fashion to further emphasize their importance in subsequent similarity analyses; weighting can be accomplished in a number of ways (*vide infra*).

This approach to target class-specific similarity searching, called *reverse fingerprinting* by Williams [234], has also been carried out in a number of other laboratories [235–242]. The application of methods utilizing "nontraditional" structural fragments [234, 237, 239] have shown promise, but none of the earlier methods including these have addressed the issue of interdependencies among structural descriptors. Two papers from the Bajorath group [240, 241] that show promise have taken steps in this direction.

Based on a growing amount of data that show that *compound* and *target promiscuity* is more ubiquitous than had earlier been suspected may present significant challenges to the development of robust target class-specific similarity searching that is difficult to overcome (See Sect. 1.3.1 for further discussion).

1.4 Summary and Conclusions

Over the past two decades, computational methods have been playing an ever-increasing role in drug discovery research due especially to the burgeoning amount of data being generated by ever faster and more powerful experimental techniques. Three concepts, molecular similarity, CS, and activity/property landscapes, in some fashion underlie all of these methods—the current work addresses molecular/structural similarity and CS, two important pillars supporting the edifice of chemical informatics.

---

[23] In order to simplify discussion, the terminology "target class specific" will be used in the remainder of this section.

Similarity is probably one of the most ubiquitous concepts in many human endeavors. Hence, it is no surprise that it also plays a significant role in many aspects of chemical informatics. And, as is essentially true in all conscious and subconscious applications of the concept, however, what precisely it is remains somewhat a mystery since "similarity like pornography is difficult to define but you know it when you see it" [10]. The inherent subjectivity of similarity poses significant problems in chemical informatics since its application in this field is, in many cases, carried out computationally. Two key issues that then must be addressed are how to represent the relevant chemical or molecular information and how to compute an effective measure of similarity from that information. This has been covered extensively for a variety of 2-D similarity measures in Sect. 1.2 that, due primarily to their generally higher computational speeds, are by far the most popular similarity measures in use today. Surprisingly, perhaps, 2-D similarity measures perform comparably or better than many 3-D measures in a variety of cheminformatics tasks, one reason along with their higher computational speeds that accounts for their popularity.

An interesting extension of similarity-based methods that shows promise involves combining similarity values using data fusion techniques that have been applied in many engineering applications. In some cases, fused similarity values have been shown to yield significantly improved results. This is especially true of an approach called group fusion, which is based on computing the similarity of compounds in a large DB with respect to a number of reference compounds using a single similarity measure. The similarity or rank values for each DB compound are then fused to yield a single similarity score or ranking. The resulting list provides a set of compounds such that those of higher rank can be selected, for example, for follow-on screening.

A discussion presented in Sect. 1.2.4 suggests a rationale, based on the surprising prevalence of similarity cliffs, as to why group fusion appears to perform better in similarity searches than the use of a single similarity measure or the fusion of multiple similarity measures, both carried out with respect to a single reference compound. This is understandable since the relatively common occurrence of similarity cliffs, which arise when two structurally dissimilar compounds have similar activities in a given assay, suggests that active compounds may in many cases be more widely dispersed through CSs than heretofore had been suspected. Moreover, the fact that the more dissimilar the set of reference compounds the better the results of group fusion similarity searches supports this contention. An unresolved issue with this approach to similarity searching is the need for multiple active reference compounds, a situation that may not be realized in the initial phase of an HTS campaign.

Aside from its computational uses in chemical informatics, similarity also plays a significant *perceptual* role in many aspects of chemistry. This clearly is the case in medicinal chemistry where chemists address the question of "what to make next" by inferring new structures for synthesis based on the structures of active and inactive compounds considered earlier. There are, of course, many other such examples one can think of, all of which raise the issue as to whether computed similarities are comparable to those perceived by chemists.

As discussed in Sect. 1.2.5, the similarity scale, which generally is taken to lie on the unit interval [0,1] of the real line, is not uniform in terms of human perception.

Humans excel at comparing very similar objects, just as chemists excel at recognizing very similar molecules. However, at some point, as objects become less and less similar, humans can no longer discern how dissimilar they are to one another, only that they are very dissimilar. This is not entirely the case computationally since computers make no value judgments; they implement specific algorithms, although a caveat discussed in Sect. 1.2.1.4 shows that computational algorithms can also exhibit idiosyncratic behaviors such as the size-dependent behavior of FP-based similarity coefficients.

A possible reason for this disparity between chemists' perceived similarity values and those obtained computationally is seen in the expressions for Tanimoto similarity and dissimilarity given in Eqs. (1.8) and (1.21), respectively. Since the denominators in both equations are identical, it is their respective numerators that determined the difference in these two coefficients. In the case of similarity, the numerator is based on the number of features in *common* in the two molecules, while in the case of dissimilarity, the numerator is based on the number of features *unique* to each molecule. Unique features, that is, features in one molecule but not in the other, are more difficult for humans to perceive than features common to both molecules. Thus, cases of low similarity (few features in common) or high dissimilarity (more unique features) are difficult for humans to perceive. Clearly, the perceptual issue goes beyond the mathematical complementarity exhibited by Eq. (1.19). Importantly, these arguments provide a mechanism that may account for the limited correspondence between computed and perceived similarities and dissimilarities.

The notion of CS is closely related to that of similarity. Section 1.3 provides a discussion of three possible representations of CSs, namely, coordinate based (Sect. 1.3.2), cell based (Sect. 1.3.3), and graph or network based (Sect. 1.3.4). The first two are well known in the chemical informatics field. The last is not, although networks are being employed to describe a growing number of chemically related systems such as those, for example, describing protein–protein interactions, drug target relationships, and pharmacological space. The network-based approach, which opens up new ways to investigate the nature of CSs, has two distinct advantages, namely, it is inherently discrete and it provides an intuitive representation of these spaces. Unfortunately, very few papers describing network-based representations of CSs have been published, but the power of this approach would seem to auger well for its future application in chemical informatics. In this regard, a new graph-based DB scheme that may provide a powerful approach for treating CSs, is gaining recognition in the computer field.

Each of the three CS representations has its strengths and weaknesses with regard to the types of applications for which they are best suited. A number of examples such as:

- Comparing compound DBs
- Selecting chemically diverse subsets
- Augmenting DBs through compound acquisition
- Similarity searching—2-D LBVS

are presented in Sect. 1.3.5 to illustrate this point.

The need for computational methods that can characterize relationships among sets of molecules is clearly manifest, especially in this age of massive and rapidly growing compound DBs. And although imperfect almost by their very nature, similarity-based methods provide the means for addressing this critical need. These methods also provide the means for constructing CSs that help to unify the chemical universe in an intuitive and computationally powerful way. Both notions are now beginning to be applied in fields outside of chemical informatics such as materials science and engineering laying the groundwork for future applications in food science and related fields.

Acknowledgments Many thanks to Prof. Dr. Jurgen Bajorath and his Group, especially Dr. Martin Vogt, for numerous helpful discussions. Thanks also to Dr. Jose Medina-Franco for providing Figs. 1.9 and 1.10 and to Drs. Mic Lajiness and Veer Shanmugasundaram for providing the data and for constructing Figs. 1.8 and 1.17, and lastly to Dr. Vijay Gokhale for reading and commenting on the entire manuscript.

References

1. Ruddigkeit L, van Deursen R, Blum LC, Reymond J-L (2012) Enumeration of 166 billion organic small molecules in the chemical universe database GBD-17. J Chem Inf Model 52:2864–2875
2. Bohacek RS, McMartin C, Guida WC (1996) The art and practice of structure-based drug design: a molecular modeling perspective. Med Res Rev 16:3–50
3. Virshup AM, Contreras-Garcia J, Wipf P, Yang W, Beratan DN (2013) Stochastic voyages into uncharted chemical space produce a representative library of all possible drug-like compounds. J Amer Chem Soc 135:7296–7303
4. Wassermann AM, Wawer M, Bajorath J (2010) Activity landscape representations for structure-activity relationship analysis. J Med Chem 53:8209–8223
5. Iyer P, Wawer M, Bajorath J (2011) Comparison of two- and three-dimensional activity landscape representations for different compound sets. MedChemComm 2:113–118
6. Bajorath J (2012) Modeling activity landscapes for drug discovery. Expert Opin Drug Discov 7:463–473
7. Iyer P, Stumpfe D, Vogt M, Bajorath J, Maggiora GM (2013) Activity landscapes, information theory, and structure-activity relationships. Mol Inf 32:421–430
8. Vogt M, Iyer P, Maggiora GM, Bajorath J (2013) Conditional probabilities of activity landscape features for individual compounds. J Chem Inf Model 53:1602–1612
9. Rouvray DH (1990) The evolution of the concept of molecular similarity. In: Johnson MA, Maggiora GM (eds) Concepts and applications of molecular similarity, chapter 2. Wiley, New York
10. Medina-Franco JL, Maggiora GM (2014) Molecular similarity analysis. In: Bajorath J (ed) Chemoinformatics in drug discovery: concepts, methods, and tools for drug discovery, chapter 15. Wiley, New York
11. Mendeleev D (1869) J Russ Phys Chem Soc 1:60
12. Meyer L (1870) Ann Suppl 7:354
13. Wilkins CL, Randic M (1980) A graph theoretical approach to structure-property and structure-activity correlation. Theoret Chim Acta 58:45–68
14. Johnson M, Basak S, Maggiora G (1988) A characterization of molecular similarity methods for property prediction. Mathl Comput Model 11:630–634
15. Johnson MA, Maggiora GM (eds) (1990) Concepts and applications of molecular similarity. Wiley, New York
16. Trinajstic N (1992) Chemical graph theory, 2nd edn. CRC, Baca Raton

17. Brown RD, Martin YC (1996) Use of structure-activity data to compare structure-based clustering methods and descriptors for use in compound selection. J Chem Inf Comput Sci 36:572–584
18. Brown RD, Martin YC (1997) The information content of 2D and 3D structural descriptors relevant to ligand-receptor binding. J Chem Inf Comput Sci 37:1–9
19. ChEMBL https://www.ebi.ac.uk/chembldb/. Accessed 1 Feb 2014
20. PubChem http://pubchem.ncbi.nlm.nih.gov. Accessed 1 Feb 2014
21. Chen J, Swamidass SJ, Dou Y, Bruand J, Baldi P (2005) ChemBD: a public database of small molecules and related chemoinformatics resources. Bioinformatics 21:4133–4139
22. DrugBank http://www.drugbank.ca. Accessed 1 Feb 2014
23. WOMBAT http://www.sunsetmolecular.com/. Accessed 1 Feb 2014
24. MDDR http://accelrys.com/products/databases/bioactivity/mddr.html. Accessed 1 Feb 2014
25. Scior JT, Bernard P, Medina-Franco JL, Maggiora GM (2007) Large compound databases for structure-activity relationships studies in drug discovery. Mini Rev Med Chem 7:851–860
26. Leach AR, Gillet VJ (2003) An introduction to chemoinformatics. Kluwer Academic, Dordrecht
27. Gasteiger J, Engel T (eds) (2003) Chemoinformatics—a textbook. Wiley-VCH, Weinheim
28. Bajorath J (ed) (2004) Chemoinformatics—concepts, methods, and tools for drug discovery. Humana, Totowa
29. Bunin BA, Siesel B, Morales G, Bajorath J (2006) Chemoinformatics: theory, practice, and products. Springer, New York
30. Bajorath J (ed) (2011) Chemoinformatics and computational chemical biology. Humana, New York
31. Willett P, Barnard JM, Downs GM (1998) Chemical similarity searching. J Chem Inf Comput Sci 38:983–986
32. Bender A, Glen RC (2004) Molecular similarity: a key technique in molecular informatics. Org Biomol Chem 2:3204–3218
33. Willett P (2009) Similarity methods in chemoinformatics. Annu Rev Inf Sci Technol 43:3–71
34. Maggiora GM, Vogt M, Stumpfe D, Bajorath J (2014) Molecular similarity in medicinal chemistry. J Med Chem 57:3186–3204
35. Lipinski C, Hopkins A (2004) Navigating chemical space for biology and medicine. Nature 432:855–861
36. Dobson CM (2004) Chemical space and biology. Nature 432:424–428
37. Koch MA, Schuffenhauer A, Scheck M, Wetzel S, Casaulta M, Odermatt A, Ertl P, Waldman H (2005) Charting biologically relevant chemical space: a structural classification of natural products (SCONP). Proc Nat Acad Sci U S A 102:17272–17277
38. Reymond J-L, van Deursen R, Blum LC, Ruddigkeit L (2010) Chemical space as a source for new drugs. Med Chem Comm 1:30–38
39. Reymond J-L, Awale M (2012) Exploring chemical space for drug discovery using the chemical universe database. ACS Chem Neurosci 3:649–657
40. Yu MJ (2013) Druggable chemical space and enumerative combinatorics. J Cheminformatics 5:19. doi:10.1186/1758–2964-5–19
41. Maggiora GM, Shanmugasundaram V (2011) Molecular similarity measures. In: Bajorath J (ed) Chemoinformatics and computational chemical biology, Chapter 2. Humana, New York
42. Baldi P, Benz RW, Hirschberg DS, Swamidass SJ (2007) Lossless compression of chemical FPs using integer entropy codes improves storage and retrieval. J Chem Inf Model 47:2098–2109
43. MACCS structural keys. Symyx software: San Ramon2005
44. Barnard JM, Downs GM (1997) Chemical fragment generation and clustering software. J Chem Inf Comput Sci 37:141–142
45. Carhart RE, Smith DH, Venkataraghaven R (1985) Atom pairs as molecular features in structure-activity studies. J Chem Inf Comput Sci 25:64–73
46. Rogers D, Hahn M (2010) Extended-connectivity FPs. J Chem Inf Model 50:742–754

47. Daylight IS (2014) Fingerprints—screening and similarity. http://www.daylight.com/dayhtml/doc/theory/theory.finger.html. Accessed 2 Feb 2014
48. ChemAxon (2014) ECFP—extended connectivity fingerprints. http://www.chemaxon.com/jchem/doc/user/ECFP.html. Accessed 3 Feb 2014
49. Hu Y, Lounkine E, Bajorath J (2009) Improving the search performance of extended connectivity fingerprints through activity-oriented feature filtering and application of a bit-density-dependent similarity function. ChemMedChem 4:540–548
50. Glen RC, Bender A, Arnby CH, Carlsson L, Boyer S, Smith J (2006) Circular fingerprints: flexible molecular descriptors with applications from physical chemistry to ADME. IDrugs 9:199–204
51. Arif SM, Holiday JD, Willett P (2009) Analysis and use of fragment-occurrence data in similarity-based virtual screening. J Comput Aided Mol Des 23:6655–6668
52. Arif SM, Hert J, Holliday JD, Malim N, Willett P (2009) Enhancing the effectiveness of FP-based virtual screening: Use of turbo similarity searching and of fragment frequencies of occurrence. In: Kadirkamanathan V, Sanguinetti G, Girolami M, Niranjan M, Noirel J (eds) Pattern recognition in bioinformatics—Proceedings 4th IAPR international conference, Springer, Berlin, pp 404–414
53. Arif SM, Holiday JD, Willett P (2010) Inverse frequency weighting of fragments for similarity-based virtual screening. J Chem Inf Model 50:1340–1349
54. Willett P, Winterman V (1986) A comparison of some measures for the determination of inter-molecular structural similarity measures of inter-molecular structural similarity. Quant Struct Act Relat 5:18–25
55. Tversky A (1977) Features of similarity. Psychol Rev 84:327–352
56. Maggiora GM, Petke JD, Mestres J (2002) A general analysis of field-based molecular similarity indices. J Math Chem 31:251–270
57. Chen X, Brown F (2007) Asymmetry of chemical similarity. ChemMedChem 2:180–182
58. Wang Y, Eckert H, Bajorath J (2007) Apparent asymmetry in fingerprint similarity searching is a direct consequence of differences in bit densities and molecular size. ChemMedChem 2:1037–1042
59. Lipkus AH (1999) A proof of the triangle inequality for the Tanimoto distance. J Math Chem 26:263–265
60. Hankerson D, Harris GA, Johnson Jr PD (1998) Introduction to information theory and data compression. CRC, Boca Raton
61. Flower DR (1988) On the properties of bit string based measures of chemical similarity. J Chem inf Comput Sci 38:379–386
62. Lajiness M (1990) Molecular similarity–based methods for selecting compounds for screening. In: Rouvray D (ed) Computational chemical graph theory. Nova Science, pp 299–316
63. Lajiness MS (1997) Dissimilarity-based compound selection techniques. Perspect Drug Disc Design 7/8:65–84
64. Dixon SL, Koehler RT (1999) The hidden component of size in two-dimensional fragment descriptors: side effects on sampling in bioactive libraries. J Med Chem 42:2887–2900
65. Fligner MA, Verducci JS, Blower PE (2002) A modification of the Jaccard–Tanimoto similarity index for diverse selection of chemical compounds using binary strings. Technometrics 44:110–119
66. Godden WJ, Xue L, Bajorath J (2000) Combinatorial preferences affect molecular similarity/diversity calculations using binary fingerprints and Tanimoto coefficients. J Chem Inf Comput Sci 40:163–166
67. Holliday JD, Salim N, Whittle M, Willett P (2003) Analysis of size dependence of chemical similarity coefficients. J Chem Inf Comput Sci 43:819–828
68. Marshall AG (1978) Biophysical chemistry. Wiley, New York
69. Hehre WJ, Radom L, Schleyer PvR, Pople JA (1986) Ab initio molecular orbital theory. Wiley, New York
70. Devillers J, Balaban AT (eds) (1999) Topological indices and related descriptors in QSAR and QSPR. Gordon and Breach Science, New York

71. Martin Y (2010) Quantitative drug design–a critical introduction, 2nd edn. CRC, New York
72. Todeschini R, Consonni V (2009) Molecular descriptors for chemoinformatics, vol 1, 2nd edn. Wiley-VCH, Weinheim
73. Guha R, Willighagen E (2010) A survey of quantitative descriptions of molecular structure. Curr Top Med Chem 12:1946–1956
74. Labute P (2000) A widely applicable set of descriptors. J Mol Graph Model 18:464–467
75. Labute P (2004) Derivation and application of molecular descriptors based on approximate surface area. In: Bajorath J (ed) Chemoinformatics: concepts, methods, and tools for drug discovery, Chapter 8. Humana, Totowa
76. Pearlman RS, Smith KS (2002) Novel software tools for chemical diversity. 3D QSAR in drug design: three-dimensional quantitative structure-activity relationships 2:339–353
77. Pearlman RS, Smith KM (1999) Metric validation and the receptor-relevant subspace concept. J Chem Inf Comput Sci 39:28–35
78. Pearlman RS (1995) Diverse solutions user's manual. University of Texas, Austin
79. Burden F (1989) Molecular identification number for substructure searches. J Chem Inf Comput Sci 29:225–227
80. Menard PR, Mason JS, Morize I, Bauerschmidt S (1998) Chemistry space metrics in diversity analysis. J Chem Inf Comput Sci 38:1204–1213
81. Schnur D (1999) Design and diversity analysis of large combinatorial libraries using cell-based methods. J Chem Inf Comput Sci 39:36–45
82. Mason JS, Beno BR (2000) Library design using BCUT chemistry-space descriptors and multiple four-point pharmacophore fingerprints: simultaneous optimization and structure-based diversity. J Mol Graphics Model 18:438–451
83. Stanton DT (1999) Evaluation and use of BCUT descriptors in QSAR and QSPR studies. J Chem Inf Comput Sci 39:11–20
84. Pirard B, Pickett SD (2000) Classification of kinase inhibitors using BCUT descriptors. J Chem Inf Comput Sci 40:1431–1440
85. González MP, Terán C, Besada TM, González-Moa MJ (2005) BCUT descriptors to predicting affinity toward A3 adenosine receptors. Bioorg Med Chem Lett 15:3491–3495
86. Shanmugasundaram V, Maggiora GM, Lajiness MS (2005) Hit-directed nearest neighbor searching. J Med Chem 48:240–248
87. Hodgkin EE, Richards WG (1987) Molecular similarity based on electrostatic potential and electric field. Int J Quantum Chem Quantum boil Symp 14:105–110
88. Sheridan RP, Kearsely SK (2002) Why do we need so many chemical similarity search methods? Drug Discov Today 7:903–911
89. Kearsley SK, Sallamack S, Fluder EM, Andose JD, Mosley RT, Sheridan RP (1996) Chemical similarity using physicochemical property descriptors. J Chem Inf Comput Sci 36:11–127
90. Sheridan RP, Miller MD, Underwood DJ, Kearsley SK (1996) Chemical similarity using geometric atom pair descriptors. J Chem Inf Comput Sci 36:128–136
91. Hert J, Willett P, Wilton DJ, Acklin P, Azzaoui K, Jacoby E, Schuffenhauer A (2004) Comparison of FP-based for virtual screening using multiple bioactive structures. J Chem Inf Comput Sci 44:1177–1185
92. Whittle M, Gillet VJ, Willett P, Alex A, Loesel J (2004) Enhancing the effectiveness of virtual screening by fusing nearest neighbor lists: a comparison of similarity coefficients. J Chem Inf Comput Sci 44:1840–1848
93. Willett P (2006) Enhancing the effectiveness of ligand-based virtual screening using data fusion. QSAR Combin Sci 25:1143–1152
94. Willett P (2013) Combination of similarity rankings using data fusion. J Chem Inf Model 53:1–10
95. Joshi R, Sanderson AC (1999) Multisensor fusion: a minimal representation framework. World Scientific, Singapore
96. Hall DL, McMullen SAH (2004) Mathematical techniques in multisensory data fusion. Artech House, Boston

97. Beliakov G, Pradera A, Tomasa C (2010) Aggregation functions: a guide for practitioners. Springer, Berlin
98. Harmonic mean (2014) Wikipedia. http://en.wikipedia.org/wiki/Harmonic_mean. Accessed 7 Jan 2014
99. Cormack GV, Clark CLA, Buettcher S (2009) Reciprocal rank fusion outperforms condorcet and individual rank learning methods. In: Proceedings of the 32nd international ACM SIGIR conference on research and development in information retrieval, Boston, 19–23 July 2009, pp 758–759
100. Chen B, Meuller C, Willett P (2010) Combination rules for group fusion in similarity based virtual screening. Mol Inf 29:533–541
101. Critchlow DE (1980) Metric methods for analyzing partially ranked data. Springer, New York
102. Nasr RJ, Swamidass SJ, Baldi PF (2009) Large scale study of multiple molecule queries. J Cheminform 1:7. http://www.jcheminf.com/content/1/1/7. Accessed 7 Jan 2014. doi:10.1186/1758-2946-1-7
103. Stumpf D, Bajorath J (2011) Similarity searching. WIRES Comput Mol Sci 1:260–282
104. Willett P (2006) Similarity-based virtual screening using 2D fingerprints. Drug Discov Today 11:1046–1053
105. Gardiner EJ, Gillet VJ, Haranczyk M, Hert J, Holliday JD, Malim N, Patel Y, Willett P (2009) Turbo similarity searching: effect of FP and dataset on virtual-screening performance. Stat Anal Data Mining 2:103–114
106. Hert J, Willett P, Wilton DJ, Acklin P, Azzaoui K, Jacoby E, Schuffenhauer A (2006) New methods for ligand-based virtual screening:use of data-fusion and machine-learning techniques to enhance the effectiveness of similarity searching. J Chem Inf Model 46:462–470
107. Miyamoto S (1990) Fuzzy sets in information retrieval and cluster analysis. Kluwer Academic, Dordrecht
108. Edgar SJ, Holliday JD, Willett P (2000) Effectiveness of retrieval in similarity searches of chemical databases: a review of performance measures. J Mol Graph Model 18:343–357
109. Willett P (2004) Evaluation of molecular similarity and molecular diversity methods using biological data. In: Bajorath J (ed) Chemoinformatics-Concepts, methods and tools for drug discovery, Chapter 2. Humana, Towata
110. Truchon J-F, Bayly CI (2007) Evaluating virtual screening: good and bad metrics for the "early recognition" problem. J Chem Inf Model 47:488–508
111. Maggiora GM (2006) On outliers and activity cliffs—why QSAR often disappoints (Editorial). J Chem Inf Model 46:1535
112. Guha R, Van Drie J (2008) Structure-activity landscape index: identifying and quantifying activity cliffs. J Chem Inf Model 48:646–658
113. Stumpfe D, Bajorath J (2012) Exploring activity cliffs in medicinal chemistry. J Med Chem 55:2932–2942
114. Stahura FL, Bajorath J (2002) Bio- and chemo-informatics beyond data management: crucial challenges and future opportunities. Drug Discov Today 7:S41–S47
115. Hu Y, Maggiora GM, Bajorath J (2013) Activity cliffs in PubChem confirmatory bioassays taking inactive compounds into account. J Comput Aided Mol Des 27:115–124
116. Lajiness MS, Maggiora GM, Shanmugasundaram V (2004) An assessment of the consistency of medicinal chemists in reviewing compound lists. J Med Chem 47:4891–4896
117. Takaoka Y, Endo Y, Yamanobe S, Kakinuma H, Okubo T, Shimazaki Y, Ota T, Sumiya S, Yoshikawa K (2003) Development of a method for evaluating drug-likeness and ease of synthesis using a data set in which compounds are assigned scores based on chemists' intuition. J Chem Inf Comput Sci 43(4)1269–1275
118. Kutchukian PS, Vasilyeva NY, Xu J, Lindvall MK, Dillon MP, Glick M, Coley JD, Brooijmans N (2012) Inside the mind of a medicinal chemist: the role of human bias in compound prioritization during drug discovery. PLoS ONE 7:e48476
119. Hawkins DM, Young SS, Rusinko A III (1997) Analysis of a large structure-activity data set using recursive partitioning. Mol Inf 16:296–302

120. Chen X, Rusinko A III, Young S (1998) Recursive partitioning analysis of a large scale structure-activity data set using three-dimensional descriptors. J Chem Inf Comput Sci 38:1054–1062
121. Rusinko A III, Farmen MW, Lambert CG, Brown PL, Young SS (1999) Analysis of a large structure/biological activity data set using recursive partitioning. J Chem Inf Comput Sci 39:1017–1026
122. Wasserman S, Faust K (1997) Social network analysis. Cambridge University , New York
123. Paolini GV, Shapland RHB, van Hoorn WP, Mason JS, Hopkins AL (2006) Global mapping of pharmacological space. Nature Biotech 24:805–815
124. Hopkins AL (2008) Network pharmacology: the next paradigm in drug discovery. Nat Chem Biol 4:682–690
125. Kesier MJ, Roth BL, Armruster BN, Ernsberger P, Irwin JJ, Shoichet BK (2007) Relating protein pharmacology by ligand chemistry. Nat Biotechnol 25:197–206
126. Yildirim MA, Goh K-I, Cusick ME, Barabási A-L, Vidal M (2007) Drug-target network. Nat Biotechnol 25:1119–1126
127. Watts DJ (2003) Six Degrees—the science of a connected age. WW Norton, New York
128. Barbási A-L (2003) Linked: how everything is connected to everything else, and what it means for business, science, and everyday life. Penguin, New York
129. Newman MEJ (2010) Networks an introduction. Oxford University, New York
130. Robinson I, Webber J, Eifrém E (2013) Graph databases. O'Reilly Media, Sebastopol, CA 95472
131. Peltason L, Bajorath J (2007) SAR Index: quantifying the nature of structure-activity relationships. J Med Chem 50:5571–5578
132. Namasivayam V, Iyer P, Bajorath J (2012) Exploring SAR continuity in the vicinity of activity cliffs. Chem Biol Drug Des 79:22–29
133. Hu Y, Bajorath J (2014) Exploring compound promiscuity patterns and multi-target activity spaces. Comput Struct Biotech J 9:1003–1012. http://dx.doi.org/10.5936/csbj.201401003. Accessed 23 Feb 2014
134. Medina-Franco JL (2013) Activity cliffs: facts or artifacts? Chem Biol Drug Des 81:553–556
135. Hu Y, Bajorath J (2010) Molecular scaffolds with high propensity to form multi-target activity cliffs. J Chem Inf Model 50:500–510
136. Wassermann AM, Bajorath J (2010) Chemical substitutions that introduce activity cliffs across different compound classes and biological targets. J Chem Inf Model 50:1248–1256
137. Martin YC, Kofron JL, Traphagen LM (2002) Do structurally similar molecules have similar biological activities? J Med Chem 45:4350–4358
138. Thor and Merlin; Version 4.62; Daylight Chemical Information Systems, Inc., Irvine, CA. http://www.daylight.com. Accessed 12 Jan 2014
139. Brown RD, Martin YC (1998) An evaluation of structural descriptors and clustering methods for use in diversity selection. SAR QSAR Environ Res 8:23–39
140. Patterson DE, Cramer RD, Ferguson AM, Clark RD, Weinberger LE (1996) Neighborhood behavior: a useful concept for validation of "molecular diversity" descriptors. J Med Chem 39:3049–3059
141. Steffen A, Kogej T, Tyrchan C, Engkvist O (2009) Comparison of molecular FP methods on the basis of biological profile data. J Chem Inf Model 49:338–347
142. Wikipedia. Curse of dimensionality. http://en.wikipedia.org/wiki/Curseof_dimensionality. Accessed 19 Jan 2014
143. Hecht-Nielsen R (1990) Neurocomputing. Addison-Wesley, Reading
144. Rupp M, Proschak E, Schneider G (2007) Kernel approach to molecular similarity based on iterative graph similarity. J Chem Inf Model 47:2280–2286
145. Joliffe IT (2002) Principle component analysis, 2nd edn. Springer, New York
146. Borg I, Groenen P (1997) Modern multi-dimensional scaling. Springer, New York
147. Dominc D, Devillers J, Chastrette M, Karcher W (1993) Non-linear mapping for structure-activity and structure-property modeling. J Chemometr 7:227–242

148. Malinowski ER (1991) Factor analysis in chemistry, 2nd edn. Wiley, New York
149. Raghavendra AS, Maggiora GM (2007) Molecular basis sets—a general similarity-based approach for representing CSs. J Chem Inf Model 47:1328–1340
150. Kruskal J (1977) The relationship between multidimensional scaling and clustering. In: Van Ryzin J (ed) Classification and clustering. Academic, New York, pp 17–44
151. Diamantaras KI, Kung SY (1996) Principal component neural networks: theory and applications. Wiley, New York
152. Molecular Operating Environment (MOE). Chemical computing group, Montreal, Quebec, Canada. http://www.chemcomp.com. Accessed 26 Feb 2014
153. Mason JS, Good AC, Martin EJ (2001) 3-D pharmacophores in drug discovery. Curr Pharm Des 7:567–597
154. Agrafiotis DK, Xu H (2003) A geodesic framework for analyzing molecular similarities. J Chem Inf Model 43:475–484
155. Agrafiotis DK, Xu H (2002) A self-organizing principle for learning non-linear manifolds. Proc Nat Acad Sci U S A 99:15869–15872
156. Agrafiotis DK (2003) Stochastic proximity embedding. J Comput Chem 24:1215–1221
157. Xue L, Stahura FL, Bajorath J (2004) Cell-based partitioning. In: Chemoinformatics: concepts, methods, and tools for drug discovery, Chapter 9. Humana , Totowa
158. Wickens TD (2009) Multiway contingency tables analysis for the social sciences. Psychology, New York
159. Bayley MJ, Willett P (1999) Binning schemes for partition-based compound selection. J Mol Graphics Model 17:10–18
160. Rush JA (1999) Cell-based methods for sampling in high-dimensional spaces. In: Truhlar DG, Howe WJ, Hopfinger AJ, Blaney J, Dammkoehler RA (eds) Rational drug design. Springer, New York, pp 73–79
161. Jain AK, Dubes RC (1988) Algorithms for clustering data. Prentice-Hall, Englewood Cliffs
162. Maggiora GM, Shanmugasundaram V, Lajiness MS, Doman TN, Schultz MW (2004) A practical strategy for directed compound acquisition. In: Oprea TI (ed) Chemoinformatics in drug discovery. Wiley-VCH, Weinheim
163. Hassan M, Bielawski JP, Hempel JC, Waldman M (1996) Optimization and visualization of molecular diversity of combinatorial libraries. Mol Divers 2:64–74
164. Sotriffer C, Manhold R, Kubinyi H, Folkers G (2011) Virutal screening—principles, challenges, and practical guidelines. Wiley, New York
165. Bork P, Jensen LJ, von Mering C, Ramani AK, Lee I, Marcotte EM (2004) Protein interaction networks from yeast to human. Curr Opin Struct Biol 14:292–299
166. Yamanishi Y, Araki M, Gutteridge A, Honda W, Kanehisa M (2008) Prediction of drug-target networks from the integration of chemical and genomic spaces. Bioinformatics 24:1232–1240
167. Zhao S, Li S (2010) Network-based relating pharmacological and genomic spaces for drug target identification. PLoS ONE 5(7):e11764. doi:10.1371/journal.pone.0011764
168. Tanaka N, Ohno K, Niimi T, Moritomo A, Mori K, Orita M (2009) Small-world phenomena in chemical library networks: application to fragment-based drug discovery. J Chem Inf Model 49:2677–2686
169. Krein MP, Sukumar N (2011) Exploration of the topology of chemical spaces with network measures. J Phys Chem A 115:12905–12918
170. Wawer M, Peltason L, Weskamp N, Teckentrup A, Bajorath J (2008) Structure-activity relationship anatomy by network-like similarity graphs and local structure-activity relationship indices. J Med Chem 51:6075–6084
171. Ripphausen P, Nisius B, Wawer M, Bajorath J (2011) Rationalizing the role of SAR tolerance for ligand-based virtual screening. J Chem Inf Model 51:837–842
172. Stumpfe D, Dimova D, Bajorath J (2014) Composition and topology of chemical spaces with network measures. J Chem Inf Model 54:451–461
173. Benz RW, Swamidass SJ, Baldi P (2008) Discovery of power-laws in chemical space. J Chem Inf Model 48:1138–1151

174. Oprea TI, Gottfries J (2001) Chemography: the art of navigating in chemical space. J Comb Chem 3:157–166
175. Albert R, Barabási A-L (2002) Statistical mechanics of complex networks. Rev Mod Phys 74:47–97
176. Harary F (1969) Graph theory. Addison-Wesley, Reading
177. Bolla M (2013) Spectral clustering and biclustering—learning large graphs and contingency tables. Wiley, New York
178. Kolaczyk ED (2009) Statistical analysis of network data—methods and models. Springer, New York
179. Liu B (2011) Web data mining: exploring hyperlinks, contents, and usage data. Springer, Heidelberg
180. van Steen M (2010) Graph theory and complex networks—an introduction. Maarten van Steen
181. Amaral LAN, Scala A, Barthélémy M, Stanley HE (2000) Classes of small-world networks. Proc Nat Acad Sci U S A 97:11149–11152
182. Barabási A, Albert R (1999) Emergence of scaling in random networks. Science 286:509–512
183. Devore JL, Berk KN (2011) Modern mathematical statistics with applications. Springer, New York
184. Irwin JJ, Shoichet BK (2005) ZINC—a free database of commercially available compoundsfor virtual screening. J Chem Inf Model 45:177–182
185. Rajan K (ed) (2013) Informatics for materials science and engineering: data-driven discovery for accelerated experimentation and applications. Elsevier, New York
186. Hudson BD, Hyde RM, Rahr E, Wood J, Osman J (1996) Parameter based methods for compound selection from chemical databases. Quant Struct-Act Relat 15:285–289
187. Holliday JD, Willett P (1996) Definitions of "dissimilarity" for dissimilarity-based compound selection. J Biomolec Screen 1:145–151
188. Menard PR, Lewis RA, Mason JS (1998) Rational screening set design and compound selection: cascaded clustering. J Chem Inf Comput Sci 38:497–505
189. Young SS, Lam RLH, Welch WJ (2002) Initial compound selection for sequential screening. Curr Opin Drug Discov Devel 5:422–427
190. Waldman M, Li H, Hassan M (2000) Novel algorithms for the optimization of molecular diversity of combinatorial libraries. J Mol Graph Model 18:412–426
191. Agrafiotis DK (1998) Diversity in chemical libraries. In Schleyer PvR, Allinger NL, Clark T, Gasteiger J, Kollman PA, Schaefer HF III, Schreiner PR (eds) The Encyclopedia of Computational Chemistry, pp 742–761, John Wiley & Sons, Chichester
192. Shanmugasundaram V, Maggiora G (2011) Application of Shannon-like diversity measures to cell-based chemistry spaces. J Math Chem 49:342–355
193. Willett P (2000) Chemoinformatics—similarity and diversity in chemical libraries. Curr Opin Biotechnol 11:85–88
194. Willett P (2004) Evaluation of molecular similarity and molecular diversity methods using biological activity data. In: Bajorath J (ed) Chemoinformatics: concepts, methods, and tools for drug discovery, Chapter 2. Springer, New York
195. Martin Y (ed) (2001) Diverse viewpoints on computational aspects of molecular diversity. J Comb Chem 3:231–250
196. Matter H (1997) Selecting optimally diverse compounds from structure databases: a validation study of two-dimensional and three-dimensional molecular descriptors. J Med Chem 40:1219–1229
197. Dunbar JB (2000) Compound acquisition strategies. Pac Symp Biocomput 5:552–562
198. Olah MM, Bologa CG, Oprea TI (2004) Strategies for compound selection. Curr Drug Discov Technol 1:211–220
199. Ma C, Lazo JS, Xie X-Q (2011) Compound acquisition and prioritization algorithm for constructing structurally diverse compound libraries. ACS Comb Sci 13:223–231

200. Lipinski CA, Lombardo F, Dominy BW, Feeney PJ (2001) Experimental and computational approaches to estimates solubility and permeability in drug discovery and development settings. Adv Drug Deliv Rev 46:3–26
201. Petit J, Meurice N, Kaiser C, Maggiora G (2012) Softening the rule of five—where to draw the line? Bioorg Med Chem 20:5343–5351
202. Bickerton GR, Pailini GV, Besnard J, Muresan S, Hopkins AL (2012) Quantifying the chemical beauty of drugs. Nat Chem 4:90–98
203. Klebe G (ed) (2000) Virtual screening: an alternative or complement to high throughput screening? Kluwer Academic, Dordrecht
204. Varnek A, Tropsha A (eds) (2008) Chemoinformatics approaches to virtual screening. RSC Publishing, Cambridge
205. Böhm H-J, Schneider G, Kubinyi H, Manhold R, Timmerman H (eds) (2008) Virtual screening for bioactive molecules. Wiley, New York
206. Bajorath J (2002) Integration of virtual and high-throughput screening. Nat Rev Drug Discov 1:882–894
207. Glen RC, Adams SE (2006) Similarity metrics and descriptor spaces—which combinations to choose? QSAR Combin Sci 25:1133–1142
208. Eckert H, Bajorath J (2007) Molecular similarity analysis in virtual screening: foundations, limitations and novel approaches. Drug Discov Today 12:225–233
209. Rester U (2008) From virtual reality—virtual screening in lead discovery and lead optimization: a medicinal chemistry perspective. Curr Opin Drug Discov Devel 11:559–568
210. Bajorath J (2009) Methods for ligand-based virtual screening. Frontiers Med Chem 4:1–22
211. Schneider G (2010) Virtual screening: an endless staircase? Nat Rev Drug Discov 9:273–276
212. Geppert H, Vogt M, Bajorath J (2010) Current trends in ligand-based virtual screening: molecular representations, data mining methods, new application areas, and performance evaluation. J Chem Inf Model 50:205–216
213. Stumpfe D, Bajorath J (2011) Similarity searching. WIREs Comput Mol Sci 1:260–282
214. Scior T, Bender A, Tresadern G, Medina-Franco JL, Mayorga KM, Langer T, Cuanalo-Contreras K, Agrafiotis DK (2012) Recognizing pitfalls in virtual screening: a critical review. J Chem Inf Model 52:867–881
215. Lavecchia A, Di Giovanni C (2013) Virtual screening strategies in drug discovery: a critical review. Curr Med Chem 20:2839–2860
216. Parker CN, Bajorath J (2006) Towards unified compound screening strategies: a critical evaluation of error sources in experimental and virtual high-throughput screening. QSAR Combin Sci 25:1153–1161
217. Yuriev E, Agostino M, Ramsland PA (2010) Challenges and advances in computational docking: 2009 in review. J Mol Recognit 24:149–164
218. Huang S-Y, Zou X (2010) Advances and challenges in protein-ligand docking. Int J Mol Sci 11:3016–3034
219. Waszkowycz B, Clark DE, Gancia E (2011) Outstanding challenges in protein-ligand docking and structure-based virtual screening. WIREs Comput Mol Sci 1:229–259
220. Mestres J, Rohrer DC, Maggiora GM (1997) A molecular field-based similarity approach to pharmacophoric pattern recognition. J Mol Graphics Model 15:114–121
221. Putta S, Lemmen l, Beroza P, Greene J (2002) A novel shape-feature based approach to virtual library screening. J Chem Inf Comput Sci 42:1230–1240
222. Koes DR, Camacho CJ (2011) Pharmer: efficient and exact pharmacophore search. J Chem Inf Model 51:1307–1314
223. Langer T (2010) Pharmacophores in drug research. Mol Inf 29:470–475
224. Mestres J, Rohrer DC, Maggiora GM (1997) MIMIC: a molecular-field matching program: exploiting applicability of molecular similarity approaches. J Comp Chem 18:934–954
225. Ballester PJ, Richards WG (2007) Ultrafast shape recognition for similarity search in molecular databases. Proc Roy Soc A 463:1307–1321

226. Hawkins P, Skillman A, Nicholls A (2007) A comparison of shape-matching and docking as virtual screening tools. J Med Chem 50:74–82
227. McGaughey GB, Sheridan RP, Baylly CI et al (2007) Comparison of topological shape and docking methods in virtual screening. J Chem Inf Model 47:1504–1519
228. Ebalunode JO, Zheng W (2009) Unconventional 2D shape similarity method affords comparable enrichment as a 3D shape method in virtual screening experiments. J Chem Inf Model 49:1313–1320
229. Yongye AB, Bender A, Martinez-Mayorga (2010) Dynamic clustering threshold reduces conformer ensemble size while maintaining a biologically relevant ensemble. J Comput Aided Mol Des 24:675–686
230. Stanton DT, Morris TW, Siddhartha R, Parker C (1999) Application of nearest-neighbor and cluster analyses in pharmaceutical lead discovery. J Chem Inf Comput Sci 39:21–27
231. Muchmore SW, Debe DA, Metz JT, Brown SP, Martin YC, Hajduk PJ (2008) Application of belief theory to similarity data fusion for use in analog searching and lead hopping. J Chem Inf Model 48:941–948
232. Swann SL, Brown SP, Muchmore SW, Patel H, Merta P, Locklear J, Hajduk PJ (2011) A unified, probabilistic framework for structure- and ligand-based virtual screening. J Med Chem 54:1223–1232
233. Sharma R, Lawrenson AS, Fisher NE et al (2012) Compound selection methods for a high-throughput screening program against a novel malarial target, PfNDH2: increasing hit rate via virtual screening methods. J Med Chem 55:3144–3154
234. Williams C (2006) Reverse fingerprinting, similarity searching by group fusion and fingerprint bit importance. Mol Divers 10:311–332
235. Xue L, Stahura FL, Godden JW, Bajorath J (2001) Fingerprint scaling increases the probability if identifying molecules with similar activity in virtual screening callculations. J Chem Inf Comput Sci 41:746–753
236. Xue L, Godden JW, Stahura FL, Bajorath J (2003) Profile scaling increases the similarity search performance of molecular fingerprints containing numerical descriptors and structural keys. J Chem Inf Comput Sci 43:1218–1225
237. Schuffenhauer A, Floersheim P, Acklin P, Jacoby E (2003) Similarity metrics for ligands reflecting the similarity of the target proteins. J Chem Inf Comput Sci 43:391–405
238. Kogej T, Engkvist Blomberg N, Muresan S (2006) Multifingerprint based similarity searches for targeted class compound selection. J Chem Inf Model 46:1201–1213
239. Batista J, Bajorath J (2008) Distribution of randomly generated activity class characteristic substructures in diverse active and database molecules. Mol Divers 12:77–83
240. Lounkine E, Auer J, Bajorath J (2008) Formal concept analysis for the identification of molecular fragment combinations specific for active and highly potent compounds. J Med Chem 51:5342–5348
241. Lounkine E, Hu Y, Batista J, Bajorath J (2009) Relevance of feature combinations for similarity searching using general or activity class-directed molecular fingerprints. J Chem Inf Model 49:561–570
242. Wassermann AM, Nisius B, Vogt M, Bajorath J (2010) Identification of descriptors capturing compound class-specific features by mutual information analysis. J Chem Inf Model 50:1935–1940

Chapter 2
The Chemical Space of Flavours

Lars Ruddigkeit and Jean-Louis Reymond

2.1 Introduction

In the complex array of molecules composing foods, flavourant molecules, although present in relatively small amounts, play a central role in determining the food flavour in terms of taste and smell. Taste molecules, which have very diverse chemical structures and properties, interact directly with receptors in the mouth to trigger taste perceptions of bitter, sweet, sour, acidic, salty and umami [1]. Fragrances are generally small, apolar and volatile compounds, which must reach olfactory receptor neurons in the upper part of the nose to trigger the complex perception of smell through interactions with approximately 900 genetically distinct G-protein-coupled olfactory receptors [2–6]. Fragrances are also used as ingredients in perfumes, soaps, shampoos or lotions. Classifications of fragrances, according to their perceived smell, produce tens to hundreds of fragrance families, although a general characterization system of smell is still difficult due to perceptual qualities [7]. The relationship between structural types and odour types is very diverse. Herein, we discuss flavourant molecules collected from the open-access databases, SuperScent [8], Flavornet [9], BitterDB [10] and SuperSweet [11], in an overall perspective of the chemical space classification of molecules to convey a global understanding of this molecular class independent of detailed structure–activity relationships [12]. This global view provides a conceptual framework to understand the chemical structural diversity of taste and smell and suggests approaches to discover new flavours through chemical space exploration.

J.-L. Reymond (✉) · L. Ruddigkeit
Department of Chemistry and Biochemistry, University of Bern, Freiestrasse 3, 3012 Bern, Switzerland
e-mail: jean-louis.reymond@dcb.unibe.ch

K. Martinez-Mayorga, J. L. Medina-Franco (eds.), *Foodinformatics*,
DOI 10.1007/978-3-319-10226-9_2

2.2 Flavour Molecules

2.2.1 Databases of Organic Molecules

Organic molecules consist of a few tens of atoms of various types (carbon, hydrogen, nitrogen, oxygen, sulphur, halogens and a few others) linked together via kinetically stable covalent single or multiple bonds. The atoms and their connectivity pattern including their three-dimensional relative positions define the molecule's identity, its molecular shape and its physicochemical and biological properties. Since the discovery of organic molecules, as the elementary building blocks of living matter, many millions of different organic molecules have been reported in the literature either as naturally occurring compounds or as the products of chemical syntheses.

Most efforts have been devoted to the area of medicinal chemistry where molecules are investigated for their drug properties. The cumulated knowledge acquired there has been placed, in part, in the public domain thanks to open-access initiative, such as the US National Institute of Health PubChem database, in which the structure and possible biological evaluation of more than 30 million of organic molecules are freely accessible [13]. The Royal Society of Chemistry runs a similar but broader open-access archive in the form of ChemSpider, a repository in which authors are encouraged to deposit their structures [14]. Additional public databases of molecules of medicinal interest are listed in Table 2.1, including collections of commercially available compounds in ZINC [15], annotated database of bioactive molecules such as ChEMBL [16] and DrugBank [17], and very large databases of theoretically possible molecules covering the entire range of what is feasible with organic chemistry, such as the chemical universe databases GDB-11 [18], GDB-13 [19] and GDB-17 [20], which list all organic molecules possible up to 11, 13 and 17 atoms obeying simple rules for chemical stability and synthetic feasibility [21].

When considering flavourants, hundreds of thousands of molecules have been investigated for their fragrant properties by various fragrance companies worldwide. However, there has been only very limited effort to establish a broad repository of flavour molecules. Nevertheless, several relatively small databases have been made accessible online in the last few years: SuperScent [8] and Flavornet [1], which list almost 2000 documented fragrances and their properties; BitterDB [10], which lists 606 molecules with documented bitter taste, containing many alkaloids; and SuperSweet [11], which list 342 molecules with proven or likely sweet taste, containing, in particular, a broad range of glycosides. When combined together, SuperScent and Flavornet assemble to a collection of 1760 different fragrance molecules, here named FragranceDB. BitterDB and SuperSweet similarly combine to 806 taste molecules, here named TasteDB.

2.2.2 Property Profiles

The properties of drug-like molecules have been extensively discussed in the literature focussing on the characteristics necessary for oral bioavailability in the form of

Table 2.1 Databases of organic molecules as of December 2013

Database	Description	Size	Web address
PubChem	Database of known molecules form various public sources	38.8 M	http://pubchem.ncbi.nlm.nih.gov
ChemSpider	Integrated resource of Royal Society of Chemistry	28.0 M	http://www.chemspider.com/
ZINC	Commercial small molecules	13.5 M	http://zinc.docking.org
ChEMBL	Bioactive drug-like small molecules annotated with experimental data	1.5 M	https://www.ebi.ac.uk/chembldb
DrugBank	Experimental and approved small-molecule drugs	6825 M	http://www.drugbank.ca
SuperScent	Database of scents from literature	1591 M	http://bioinf-applied.charite.de/superscent/
Flavornet	Volatile compounds from the literature based on GC–MS	738 M	http://flavornet.org
FragranceDB	SuperScent + Flavornet	1760 M	–
SuperSweet	Database of carbohydrates and artificial sweeteners	342 M	http://bioinf-applied.charite.de/sweet/index.php?site=home
BitterDB	Database of bitter Cpds from literature and Merck index	606 M	http://bitterdb.agri.huji.ac.il/bitterdb/
TasteDB	SuperSweet + BitterDB	806 M	–
GDB-11	Possible small molecules up to 11 atoms of C, N, O, F	26.4 M	http://www.gdb.unibe.ch
GDB-13	Possible small molecules up to 13 atoms of C, N, O, S, Cl	980 M	http://www.gdb.unibe.ch
GDB-17	Possible small molecules up to 17 atoms of C, N, O, S, halogen	166.4 G	http://www.gdb.unibe.ch

Lipinski's "rule of five", which sets boundaries to molecular weight (MW ≤ 500 Da), the octanol–water partition coefficient P (log$P$ ≤ 5), and the number of hydrogen-bond-donor atoms (HBD ≤ 5) and hydrogen-bond-acceptor atoms (HBA ≤ 10) [9]. A narrower definition with tighter boundaries on molecular weight (MW ≤ 300 Da), polarity (log$P$ ≤ 3) and flexibility in terms of rotatable bonds (RBC ≤ 3) have also been defined to select molecules suitable as "fragments", which are generally smaller molecules showing weak activities, but which can be optimized by adding substituents [22].

A similar set of boundaries has not been proposed for flavours. While the property ranges necessary for taste molecules is a priori rather large, one can guess that for fragrant molecules, upper values in terms of molecular weight and polarity are necessary to enable a minimum amount of volatility, which is the key feature necessary for fragrances to reach their site of action. To understand which boundaries are suitable, we present herein the property profiles of the flavour collections, FragranceDB and TasteDB, and compare them with those of drug-like molecules in ChEMBL (bioactive molecules) [16], ZINC (commercial compounds for bioactivity screening) [15] and GDB-13 (possible molecules up to 13 atoms) [19].

The heavy-atom count (HAC, heavy atoms = all non-hydrogen atoms) profile shows that FragranceDB contains predominantly very small molecules with an upper boundary at approximately 21 atoms (Fig. 2.1a). A frequency peak appears at

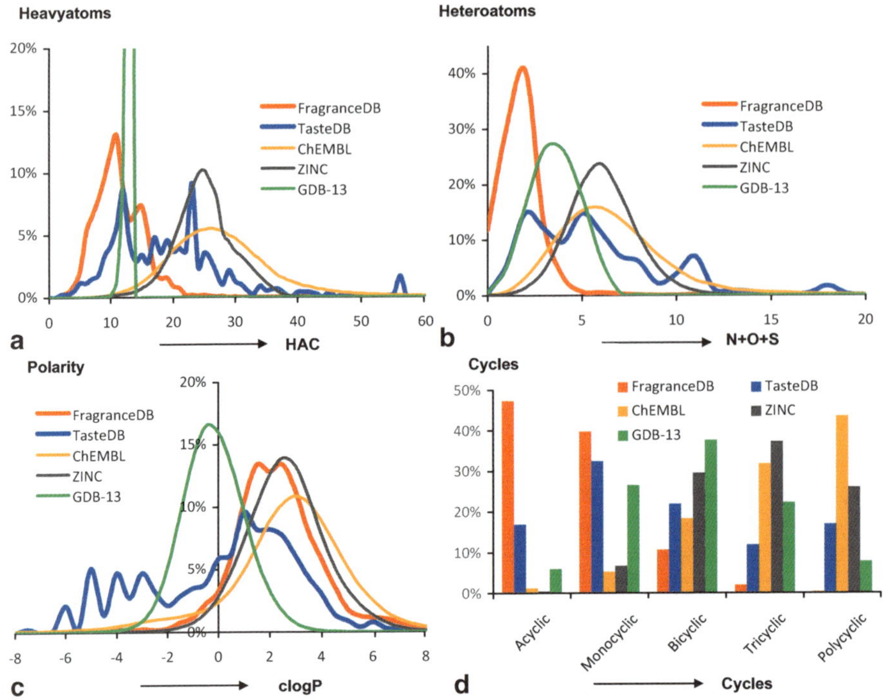

Fig. 2.1 Property histograms of fragrance and taste databases in comparison to ChEMBL, ZINC and GDB-13

9–11 heavy atoms corresponding to a diverse constellation comprising aliphatic linear and branched alkenes, aldehydes, alcohols, ketones and esters, various simple benzene, phenol and benzaldehyde analogues, furanones and monoterpenes. FragranceDB shows only very limited size overlap with drugs (ChEMBL) and commercial drug-like compounds (ZINC), which peak at the size of 20–30 heavy atoms. The chemical universe database GDB-13 falls within the size boundary of FragranceDB and offers a very large diversity of potential fragrances, including, in particular, analogues of monoterpenes with 10–11 atoms. TasteDB, on the other hand, covers a much broader size range, in agreement with the fact that flavours do not require volatility to reach their site of action. An abundance peak is nevertheless visible at 10–12 atoms and corresponds to various hexoses and their reduced hexitols, together with monoterpenes (menthone, camphor, citronellol), coumarins, anisols and some amino acids. Taste molecules in the size range of drugs (20–30 atoms) correspond to simple di-glycosides as well as various alkaloids and aromatic compounds and peptides. The frequency peak at HAC =56 corresponds to steviol glycosides listed in the database SuperSweet [23].

The heteroatom composition of flavours versus drugs is best compared by considering the sum of oxygen, nitrogen and sulphur atoms (Fig. 2.1b). Halogens are rather rare in flavours, although organochlorine compounds such as sucralose have a sweet taste. FragranceDB stands out with a very low number of heteroatoms peaking at just two heteroatoms, which are mostly oxygen atoms as found in volatiles aldehydes and ketones, alcohols, carboxylic esters and acids. As for the HAC profile, the overlap with drug molecules in ChEMBL and drug-like compounds in ZINC, in terms of heteroatom numbers, is small because drug molecules generally have a larger number of functional groups due to their larger size. Note that drug molecules very often contain multiple nitrogen atoms as well as amide bonds which are almost entirely absent in fragrances. The GDB-13 database displays relatively more heteroatoms despite of the small molecular size due to a combinatorial enumeration favouring highly functionalized molecules. The heteroatom profile of TasteDB is much broader in line with the broader range of molecular weights, again a consequence of the abundance of sweet-tasting oligosaccharides, including the steviol glycosides with a high density of hydroxyl groups.

A further insight into global properties can be gained by considering the logarithm of the calculated octanol/water partition coefficient clog*P* as a measure of polarity (Fig. 2.1c). Clog*P* indicates lipophilic molecules at high positive values, water-soluble molecules at strongly negative values and amphiphilic molecules around zero. Here, FragranceDB overlaps nicely with the drug and drug-like molecules in ChEMBL and ZINC by covering the range $0 < \mathrm{clog}P < 5$, which is a polarity range well suitable for rapid diffusion in biological media. This probably reflects the necessity of fragrances to diffuse from the gas phase to the olfactory neurons to reach their receptors, which requires properties similar to those necessary for drugs to reach their site of action. This property is also shared by the majority of TasteDB; however, in this case a significant fraction of the database extends into negative clog*P* values, comprising monosaccharides, disaccharides and related polyols, steviol glycosides, and amino acids and peptides such as aspartame. It should be noted that GDB-13, which reflects the combinatorial enumeration of the entire chemical space, peaks at $\mathrm{clog}P = 0$ due to the large fraction of cationic polyamines in the database which extend into negative clog*P* values. Due to the large size of GDB-13, however (almost one billion molecules), the database still contains an extremely large number of molecules in the polarity range of fragrances compared to the other databases.

Structural rigidity is a defining molecular property in drugs because conformational entropy strongly reduces binding affinity. Generally, molecules with large number of cycles are more rigid and have a better chance to bind strongly and selectively to their target. Remarkably, FragranceDB is predominantly a collection of acyclic compounds, with an abundance of acyclic aliphatic alcohols, aldehydes, acids and esters, such as butter and fruit aroma (Fig. 2.1d). Monocyclic molecules are also abundant, in particular cyclic terpenes, such as limonene or menthol; and monocyclic aromatic molecules, such as cinnamaldehyde. The abundance of acyclic and monocyclic compounds in FragranceDB contrasts with the typical drug molecules in ChEMBL and ZINC, which tend to be polycyclic, also as a consequence of their size. The combinatorial enumeration of molecules in GDB-13 corresponding to the size range of fragrances favours bicyclic molecules as the most abundant topology. TasteDB contains mostly monocyclic molecules, many of which are monosaccharides, but also extends into polycyclic molecules due to the presence of oligosaccharides and steroids in the collection.

2.3 Visualizing the Chemical Space of Flavours

2.3.1 The Chemical Space

In the context of organic chemistry, the term "chemical space" describes the ensemble of all known and/or possible molecules, but also the various multidimensional "property spaces" that can be defined by assigning dimensions to numerical descriptors of molecular structures [24, 25]. Such property spaces provide a general organization principle, which helps understand the molecular diversity available in large databases often containing many millions of molecules (Table 2.1). To obtain visual representations of property spaces, one usually performs principal component analysis (PCA) and representation of the (PC1, PC2)-plane containing the largest variance. This mathematical procedure is equivalent to taking a picture of the multidimensional space from the angle showing the largest diversity (Fig. 2.2) [26–32].

Thousands of numerical descriptors of molecular structure are known, and the number of possible property spaces is therefore unlimited. Recently, we showed that the chemical space of molecular quantum numbers (MQN), a set of 42 simple integer value descriptors counting atoms, bonds, polar groups and topological fea-

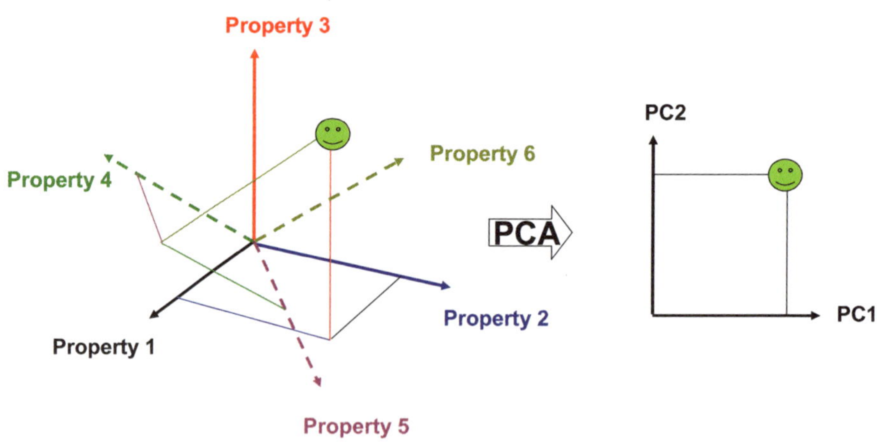

Fig. 2.2 Principal component analysis (*PCA*) projects a multidimensional property space into the plane of the largest variance

tures, such as cycles, provides a simple classification system of large databases and produces insightful (PC1, PC2)-maps for a variety of databases [33]. These PC-maps separate molecules by their mass, the number of cycles and rotatable bonds and their polarity, as can be illustrated by colour coding with property values. We have used such MQN-space maps to design interactive searchable maps of various public databases including zoom-in function and visualization of the molecules with links to their source database in the form of a "Google-map"-type application freely available from www.gdb.unibe.ch [34]. A related classification system and interactive visualization system were also realized using a simplified molecular-input line-entry system (SMILES) fingerprint (SMIfp), counting the occurrences of characters occurring in the SMILES representation of molecules [35]. One of the most striking features of these classification systems is that they group molecules by their pharmacophoric features and biological activities, and thus enable virtual screening in prospective searches [36].

2.3.2 Maps of the Flavours—Chemical Space

To gain an overview of the chemical space of flavours, we have performed a PCA visualization of the merged database containing FragranceDB and TasteDB, totalling 2517 compounds. These databases are represented in their (PC1, PC2)-plane which can be considered as a general 2-D map of their chemical space.

For the case of the MQN-space representation shown in Fig. 2.3a–d, the molecules spread by increasing size in the horizontal PC1-axis covering 67.97% of data variability. The vertical PC2-axis separates molecules by structural rigidity covering 15.54% of data variability. The total data variability represented by the (PC1, PC2)-plane amounts to 83.51%, which is typical for the projection of large databases from MQN-space. The molecules are grouped in descending diagonal stripes grouping molecules with an increasing number of cycles and ring atoms. Acyclic and monocyclic compounds are the most abundant category in FragranceDB, respectively, TasteDB. The category map in Fig. 2.3d shows that FragranceDB is essentially an acyclic/monocyclic compound database of small molecules, while TasteDB extends in large and polycyclic molecules.

In the maps of the SMIfp-space shown in Fig. 2.3e–h, the PC1-axis covers 66.9% of data variability and the PC2-axis covers 18.97%, totalling to 85.87% of data variability visible in the (PC1, PC2)-plane. Molecules spread by increasing size along the descending diagonal (Fig. 2.3e). The horizontal PC1-axis separates molecules according to the number of nonaromatic carbons (Fig. 2.3g), and the vertical axis according to the number of aromatic carbons (Fig. 2.3f). When comparing the category map in Fig. 2.3h with the property values in Fig. 2.3e–h, one can appreciate that FragranceDB contains mostly nonaromatic molecules, which correspond, in large part, to the acyclic molecules seen in the MQN-map of Fig. 2.3b. On the

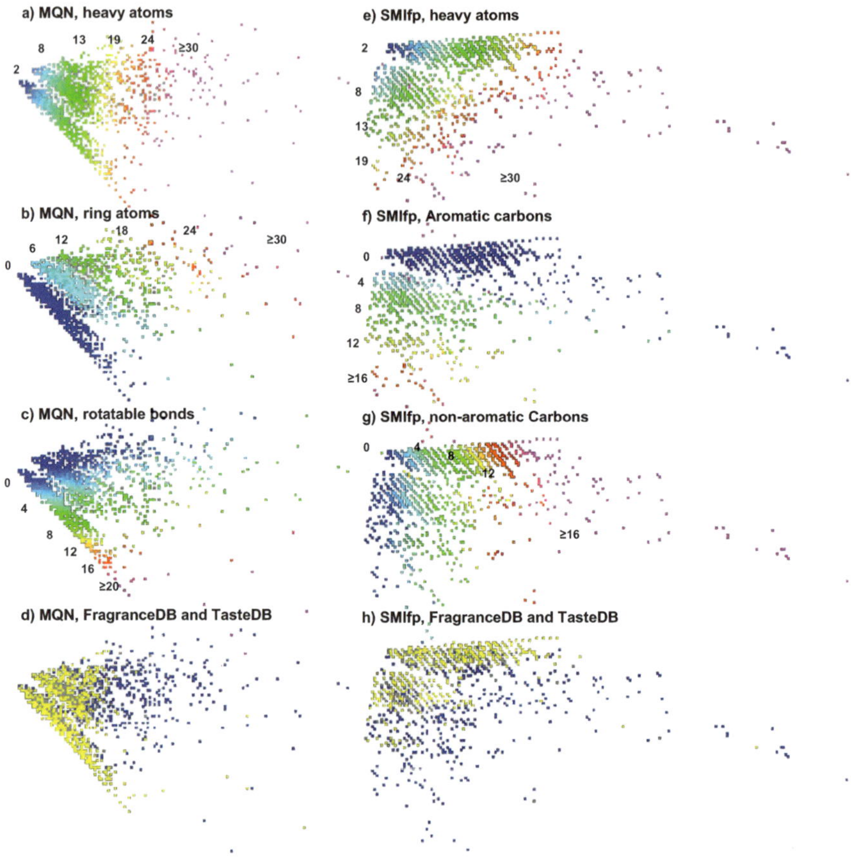

Fig. 2.3 Colour-coded maps of the flavours and taste chemical space. (PC1, PC2)-maps for PCA of the 42-dimensional MQN-space (**a–d**) and 34-dimensional SMIfp-space (**e–h**) are colour-coded by increasing value of the indicated property in the scale *blue–cyan–green–yellow–orange–red–magenta* with the corresponding value indicated on the map, for (**d, h**) *yellow* = flavour, *blue* = taste, and *grey* = pixel with mixed categories

other hand, TasteDB spans a broader range of SMIfp values, in particular, many taste molecules contain a large number of aromatic carbon atoms.

Overall, the MQN- and SMIfp-maps of the combined FragranceDB and TasteDB illustrate the broad range of structural types encountered in flavours. Note that the (PC1, PC2)-plane does not reflect any distribution of polarity properties. These are generally to be found in the PC3-dimension which requires additional representations not discussed here.

2.4 Fragrance Analogues in Chemical Space

2.4.1 *Similarity Searching by City-Block Distance*

The MQN- and SMIfp-spaces discussed in the previous section allow not only simple PCA-mapping of chemical space but also an extremely fast search for analogues using dedicated online browsers, which are freely accessible for use at www.gdb.unibe.ch. The browsers search for analogues of any query molecule as drawn in the query window using the principle of nearest neighbours in the multidimensional property space by measuring the city-block distance (CBD) between molecules. The CBD separating two molecules is the sum of the absolute differences between descriptor pairs across the 42 MQN and the 34 SMIfp descriptors. By pre-organizing databases according to file systems named X-MQN and X-SMIfp, databases of many millions of compounds can be searched within seconds for CBD_{MQN} and CBD_{SMIfp} neighbours, respectively, of any query molecule [37].

We have performed extensive comparisons between CBD and the more common Tanimoto coefficient as pairwise similarity measured between molecules and found the performance of both methods to be largely comparable, in particular, for the high-similarity pairs, i.e. both similarity measures will indicate the same molecules as the most similar, but differ substantially when considering very dissimilar compounds. On the other hand, searching according to the Tanimoto similarity is much slower than searching by CBD. The X-MQN and X-SMIfp systems incorporate additional options to direct any analogue search by restricting certain parameters in the analogues shown to certain subclasses (charges, HBD, HBA, elemental formula or compliance with drug-likeness rules), as visible in the search-window interface for the database ZINC using MQN-similarity searching (Fig. 2.4).

2.4.2 *Fragrance Analogues from MQN-Space*

The chemical space neighbourhood search gives particularly interesting results when considering fragrances. In the context of an analogue search within databases of commercially available compounds such as ZINC, one can identify interesting analogues by MQN- or SMIfp-similarity searching by preserving the number of HBD and HBA atoms, the electrostatic charges and optionally the elemental formula to avoid the selection of analogues with multiple heteroatoms, in particular nitrogen-rich heterocycles which are particularly abundant due to their importance in drug-discovery applications. Only the MQN-similarity search is exemplified here, but the SMIfp-similarity gives comparable results.

In Fig. 2.5, the MQN neighbours of the peppermint fragrance component, menthone, are shown. There are 27 commercially available compounds within $CBD_{MQN} \leq 12$, which is a useful distance boundary in the MQN-space [37]. These commercial analogues not only contain menthone itself (hit no. 1), a regioisomer (hit no. 2), but also various other cyclohexanones with the same number of acyclic

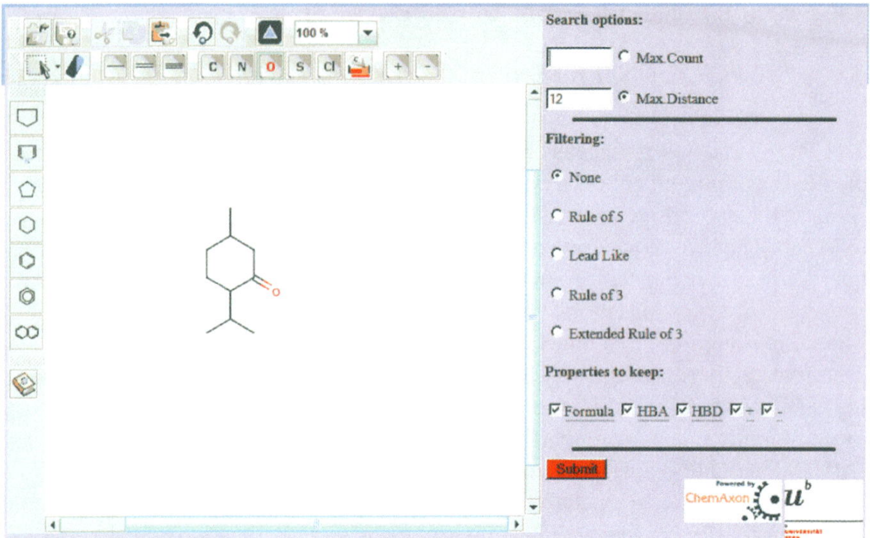

Fig. 2.4 The search-window option to identify the nearest neighbours of menthone in the MQN-space of the database ZINC (see also Sect. 1.4.2)

carbon atom substituents (hit nos. 3–9). Cycloheptanones (hit nos. 13–15) and cyclopentanones (hit nos. 26–27) are also proposed by the MQN-similarity search.

One can also extend the search to other databases containing a larger diversity of molecules. The chemical universe database GDB-13, which lists 977 million molecules of up to 13 atoms of C, N, O, S and Cl possible following simple rules of chemical stability and synthetic feasibility, is the largest database of small molecules to date [19]. GDB-13 is particularly relevant for fragrance analogue searches since it contains molecules in the size range most populated by fragrances; in particular, the majority of monoterpenes have less than 13 atoms. When applying the MQN-similarity search to typical fragrances, one can appreciate the very large number of high-similarity fragrance analogues that are possible, including isomers (Table 2.2). The vast majority of these molecules are presently unknown, and many do not pose any particular synthetic challenge, suggesting that large numbers of fragrant molecules remain to be explored.

2.5 Conclusion and Outlook

The general properties of flavour molecules, comprising fragrances which are relatively small organic compounds with few polar functional groups, such as to be volatile, and the more polar and diverse taste molecules, define a subset of the chemical space that is clearly separated from the well-known drug-like molecules. A global understanding of chemical space aided by representations such as the

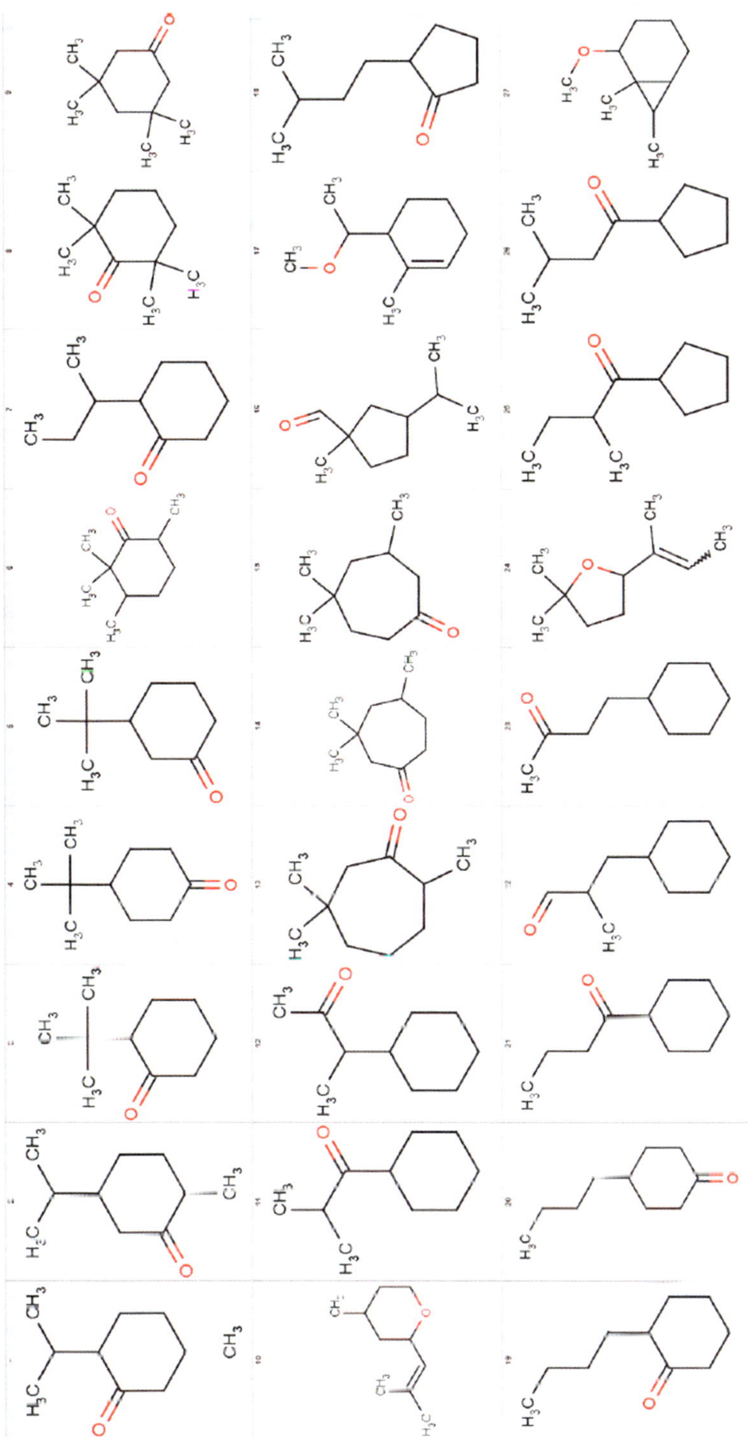

Fig. 2.5 MQN-nearest neighbour isomers of menthone (hit no. 1) in the ZINC database preserving the same number of H-bond donor atoms (0) and H-bond acceptor atoms (1)

Table 2.2 Number of fragrance analogues found by nearest-neighbour searching in the MQN-space of ZINC and GDB-13 within the distance boundary $CBD_{MQN} \leq 12$

Fragrance	Formula	ZINC $CBD_{MQN} \leq 12$		GDB-13 $CBD_{MQN} \leq 12$	
		All	Isomers	All	Isomers
Furaneol	$C_6H_8O_3$	200	3	14,412	54
Isoamyl acetate	$C_7H_{14}O_2$	3025	42	164,151	1025
Caprylic acid	$C_8H_{16}O_2$	1437	14	427,990	28
Vanillin	$C_8H_8O_3$	4771	34	397,263	2041
Cinnamaldehyde	C_9H_8O	1403	13	26,249	337
Limonene	$C_{10}H_{16}$	773	18	112,817	2141
α-Pinene	$C_{10}H_{16}$	64	9	65,614	1637
Camphor	$C_{10}H_{16}O$	200	11	243,162	9733
Menthone	$C_{10}H_{18}O$	1147	43	605,667	6858
Rose oxide	$C_{10}H_{18}O$	889	44	624,293	10,574
Menthol	$C_{10}H_{20}O$	734	26	383,641	1460
Citronellol	$C_{10}H_{20}O$	1642	38	2,927,465	5429
Lauraldehyde	$C_{12}H_{24}O$	260	4	93,700	5878

PC-maps of the MQN- and SMIfp-chemical spaces presented here, illustrate the extent of the structural diversity at hand. This chemical space is currently relatively sparsely populated compared to its potential, implying that many millions of additional flavour molecules remain to be discovered. Proximity searches in these chemical spaces can greatly facilitate the identification of flavour analogues.

The graphical and global understanding of flavour–chemical diversity presented in this chapter will probably serve as a confirmatory illustration of expert knowledge to fragrance chemists. On the other hand, such overviews are excellent tools to help in the dissemination of flavour chemistry to the broader scientific community and the definition of further goals in terms of exploring the flavour–chemical space. In particular, one can hypothesize that a thorough analysis of structure–activity relationships in a chemical space perspective could lead to a better understanding of the diversity of odour and taste perception and reveal the general principles underlying the genetic diversity of the olfactory system.

Acknowledgements This work was supported financially by the University of Bern and the Swiss National Science Foundation.

References

1. Cygankiewicz AI, Maslowska A, Krajewska WM (2013) Molecular basis of taste sense: involvement of GPCR receptors. Crit Rev Food Sci Nutr 54(6):771–780. doi:10.1080/10408398.2011.606929
2. Buck L, Axel R (1991) A novel multigene family may encode odorant receptors: a molecular basis for odor recognition. Cell 65(1):175–187. doi:http://dx.doi.org/10.1016/0092-8674(91)90418-X

3. Malnic B, Hirono J, Sato T, Buck LB (1999) Combinatorial receptor codes for odors. Cell 96(5):713–723. doi:http://dx.doi.org/10.1016/S0092-8674(00)80581-4
4. Shepherd GM (2004) The human sense of smell: are we better than we think? PLoS Biol 2(5):e146. doi:10.1371/journal.pbio.0020146
5. Mason JR, Clark L, Morton TH (1984) Selective deficits in the sense of smell caused by chemical modification of the olfactory epithelium. Science 226(4678):1092–1094
6. Briggs MH, Duncan RB (1961) Odour receptors. Nature 191:1310–1311
7. Kaeppler K, Mueller F (2013) Odor classification: a review of factors influencing perception-based odor arrangements. Chem Senses 38(3):189–209. doi:10.1093/chemse/bjs141
8. Dunkel M, Schmidt U, Struck S, Berger L, Gruening B, Hossbach J, Jaeger IS, Effmert U, Piechulla B, Eriksson R, Knudsen J, Preissner R (2009) SuperScent—a database of flavors and scents. Nucleic Acids Res 37(Suppl 1):D291–294. doi:10.1093/nar/gkn695
9. Lipinski CA, Lombardo F, Dominy BW, Feeney PJ (1997) Experimental and computational approaches to estimate solubility and permeability in drug discovery and development settings. Adv Drug Deliv Rev 23(1–3):3–25
10. Wiener A, Shudler M, Levit A, Niv MY (2012) BitterDB: a database of bitter compounds. Nucleic Acids Res 40(Database issue):D413–419
11. Ahmed J, Preissner S, Dunkel M, Worth CL, Eckert A, Preissner R (2011) SuperSweet—a resource on natural and artificial sweetening agents. Nucleic Acids Res 39(Suppl 1):D377–382. doi:10.1093/nar/gkq917
12. Kovatcheva A, Golbraikh A, Oloff S, Xiao Y-D, Zheng W, Wolschann P, Buchbauer G, Tropsha A (2004) Combinatorial QSAR of ambergris fragrance compounds. J Chem Inf Comp Sci 44(2):582–595. doi:10.1021/ci034203t
13. Wang Y, Xiao J, Suzek TO, Zhang J, Wang J, Bryant SH (2009) PubChem: a public information system for analyzing bioactivities of small molecules. Nucleic Acids Res 37(Web Server issue):W623–633
14. Williams AJ (2008) Public chemical compound databases. Curr Opin Drug Discov Devel 11(3):393–404
15. Irwin JJ, Sterling T, Mysinger MM, Bolstad ES, Coleman RG (2012) ZINC: a free tool to discover chemistry for biology. J Chem Inf Model 52(7):1757–1768. doi:10.1021/ci3001277
16. Gaulton A, Bellis LJ, Bento AP, Chambers J, Davies M, Hersey A, Light Y, McGlinchey S, Michalovich D, Al-Lazikani B, Overington JP (2012) ChEMBL: a large-scale bioactivity database for drug discovery. Nucleic Acids Res 40(Database issue):D1100–1107. doi:10.1093/nar/gkr777
17. Knox C, Law V, Jewison T, Liu P, Ly S, Frolkis A, Pon A, Banco K, Mak C, Neveu V, Djoumbou Y, Eisner R, Guo AC, Wishart DS (2011) DrugBank 3.0: a comprehensive resource for 'Omics' research on drugs. Nucleic Acids Res 39(Suppl 1):D1035–1041. doi:10.1093/nar/gkq1126
18. Fink T, Reymond JL (2007) Virtual exploration of the chemical universe up to 11 atoms of C, N, O, F: assembly of 26.4 million structures (110.9 million stereoisomers) and analysis for new ring systems, stereochemistry, physicochemical properties, compound classes, and drug discovery. J Chem Inf Model 47(2):342–353
19. Blum LC, Reymond JL (2009) 970 million druglike small molecules for virtual screening in the chemical universe database GDB-13. J Am Chem Soc 131(25):8732–8733
20. Ruddigkeit L, van Deursen R, Blum LC, Reymond JL (2012) Enumeration of 166 billion organic small molecules in the chemical universe database GDB-17. J Chem Inf Model 52(11):2864–2875. doi:10.1021/ci300415d
21. Reymond JL, Awale M (2012) Exploring chemical space for drug discovery using the chemical universe database. ACS Chem Neurosci 3(9):649–657
22. Congreve M, Carr R, Murray C, Jhoti H (2003) A rule of three for fragment-based lead discovery? Drug Discov Today 8(19):876–877
23. Ruddat M, Heftmann E, Lang A (1965) Steviol glycoside biosynthesis. Arch Biochem Biophys 110(3):496–499
24. Pearlman RS, Smith KM (1998) Novel software tools for chemical diversity. Perspect Drug Discov 9–11:339–353

25. Reymond JL, Van Deursen R, Blum LC, Ruddigkeit L (2010) Chemical space as a source for new drugs. Med Chem Comm 1:30–38. doi:10.1039/c0md00020e
26. Oprea TI, Gottfries J (2001) Chemography: the art of navigating in chemical space. J Comb Chem 3(2):157–166
27. Medina-Franco JL, Martinez-Mayorga K, Giulianotti MA, Houghten RA, Pinilla C (2008) Visualization of the chemical space in drug discovery. Curr Comput-Aided Drug Des 4(4):322–333. doi:10.2174/157340908786786010
28. Medina-Franco JL, Martinez-Mayorga K, Bender A, Marin RM, Giulianotti MA, Pinilla C, Houghten RA (2009) Characterization of activity landscapes using 2D and 3D similarity methods: consensus activity cliffs. J Chem Inf Model 49(2):477–491
29. Rosen J, Gottfries J, Muresan S, Backlund A, Oprea TI (2009) Novel chemical space exploration via natural products. J Med Chem 52(7):1953–1962
30. Singh N, Guha R, Giulianotti MA, Pinilla C, Houghten RA, Medina-Franco JL (2009) Chemoinformatic analysis of combinatorial libraries, drugs, natural products, and molecular libraries small molecule repository. J Chem Inf Model 49(4):1010–1024
31. Akella LB, DeCaprio D (2010) Cheminformatics approaches to analyze diversity in compound screening libraries. Curr Opin Chem Biol 14:325–330
32. Le Guilloux V, Colliandre L, Bourg S, Guénegou G, Dubois-Chevalier J, Morin-Allory L (2011) Visual characterization and diversity quantification of chemical libraries: 1. Creation of delimited reference chemical subspaces. J Chem Inf Model 51(8):1762–1774. doi:10.1021/ci200051r
33. van Deursen R, Blum LC, Reymond JL (2010) A searchable map of PubChem. J Chem Inf Model 50(11):1924–1934
34. Awale M, van Deursen R, Reymond JL (2013) MQN-Mapplet: visualization of chemical space with interactive maps of DrugBank, ChEMBL, PubChem, GDB-11, and GDB-13. J Chem Inf Model 53(2):509–518. doi:10.1021/ci300513m
35. Schwartz J, Awale M, Reymond JL (2013) SMIfp (SMILES fingerprint) chemical space for virtual screening and visualization of large databases of organic molecules. J Chem Inf Model 53(8):1979–1989. doi:10.1021/ci400206h
36. Blum LC, van Deursen R, Bertrand S, Mayer M, Burgi JJ, Bertrand D, Reymond JL (2011) Discovery of alpha7-nicotinic receptor ligands by virtual screening of the chemical universe database GDB-13. J Chem Inf Model 51:3105–3112
37. Ruddigkeit L, Blum LC, Reymond JL (2013) Visualization and virtual screening of the chemical universe database GDB-17. J Chem Inf Model 53(1):56–65. doi:10.1021/ci300535x

Chapter 3
Chemoinformatics Analysis and Structural Similarity Studies of Food-Related Databases

Karina Martinez-Mayorga, Terry L. Peppard, Ariadna I. Ramírez-Hernández, Diana E. Terrazas-Álvarez and José L. Medina-Franco

Chemoinformatics approaches to problem solving are commonly used in both academia and industry, and while a major focus is the pharmaceutical industry, many other sectors of the chemical industry lend themselves to it equally well. The chemoinformatic concepts, thoroughly discussed in Chap. 1 of this book, are general and can also be applied to address problems frequently encountered in food chemistry. A general strategy when applying these computational methods is to replace biological activity by a food-related property, for instance, flavor character or antioxidative activity. In many cases, the representation of the chemical structure remains the same (using, for example, molecular fingerprints, physicochemical and/or structure/substructure representations). In other words, structure-activity relationships (SAR) studies commonly conducted in medicinal chemistry for the purpose of drug discovery can be generalized to the study of structure–property relationships (SPR) for virtually any chemistry-related project [1]. Herein, we discuss representative and specific applications of methods used in chemoinformatics to mine data and characterize SPR information relevant to food chemistry. The chapter is organized into two major sections. First, we discuss exemplary applications of chemoinformatic analyses and characterization of the chemical space of compound databases. In this section, we cover major related concepts such as chemical space and molecular representation. The second section is focused on the application of similarity searching to food chemical databases.

J. L. Medina-Franco (✉) · K. Martinez-Mayorga
Departamento de Fisicoquímica, Instituto de Química, Universidad Nacional Autónoma de México, Av. Universidad 3000, Mexico City 04510, Mexico

Torrey Pines Institute for Molecular Studies, 11350 SW Village Parkway, Port St. Lucie, FL 34987, USA
e-mail: medinajl@unam.mx

T. L. Peppard
Robertet Flavors, Inc., 10 Colonial Dr. Piscataway, NJ 08854, USA

A. I. Ramírez-Hernández · D. E. Terrazas-Álvarez
Departamento de Fisicoquímica, Instituto de Química, Universidad Nacional Autónoma de México, Av. Universidad 3000, Mexico City 04510, Mexico

K. Martinez-Mayorga, J. L. Medina-Franco (eds.), *Foodinformatics*,
DOI 10.1007/978-3-319-10226-9_3

3.1 Chemoinformatic Analyses

Chemoinformatics, "cheminformatics," and "chemical information science" are different terms that have been coined for the common goal of applying informatics methods to solve chemical problems [2]. Chemoinformatics has also been defined as "a scientific field based on the representation of molecules as objects (graphs or vectors) in a chemical space" [3]. Further definitions are surveyed by Varnek and Baskin [3] and Willet [4]. Major aspects of chemoinformatics include the representation of chemical compounds, storing and mining information in databases, and generating and analyzing data [2].

Representation Molecular representation is at the core of chemoinformatics. There are two major types of representation: graphs and descriptor vectors. Graph-based approaches are applied to conduct structure and substructural analysis. These methods are easy to interpret and allow relatively straightforward communication with non-computational experts. Representations employing descriptor vectors are commonly used in chemoinformatics for database processing, clustering, similarity searching, and developing descriptive and predictive models of SAR; for example, QSPR/QSAR models and activity landscape models [1]. More than 5000 descriptors of different design have been developed [5]. The choice of descriptors used to analyze compound data sets gives rise to different chemical spaces.

In the food chemistry field, it has been recognized that there is a need for standardized food descriptions [6]. Food databases such as INFOODS contain free text. Representative databases relevant to the food chemistry field are presented in more detail in Chap. 9. Such databases require curation of their chemical structures as well as of the associated descriptions. Curation then involves the standardization of vocabulary, dictionaries to homogenize terms, and deletion of unnecessary wording. This is a tedious, but an important and necessary step. Relevant food databases not involving chemical structures are also in common use in the food industry. These databases may have different purposes, involving: cooking methods, ingredients, recipes, cuisine, and preparation location. In this context, the concept "food description" is used in a broad sense and applies to chemical and non-chemical databases. These databases allow for the sharing and exchange of food composition data. Some of the aspects that affect the quality of the information are: nutrient definitions, analytical methods used, and food description. The need for a "universal system" to describe and store food information has been recognized [6].

Another important aspect of food databases is that food and some food additives are, by nature, mixtures of components. For example, flavors frequently comprise or contain extracts of plants. Such mixtures and combinations of mixtures provide fertile ground for innovation. Similarly, in the search for bioactive molecules, natural products have been and continue to be a primary source of molecules with potential therapeutic effect. In fact, traditional medicine around the world is ancestral and still in use. An interesting example of this is the medicinal herb St John's wort (*Hypericum Perforatum*) which is prescribed in some countries for the treatment for depression [7]. The chemical composition and pharmacological effect of the individual constituents have been characterized; however, the less dramatic side effects typically observed cf. standard antidepressant drugs seems to be related to the mixture's complexity.

With the aim of standardizing the description of food-related databases and its analysis, Haddad et al. [8], for example, used a structural representation consisting of 1664 odorants, and used this information for classifying odorants based on similarity measures, as explained later in this chapter.

Chemical Space The concept of chemical space has broad application not only in drug discovery but also in virtually any chemistry-related dataset. It has been pointed out that "unlike real physical space, a chemical space is not unique; each ensemble of graphs and descriptors defines its own chemical space" [3]. Chemical space has been directly compared to the cosmic universe and several definitions have been proposed in the literature [9]. For example, Virshup et al. [10] recently defined chemical space as "an M-dimensional Cartesian space in which compounds are located by a set of M physicochemical and/or chemoinformatic descriptors." Comparison of the chemical space of compound collections is important for library selection and design [11]. When designing new libraries, or screening existing libraries, it is relevant to consider the chemical space coverage of the new compounds, the structural novelty, and the pharmaceutical relevance. Systematic analysis of the chemical space of compound libraries, in particular, large collections, requires computational approaches [12]. As we recently pointed out, depending on project goals, a wide range of approaches have been developed to populate, mine, and select relevant areas of chemical space [13].

It is possible to draw a direct analogy between chemical space and flavor space. A thorough discussion of chemical space is described elsewhere [9], while a comprehensive discussion of flavor and fragrance-relevant chemical space is discussed by Reymond et al. in Chap. 2 of this book.

Chemical Databases Chemical libraries vary in nature, composition, and design, and each may serve one or more specific purposes. Compound collections used for virtual (*in silico*) screening include combinatorial libraries, commercial vendors' compounds, and natural products [14]. Molecular databases may contain hundreds, thousands, or even millions of molecules, these may be existing chemicals, or they may be hypothesized compounds, e.g., for later chemical synthesis. Libraries of existing compounds may be commercial, public domain, or proprietary.

Such chemical databases can be used for a wide variety of purposes, such as the development and systematic analysis of SAR [15] and identification of polypharmacology [16]. The constant increase in the number of molecules stored in compound databases [17] has led to the concept of chemical space (*vide supra*).

Repurposing or repositioning of chemical compounds is an approach to accelerate the identification of a new use for a compound with a pre-existing use. Repurposing can be achieved computationally or experimentally or by using a combination of the two approaches. In the pharmaceutical area, it is known as drug repurposing [18] and represents an application based on increasing evidence for the concept of *polypharmacology*, i.e., that observed clinical effects are often due to the interaction of single or multiple drugs with multiple targets [19]. Reviews and discussions are described in the literature in an integrated manner with related concepts such as polypharmacology, chemogenomics, phenotypic screening, and high-throughput *in vivo* testing [20].

A number of food phytochemicals and food-related molecular databases are available [21]. Food and food-related databases are described in more detail in Chap. 9 of this book. Major examples of public databases of chemical compounds annotated with biological activity for drug-discovery applications have been developed. Prominent examples include: BindingDB, ChEMBL, PubChem, and WOrld of Molecular BioAcTivity (WOMBAT). These databases and others described in Chap. 9 can be analyzed and compared for knowledge of chemical space coverage and potential repurposing, for example, using the concept of similarity searching.

Chemoinformatic Profiling of Chemical Databases Chemoinformatics has a fundamental role in the diversity analysis of compound collections and in the mining of chemical space. Chemoinformatic approaches designed to mine and navigate through the chemical space of compound collections is described in detail elsewhere (Chap. 1 of this book). The various approaches in conducting chemoinformatic characterization of compound libraries are mainly distinguished by the structural representations and criteria used to characterize the chemical libraries. Typically, compound databases are compared using physicochemical properties, molecular scaffolds, or structural fingerprints. Following the same or similar approaches to those used to characterize databases of interest in the pharmaceutical industry, it is possible to conduct analysis of food chemical databases.

Since these three major types of structural representation are focused on specific aspects of the structures, it is convenient to use more than one criterion for comprehensive analysis of the structural and property diversity of molecular databases. This is because each of these methods has its own strengths and weaknesses. For example, the use of whole molecule properties (holistic properties) has the advantage of being intuitive and straightforward to interpret. However, physicochemical properties do not provide information regarding structural patterns, and molecules with different chemical structures can have the same or similar physicochemical properties. Similar to physicochemical descriptors, chemotypes or scaffolds may be readily interpreted and enable easy communication with medicinal chemists and biologists. For example, scaffold analysis has led to concepts which are widely used in medicinal chemistry and drug discovery, e.g., "scaffold hopping" [22] and "privileged structures" [23]. One of the shortcomings of molecular scaffold analysis is a lack of information regarding structural similarity primarily due to the side chains cf. the inherent similarity or dissimilarity of the scaffolds themselves. An obvious solution is the analysis not only of the molecular frameworks *per se* but also of the side chains, the functional groups, and other substructural analysis strategies [24].

Molecular fingerprints are widely used and have been successfully applied to a number of chemoinformatic and computer-aided molecular applications. A challenge of some fingerprints is that they are more difficult to interpret. Also, it is well known that chemical space may be highly dependent on the types of fingerprints used to derive it. In order to reduce the dependence of chemical space on the choice of structure representation, several SAR/SPR studies have implemented consensus methods in order to combine the information encoded by different molecular representations. Use of multiple fingerprints and representations to derive consensus conclusions (e.g., *consensus activity cliffs*) has been proposed as a solution [1].

We have conducted a comprehensive chemoinformatic characterization of a subset of the Flavor and Extract Manufacturers Association (FEMA) Generally Recognized As Safe (GRAS) list of approved flavoring substances (discrete chemical entities only) [25, 26]. To this end, we employed a set of rings, atom counts (carbon, nitrogen, oxygen, sulfur, and halogen atoms), six molecular properties (octanol/water partition coefficient, polar surface area, numbers of hydrogen bond donors and acceptors, number of rotatable bonds, and molecular weight), and seven structural fingerprints of different design: MACCS keys radial fingerprints (also known as extended connectivity fingerprints), chemical hashed fingerprints (implemented in ChemAxon), atom pair (Carhart), fragment pair, pharmacophore fingerprints, and weighted Burden number. In that work, we considered a set of 2244 compounds based on the FEMA GRAS list, complete through GRAS 25 [26]. An early version of this GRAS database is briefly described in Peppard et al. [27]. This data set was compared to a database of 1713 approved drugs, two databases of natural products (with 2449 and 467 molecules, respectively) a set of 10000 commercial compounds, a database of 2116 flavors and scents, and a collection of 32357 compounds used in traditional Chinese medicine. It was concluded that the molecular size of the GRAS flavoring substances and the SuperScent database is, in general, smaller cf. members of the other databases analyzed. The lipophilicity profile of these two databases, a key property to predict human bioavailability, was similar to approved drugs. Using a visual representation of chemical space based on a principal component analysis based on the number of aromatic rings and six additional molecular properties, it was concluded that a large number of GRAS chemicals overlapped a broad region of the property space occupied by drugs. The GRAS list analyzed in that work has high structural diversity, comparable to approved drugs, natural products, and libraries of screening compounds (Table 3.1).

Table 3.1 Reference databases used to characterize and compare FEMA GRAS list (3–25) and SuperScent

Database	Content	Size
FEMA GRAS	Flavors	2244
AnalytiCon	Natural products	2449
Specs NP	Natural products	467
DrugBank	Approved drugs	1713
SpecsWD3	Approved drugs	10000
TCM	Natural products	32357
SuperScent	Flavors and fragrances	2116

3.2 Similarity Searching

Computational approaches, including those based on molecular modeling and chemoinformatics tools, are increasingly being used to help identify compounds with biological activity. In particular, *in silico* or virtual screening is a valuable means of focusing experimental efforts on filtered sets of compounds yielding a higher probability of having the desired biological activity [28]. The rationale here is that the information of the system encoded in the computational procedure will increase the probability of identifying compounds with biological activity. Hit identification using computational screening requires several interactive and iterative steps and requires a careful selection of the methods to be used. The selection of a particular approach depends on the aim of the project, the information available for the system, and the computational resources available. In addition, one needs to consider the inherent limitations of each step involved and computational cost.

Virtual screening methods can be roughly organized into two major groups, namely, ligand based and structure based [29]. Ligand-based approaches use structure-activity data from a set of known actives in order to identify candidate compounds for experimental evaluation. A common ligand-based approach is based on the molecular similarity concept, which states that structurally similar molecules are more likely to have similar biological activity [30]. Significant exceptions to this rule do occur, with so-called activity cliffs describing situations where compounds with similar structure have, unexpectedly, very different biological activity [31]. Other ligand-based methods include substructure, clustering, quantitative structure-activity relationships (QSAR), pharmacophore, and three-dimensional (3D) shape matching techniques [32].

Structure-based approaches use the 3D structure of the target, usually obtained from X-ray crystallography or nuclear magnetic resonance (NMR) spectroscopy. However, in the absence of a receptor's 3D structural information, homology modeling [32] has successfully been used in virtual screening [33]. One of the most common structure-based methods is molecular docking. If information for both the experimentally active compound(s) and the 3D structure of the target are available, then the ligand- and structure-based virtual screening methods can be combined. Indeed, combining both methods increases the possibility of identifying active compounds [34].

Similarity searching is a typical ligand-based approach. Selection of the query or reference compounds in virtual screening is one of the crucial initial steps required for a successful outcome. Depending on both the dataset and the biological activity, it is possible that one or more reference compounds are associated with activity cliffs, i.e., that each might be a potential "activity cliff generator" [35]. An activity cliff generator is defined as a molecular structure that has a high probability of forming an activity cliff with molecules tested in the same biological assay. Since activity cliffs represent significant exceptions to the similarity principle, typically leading to erroneous results in similarity searching, it has recently been proposed that activity cliff generators be identified and removed from data sets before selecting reference compounds. Moreover, removal of activity cliff generators has been proposed as a general strategy, to be employed before developing predictive models such as those obtained with traditional QSAR, or other machine learning algorithms based on the similarity property principle [36].

Selection of chemical databases for similarity searching (or any other virtual screening approach) is another major component of the searching protocol. As mentioned in the previous section, a number of compound databases from different sources can be used. Notably, similarity searching can be applied to compound collections initially assembled for a different purpose, detailed above as repurposing. For example, Méndez-Lucio et al. recently conducted a 3D similarity search of DrugBank, a database of drugs approved for clinical use, with a distinct inhibitor of DNA methyltransferases, an emerging and promising epigenetic target for the treatment of cancer and other diseases [37]. The anti-inflammatory drug olsalazine was one of the most similar molecules to the reference compound, and it indeed showed hypomethylating activity based on a well-characterized live-cell imaging assay mediated by DNMT isoforms [38].

Information contained in databases is, in almost all cases, multivariate in nature; those related to food chemicals present particular challenges. One issue frequently encountered is that the chemical information is ambiguous. For example, materials may comprise a mixture of constituents, as in the case of essential oils; a mixture of isomers; or single components, but having incomplete stereochemical information. This adds to the unavoidable problem of missing information in chemical databases, such as protonation state of amino or carboxylic acid groups, prevalence of particular tautomers, etc. Moreover, these structural characteristics change depending on environment, for instance, when bound to a biological target (or targets). Since these are unavoidable and "dynamic" structural features, the preference is to ignore protonation states and consider the most stable tautomer for a given molecule.

When geometric isomers or stereoisomers are incompletely defined, one strategy is to consider all possible isomers in the computations. Alternatively, it is possible to use structural representations that do not take into account stereochemical information, although this will, of course, convey less chemical information. In the case of mixtures comprising multiple constituents, it is not possible to perform traditional chemoinformatic studies based on chemical structure (although there are studies that can be performed based purely on the nonstructural content of the databases). For such mixtures, e.g., essential oils, oleoresins, or other natural extracts, chemoinformatic studies can be performed if the composition and property description (organoleptic, biological activity, etc.) can be obtained for each constituent. In addition, the possibility of synergistic effects cannot be dismissed or, as in the case of St. John's wort, reduce side effects (in the treatment of mood disorders) due to the composition of the herb.

Another aspect to consider when dealing with food chemical databases is the dimensionality and, often times, the non-standardized description of the chemicals. In such cases, it is necessary to first use dictionaries or lexicons to ensure the information is as homogeneous as possible. This process, which is part of the curation of the database, may require manual intervention in which case it may not be entirely unbiased. Curation also includes deletion of unnecessary wording and of duplicates.

Once these steps have been performed, the database may now have chemicals without description; these will be discarded.

A final consideration is that the cleaned-up database which contains more than one description for each chemical is multi-dimensional cf. databases of chemical compounds containing just one biological activity. A similar scenario can be seen in the case of chemical databases containing the results of multiple biological assays.

There are reports in the literature by us and also by others facing these challenges. For example, both Zarzo et al. (*vide infra*) and our group have discussed the curation and chemoinformatic description of odor and flavor databases, respectively. Regarding the analysis of chemical structures, we performed structural similarity of chemical structures based on fingerprint representations. In this arena, Sprous et al. [39], Pintore et al. [40], and Jensen et al. [41] have reported related studies.

Zarzo et al. [42] characterized an odor database; the first step consisted of encoding the odor description of the database in a dichotomic format, where 0 corresponded to the absence of a given descriptor, while 1 represented its presence. From those data, the authors were able to perform a descriptive analysis of the database and show the incidence of each descriptor in the database. They also demonstrated associations among descriptors, in other words, pairs of descriptors that repeatedly were used together in the database. Lastly, using principal component analysis on a selected subset of the database, the authors constructed the corresponding "odor space." The 2D graphical representation of this odor space organized descriptors in the same regions of the plot that are intuitively similar, such as fruity (pineapple, berry, peach, cherry, apple, etc.), floral (rose, sweet, other floral), etc. One of the outcomes of this work was the presentation of an odor space which provides useful information when training sensory panels for odor profiling.

We performed a chemoinformatic analysis of the FEMA-GRAS list (containing both chemical structures and associated sensory attributes), the first steps of which comprised the compilation and curation of the database [25]. After standardization of descriptive flavor terms using a recognized sensory lexicon (ASTM, American Society for Testing and Materials publication DS 66) and removal of unnecessary wording, the resultant database was analyzed for the incidence of descriptors and their associations using three independent methods: principal component analysis, clustering, and flavor descriptor relationships. We found that certain descriptors appear in the same region of the flavor space generated with the principal component analysis, as well as within nearby clusters when generating a clustering-based heat map, and also in a pair-wise analysis of descriptor associations. The correspondence of results obtained with these three methods gives confidence in the results.

The concept of information content, commonly used in the field of chemoinformatics, has been applied to olfactory databases by Pintore et al. [40]. The challenge of establishing a standard olfactory description of chemicals is recognized by the authors. Two olfactory databases were compared, according to the consistency of odor description. Based on 2D representations, the authors applied several classification methods, along with corresponding means of validation. The authors related this consistency to the information content of the databases, and concluded that one of the main difficulties when working with odor databases is the subjectivity used, even by experts, to describe odor perception. Not surprisingly, this led to some wide discrepancies in descriptions of the same compound in the two databases. In this study, the 2D representations of the chemical structures included in the two databases were used to explore the consistency of the odor descriptions rather than to perform structural similarity with the aim of finding either similar compounds for structure–property relationships, or compounds with similar property profiles (biological activity, odor description, etc.).

Sprous and Salemme [39] reported a comparison of the FEMA GRAS compounds with compounds contained in the Drugbank database. The study was based on determining the chemoinformatic profile of the database (*vide supra*), computing the population of structural and physicochemical features, such as molecular weight, molecular flexibility, logP, logS, and numbers of acceptor, donor, acidic and basic atoms, etc. The authors concluded that, in general, GRAS compounds occupy a different and identifiable region of chemical space relative to pharmaceuticals. However, more recent subsets of the GRAS list, which contain fewer compounds from natural sources, are more diverse, thus expanding the chemical space occupied by compounds of previous versions of the FEMA/GRAS list.

Haddad et al. [8] developed a metric for odorant comparison based on a chemical space constructed from 1664 molecular descriptors. A refined version of this metric was devised following the elimination of redundant descriptors. The study included the comparison with models previously reported for nine datasets. The final, so-called multidimensional metric, based on Euclidean distances measured in a 32-descriptor space, was more efficient at classifying odorants cf. reference models previously reported. Thus, this study demonstrated the use of structural similarity for the classification of odors in multidimensional space.

In order to identify potential bioactivity among the food-flavoring components that comprise the FEMA GRAS list, we recently conducted ligand-based virtual screening for compounds with structures similar to approved antidepressant drugs [43]. The virtual screening was performed by means of fingerprint-based similarity searching. Valproic acid turned out to be the most similar antidepressant to a small number of GRAS compounds. Guided by the hypothesis that the inhibition of histone deacetylase-1 (HDAC1) may be associated with the efficacy of valproic acid in the treatment of bipolar disorder, we screened the GRAS compounds most similar to valproic acid for HDAC1 inhibition. The GRAS chemicals nonanoic acid and 2-decenoic acid inhibited HDAC1 at the micromolar level, with potency comparable to that of valproic acid. GRAS compounds likely do not exhibit strong enzymatic inhibitory effects at the concentrations typically employed in foods and beverages. As shown in that study, GRAS chemicals are able to bind, albeit weakly, to important therapeutic targets. Additional studies on bioavailability, toxicity at higher concentrations (GRAS flavor molecules being safe when used at or below the levels approved for foods and beverages) and off-target effects are warranted. The results of that work demonstrate that similarity searching followed by experimental evaluation can be used for rapid identification of GRAS chemicals with possible biological activity, with potential application for promoting health and wellness [43].

Table 3.2 GRAS flavor chemicals with highest similarity to known analgesics

CAS #	Name	Structure
1093200-92-0	N-[(4-Amino-2,2-dioxido-1H-2,1,3-benzothiadiazin-5-yl)oxy)]-2,2-dimethyl-N-propylpropanamide	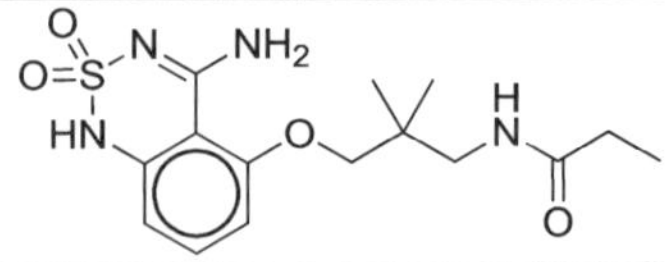
83-67-0	Theobromine	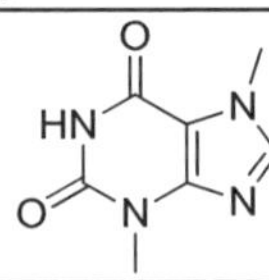

In two subsequent studies, again using structural similarity, we compared the FEMA GRAS list with analgesics and with compounds used as satiety agents. The list of analgesics comprised ten structurally diverse molecules currently used in the clinic. A total of eight satiety agents were identified in the literature, and these were used for similarity searching. The satiety agents included those currently used in the clinic, as well as those still in clinical trials.

In both studies, reference compounds were compared with the FEMA GRAS list using three software programs (MOE, ChemAxon, and PowerMV), with a total of seven structural representations. Compounds identified by different programs and representations were chosen as consensus compounds for further study. Then, a chemical space was constructed based on physicochemical properties. Nearest neighbors were identified based on Euclidian distances considering all the dimensions (properties). Based on the comparison of structural features and physicochemical properties, two FEMA GRAS compounds (listed on Table 3.2)were identified as similar to the reference analgesics. In the second study, a total of nine FEMA GRAS compounds were identified as similar to those used as reference satiety agents (see Table 3.3). For compounds having a known mode of action, *in vitro* studies using the identified GRAS chemicals could help determine whether or not they may have a satiety or analgesic effect in humans. However, it must be borne in mind that biological effects, in the large majority of cases, result from complex and multiple interactions in the body, as already described above in the area of polypharmacology.

Phytochemicals derived from eatable plants represent a remarkable source of bioactive compounds. In a recent study, Jensen et al. [41] performed a high-throughput analysis of phytochemicals in order to uncover associations between diet and health benefits using text mining and chemoinformatic methods. The first step of that study involved the extraction of associations between the terms of plants and phytochemicals, analyzing 21 million abstracts in PubMed/MEDLINE covering the period 1998–2012. This information was merged with the Chinese Natural Product Database and the Ayurveda dataset, which was also curated by the authors. The final dataset contained almost 37000 phytochemicals. A remarkable outcome

Table 3.3 GRAS flavor chemicals with highest similarity to known satiety agents

CAS #	Name	Structure
100-86-7	2-mehtyl-1-phenylpropan-2-ol	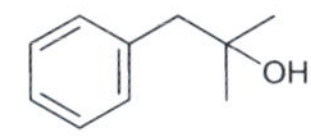
103-05-9	2-Methyl-4-phenyl-2-butanol	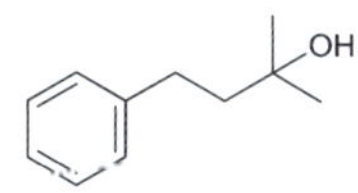
83-67-0	Theobromine	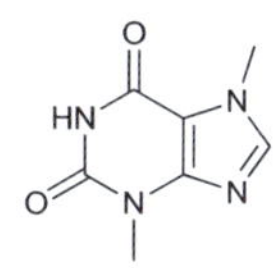
4265-16-1	2-Benzofurancarboxaldehyde	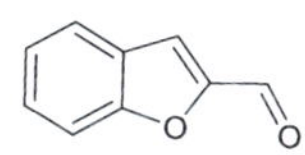
39537-23-0	L-Alanyl-L-glutamine	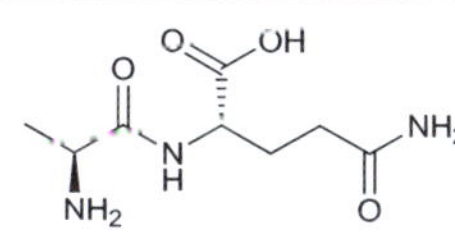
714229-20-6	Advantame	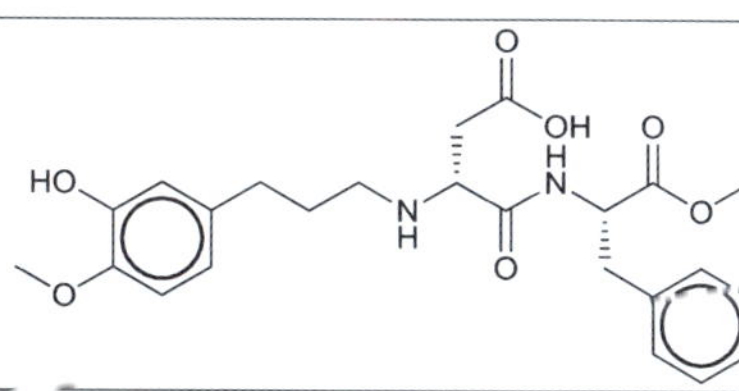
1323-75-7	(2Z)-2-Mehtyl-5-{2-methyl-3-methylidenebicyclo[2.2.1]heptan-2-yl}pent-2-en-1-yl 2-phenylacetate	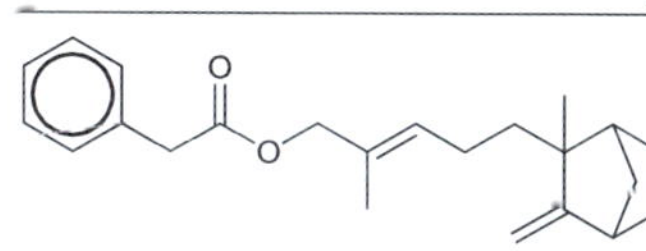
1139-30-6	(1R,4R,6R,10S)-9-Methylene-4,12,12-trimethyl-5-oxatricyclo[8.2.0.04,6]dodecane	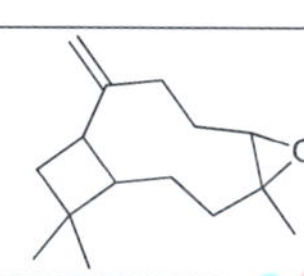
10024-57-4	(4-Methylphenyl) dodecanoate	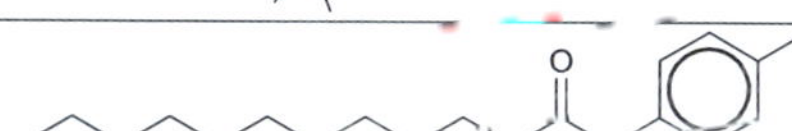

of that work is the structured and standardized database of phytochemicals associated with medicinal plants. As claimed by the authors, their approach facilitates the identification of novel bioactive compounds from natural sources, and the repurposing of medicinal plants for diseases other than those traditionally used for, with the added benefit that the information collected can help elucidate mechanism of action [41]. As a case study, the authors applied structural similarity searching in order to find molecules in their compiled database of phytochemicals with activity against a protein involved in the colon cancer pathway or a colon cancer drug target; the reference compounds were those reported in the ChEMBL database. A set of molecules from this study have not only reported health benefit against colon cancer but also verified activity against colon cancer protein targets.

The studies here described exemplify the application of the concepts and methodologies widely used in pharmaceutical settings, such as of data mining, diversity analysis, polypharmacology, repurposing, and similarity searching, in databases containing food additives and phytochemicals.

Acknowledgments K.M-M. thanks the Institute of Chemistry-UNAM and DGAPA-UNAM for funding (PAPIIT IA200513). The authors also wish to thank Robertet Flavors for permission to publish this chapter.

References

1. Medina-Franco JL, Yongye AB, López-Vallejo F (2012) Consensus models of activity landscapes. In: Matthias D, Kurt V, Danail B (eds) Statistical modeling of molecular descriptors in QSAR/QSPR. Wiley-VCH, Weinheim, pp 307–326
2. Engel T (2006) Basic overview of chemoinformatics. J Chem Inf Model 46:2267–2277
3. Varnek A, Baskin II (2011) Chemoinformatics as a theoretical chemistry discipline. Mol Inf 30:20–32
4. Willett P (2011) Chemoinformatics: a history. WIREs Comput Mol Sci 1:46–56
5. Todeschini R, Consonni V (2000) Handbook of molecular descriptors. Wiley-VCH, Weinheim
6. Pennington JT (2006) Issues of food description. Food Chem 57:145–148
7. Caccia S, Gobbi M (2009) St. John's wort components and the brain: uptake, concentrations reached and the mechanisms underlying pharmacological effects. Curr Drug Metab 10:1055–1065
8. Haddad R, Khan R, Takahashi YK, Mori K, Harel D, Sobel N (2008) A metric for odorant comparison. Nat Methods 5:425–429
9. Medina-Franco JL, Martínez-Mayorga K, Giulianotti MA, Houghten RA, Pinilla C (2008) Visualization of the chemical space in drug discovery. Curr Comput Aided Drug Des 4:322–333
10. Virshup AM, Contreras-García J, Wipf P, Yang W, Beratan DN (2013) Stochastic voyages into uncharted chemical space produce a representative library of all possible drug-like compounds. J Am Chem Soc 135:7296–7303
11. Fitzgerald SH, Sabat M, Geysen HM (2006) Diversity space and its application to library selection and design. J Chem Inf Model 46:1588–1597
12. Akella LB, DeCaprio D (2010) Cheminformatics approaches to analyze diversity in compound screening libraries. Curr Opin Chem Biol 14:325–330
13. Medina-Franco JL, Martinez-Mayorga K, Meurice N (2014) Balancing novelty with confined chemical space in modern drug discovery. Expert Opin Drug Discov 9:151–165

14. Harvey AL (2008) Natural products in drug discovery. Drug Discov Today 13:894–901
15. Scior T, Bernard P, Medina-Franco JL, Maggiora GM (2007) Large compound databases for structure-activity relationships studies in drug discovery. Mini Rev Med Chem 7:851–860
16. Hopkins AL (2008) Network pharmacology: the next paradigm in drug discovery. Nat Chem Biol 4:682–690
17. Gozalbes R (2011) Rational generation of focused chemical libraries: an update on computational approaches. Comb Chem High Throughput Screen 14:428–428
18. Ashburn TT, Thor KB (2004) Drug repositioning: Identifying and developing new uses for existing drugs. Nat Rev Drug Discov 3:673–683
19. Paolini GV, Shapland RHB, van Hoorn WP, Mason JS, Hopkins AL (2006) Global mapping of pharmacological space. Nat Biotechnol 24:805–815
20. Medina-Franco JL, Giulianotti MA, Welmaker GS, Houghten RA (2013) Shifting from the single to the multi target paradigm in drug discovery. Drug Discov Today 18:495–501
21. Scalbert A, Andres-Lacueva C, Arita M, Kroon P, Manach C, Urpi-Sarda M, Wishart D (2011) Databases on food phytochemicals and their health-promoting effects. J Agric Food Chem 59:4331–4348
22. Schneider G, Neidhart W, Giller T, Schmid G (1999) Scaffold-hopping by topological pharmacophore search: a contribution to virtual screening. Angew Chem Int Ed 38:2894–2896
23. Duarte CD, Barreiro EJ, Fraga CA (2007) Privileged structures: a useful concept for the rational design of new lead drug candidates. Mini Rev Med Chem 7:1108–1119
24. Villar HO, Hansen MR, Kho R (2007) Substructural analysis in drug discovery. Curr Comput Aided Drug Des 3:59–67
25. Martínez-Mayorga K, Peppard TL, Yongye AB, Santos R, Giulianotti M, Medina-Franco JL (2011) Characterization of a comprehensive flavor database. J Chemom 25:550–560
26. Medina-Franco JL, Martínez-Mayorga K, Peppard TL, Del Rio A (2012) Chemoinformatic analysis of GRAS (Generally Recognized as Safe) flavor chemicals and natural products. PLoS One 7:e50798
27. Peppard TL, Le M, Pandya RN (2008) Prediction tool for modern flavor development. In: Hofmann T, Meyerhof W, Schieberle P (eds) Recent Highlights in Flavor Chemistry & Biology. Proceedings of the 8th Wartburg Symposium on flavour chemistry and biology. Deutsche Forschungsanstalt für Lebensmittelchemie, Garching, pp 374–378
28. Scior T, Bender A, Tresadern G, Medina-Franco JL, Martínez-Mayorga K, Langer T, Cuanalo-Contreras K, Agrafiotis DK (2012) Recognizing pitfalls in virtual screening: a critical review. J Chem Inf Model 52:867–881
29. Alvarez J, Shoichet B (2005) Virtual screening in drug discovery. Taylor & Francis Group, LLC CRC Press, Boca Raton
30. Maldonado AG, Doucet JP, Petitjean M, Fan BT (2006) Molecular similarity and diversity in chemoinformatics. from theory to applications. Mol Divers 10:39–79
31. Maggiora GM (2006) On outliers and activity cliffs-why QSAR often disappoints. J Chem Inf Model 46:1535
32. Villoutreix BO, Renault N, Lagorce D, Sperandio O, Montes M, Miteva MA (2007) Free resources to assist structure-based virtual ligand screening experiments. Curr Protein Pept Sci 8:381–411
33. Radestock S, Weil T, Renner S (2008) Homology model-based virtual screening for GPCR ligands using docking and target-biased scoring. J Chem Inf Model 48:1104–1117
34. Kruger DM, Evers A (2010) Comparison of structure- and ligand-based virtual screening protocols considering hit list complementarity and enrichment factors. Chemmedchem 5:148–158
35. Mendez-Lucio O, Perez-Villanueva J, Castillo R, Medina-Franco JL (2012) Identifying activity cliff generators of PPAR ligands using SAS maps. Mol Inf 31:837–846
36. Cruz-Monteagudo M, Medina-Franco JL, Pérez-Castillo Y, Nicolotti O, Cordeiro MNDS, Borges F (2014) Activity cliffs in drug discovery: Dr. Jekyll or Mr. Hyde? Drug Discov Today 19:1069–1080
37. Rius M, Lyko F (2012) Epigenetic cancer therapy: rationales, targets and drugs. Oncogene 31:4257–4265

38. Méndez-Lucio O, Tran J, Medina-Franco JL, Meurice N, Muller M (2014) Towards drug repurposing in epigenetics: olsalazine as a novel hypomethylating compound active in a cellular context. ChemMedChem 9:560–565
39. Sprous DG, Salemme FR (2007) A comparison of the chemical properties of drugs and FEMA/FDA notified GRAS chemical compounds used in the food industry. Food Chem Toxicol 45:1419–1427
40. Pintore M, Wechman C, Sicard G, Chastrette M, Amaury N, Chretien JR (2006) Comparing the information content of two large olfactory databases. J Chem Inf Model 46:32–38
41. Jensen K, Panagiotou G, Kouskoumvekaki I (2014) Integrated text mining and chemoinformatics analysis associates diet to health benefit at molecular level. PLoS One 10:e1003432
42. Zarzo M, Stanton DT (2006) Identification of latent variables in a semantic odor profile database using principal component analysis. Chem Senses 31:713–724.
43. Martinez-Mayorga K, Peppard TL, López-Vallejo F, Yongye AB, Medina-Franco JL (2013) Systematic mining of generally recognized as safe (GRAS) flavor chemicals for bioactive compounds. J Agric Food Chem 61:7507–7514

Chapter 4
Reverse Pharmacognosy: A Tool to Accelerate the Discovery of New Bioactive Food Ingredients

Quoc Tuan Do, Maureen Driscoll, Angela Slitt, Navindra Seeram, Terry L. Peppard and Philippe Bernard

4.1 Introduction

In many ancient civilizations, such as the Chinese, Egyptian, Indian, and Sumarian, foods were considered as medicine and traditional medicines would usually favor prevention over cure. Hippocrates, the father of Western medicine, famously considered food as medicine and medicine as food (~500 BC). During approximately the same period, in China, the so-called Yellow Emperor's Inner Classic was compiled which represents the first codification of Chinese food therapy. So the concept of foods providing health benefits is not new. Today's functional foods may be regarded as a modern continuation of our ancestors' quest for good health. But what is a functional food? "Functional foods can be considered to be those whole, fortified, enriched or enhanced foods that provide health benefits beyond the provision of essential nutrients (e.g., vitamins and minerals), when they are consumed at efficacious levels as part of a varied diet on a regular basis" [26]. With better-informed consumers, the increase in life expectancy, and growing regulatory constraints, the food industry is today striving for constant

---

Dedication—This chapter is dedicated to the memory of John Sciré, who sadly passed away in November 2013. It was largely through his efforts and his enthusiasm that work on the flavorings was able to be undertaken.

Q. T. Do (✉) · P. Bernard
Greenpharma, S.A.S, 3, allée du Titane, 45100 Orléans, France
e-mail: quoctuan.do@greenpharma.com

M. Driscoll · A. Slitt · N. Seeram
Department of Biomedical and Pharmaceutical Sciences, College of Pharmacy, University of Rhode Island, 7 Greenhouse Road, Kingston, RI 02881, USA

T. L. Peppard
Robertet Flavors Inc., 10 Colonial Dr., Piscataway, NJ 08854, USA

K. Martinez-Mayorga, J. L. Medina-Franco (eds.), *Foodinformatics*,
DOI 10.1007/978-3-319-10226-9_4

innovation. Consequently, there are many opportunities for novel active food ingredients. Indeed, the global functional foods market is projected to reach nearly $30 billion by 2014 [44]. How can we try to fulfill the needs of this industry? We propose applying the technique of reverse pharmacognosy (RPG) to accelerate the discovery of new bioactive food ingredients and the substantiation of bioactivity in support of certain health claims. To define reverse pharmacognosy, we first define pharmacognosy.

The term pharmacognosy comes from the Greek *pharmakon* which means drug or recipe and *gnosis* which means knowledge. A simple definition could be: "Pharmacognosy is the science which studies natural compounds with therapeutic, cosmetic and agri-food applications" [6]. The workflow starts with a selection of plants based on ethnopharmacological data [1] and biodiversity [15]. Extracts are made, which are tested in biological assays. Active extracts are further fractionated and then tested again in a fraction-test iterative process until identification of the molecule(s) responsible for the biological activity.

The aim of RPG is to exploit the overwhelming amount of data generated by pharmacognosy. It was recently introduced to find new therapeutic activities among natural products and their botanical sources by means of database mining and computational tools. RPG represents a complementary approach to pharmacognosy, which makes it possible to find applications for living organisms based on the bioactive compounds they contain and the biological properties of these compounds. Inverse screening and natural compound/natural source databases are essential components of RPG. The workflow starts with a selected molecule (based on absence of toxicity, ease of sourcing, etc.). We identify putative affinity with proteins of interest, using *in silico* approaches to reduce the number of *in vitro* assays required to be performed, and then validate predicted activities with suitable *in vitro* tests. When biological activities are confirmed, we can position all extracts containing the studied compound (assuming present at sufficiently high concentration) in the applications linked by the modulation of the identified targets, provided of course that there are no adverse effects. Allergenic and other safety issues are crucial considerations in the development of future bioactive ingredients. Hence, several authors have considered food additives in the Flavor and Extract Manufacturers Association (FEMA) GRAS list of approved flavoring materials as another potential source of bioactive molecules, or promising starting points for the development of such [66, 46, 41]. (The relationship between FEMA GRAS status and GRAS status subject to Food and Drug Administration, FDA, approval is mentioned below in the Results and Discussion section.) In this work, we describe examples of studies aimed at finding new active ingredients from natural products, and from molecules in the FEMA GRAS list using an RPG strategy. In either case, it may well be that the best outcomes are obtained by merely using such molecules as starting points ("hits") for further development of functional ingredients, employing the "hit-to-lead"-type approach favored by medicinal chemists.

4.2 Materials and Methods

4.2.1 In Silico Models

4.2.1.1 Protein-Based Approach

RPG needs a database with information relating natural compounds and living organisms that produce them, e.g., plants, microorganisms, etc. In this way, when an interesting activity is identified for a compound, we have natural sources for it and can develop an extraction process to yield an extract enriched in the desired molecule. We perform our studies based on Greenpharma database, a proprietary in-house database containing 150,000 natural molecules and 160,000 organisms, with 50,000 entries for traditional uses of plants and 20,000 biological data records. It is designed with open-source tools (Linux, Apache, mySQL, Php, Sketcher, etc.) [3].

We also need a target database comprising three-dimensional (3D) structures of proteins of therapeutic interest and docking software to predict the affinity of target compounds with their putative protein partners. In our case, we have developed "Selnergy" for virtual screening [15]. It is based on Surflex-Dock in the Sybyl Molecular Modeling Package (Tripos, MO, USA) with a target database of 10,000 protein 3D structures. Proteins structures are extracted either from crystallography data in the Protein Data Bank (http://www.rcsb.org) or from homology modeling (e.g., some G-protein-coupled receptors). A procedure was set up to include or exclude protein models in the Selnergy database. It is based on how well Selnergy can reproduce the pose of a co-crystallized ligand when docked with its cognate protein partner. Furthermore, the protein model must be able to discriminate decoy from active compounds [17]. For a review of the protein database and *in silico* tools useful for RPG, refer to [3].

4.2.1.2 Ligand-Based Approach

One important prerequisite of the protein-based approach is obviously the need to have a protein 3D structure. Furthermore, molecules can have biological activities without identified targets. Yet this type of data is also of interest. Due to the existence of several databases containing small molecules and information about their biological activities, one can envisage using these information sources to identify new activities based on structure–activity relationships [33], with structurally similar compounds being likely to have similar biological activities. Below are several public domain databases of interest for the ligand-based approach:

ChEMBL [21] "ChEMBL is an Open Data database containing binding, functional and ADMET information for a large number of drug-like bioactive compounds. These data are manually abstracted from the primary published literature on a regular basis, then further curated and standardized to maximize their quality and utility across a wide range of chemical biology and drug-discovery research problems. Currently, the database contains 5.4 million bioactivity measurements for more than 1 million compounds and 5200 protein targets. Access is available through a web-based interface, data downloads and web services at: https://www.ebi.ac.uk/chembldb."

Pubchem [32] Pubchem is a database maintained by the National Center for Biotechnology Information (NCBI), which is part of the US National Institutes of Health (NIH). PubChem can be freely accessed through a web user interface or is downloadable by File Transfer Protocol (FTP) at http://pubchem.ncbi.nlm.nih.gov. Pubchem is organized into three main parts: substances (~126 million entries of compound mixtures, extracts, etc.), pure compounds (48 million unique structures), and bioassays (~740,000 records). Users can search the database by name, PubChem identifiers, structures of molecules to retrieve small molecules, calculated physicochemical data, and experimental biological data. Structure–activity relationship tools are available for further analysis of the extracted results.

Drugbank [34] "The DrugBank database is a unique bioinformatics and cheminformatics resource that combines detailed drug (i.e., chemical, pharmacological and pharmaceutical) data with comprehensive drug target (i.e., sequence, structure, and pathway) information. The database contains 6825 drug entries including 1541 FDA-approved small molecule drugs, 150 FDA-approved biotech (protein/peptide) drugs, 86 nutraceuticals and 5082 experimental drugs. Additionally, 4323 non-redundant protein (i.e., drug target/enzyme/transporter/carrier) sequences are linked to these drug entries. Each DrugCard entry contains more than 150 data fields with half of the information being devoted to drug/chemical data and the other half devoted to drug target or protein data." The database can be freely accessed and downloaded at http://www.drugbank.ca/.

BindingDB [38] "**BindingDB** is a public, web-accessible database of measured binding affinities, focusing chiefly on the interactions of proteins considered to be drug-targets with small, drug-like molecules. BindingDB contains 1,009,290 binding data, for 6589 protein targets and 427,325 small molecules. There are 2046 protein–ligand crystal structures with BindingDB affinity measurements for proteins with 100 % sequence identity, and 5815 crystal structures allowing proteins to 85 % sequence identity."

The Protein–Small-Molecule Database (PSMDB) [75] "The Protein–Small-Molecule Database (PSMDB) provides non-redundant sets of protein–small-molecule complexes that are especially suitable for structure-based drug design and protein–small-molecule interaction research." It is designed to be easily updated and to avoid redundancies in terms of ligands (by using structural similarity) and proteins (by using protein sequence homology). Ligands are considered if they have at least seven heavy atoms. The database is downloadable, proteins and ligands being in separate files. PSMDB can be accessed at http://compbio.cs.toronto.edu/psmdb/.

CREDO [61] "CREDO is a unique relational database storing all pairwise atomic interactions of inter- as well as intra-molecular contacts between small molecules and macromolecules found in experimentally determined structures from the Protein Data Bank. These interactions are integrated with further chemical and biological data. The database implements useful data structures and algorithms such as cheminformatics routines to create a comprehensive analysis platform for drug discovery. The database can be accessed through a web-based interface, downloads of data sets and web services at http://marid.bioc.cam.ac.uk/credo."

Examples of commercial database from several companies can be found such as Wombat, WDI, MDDR, CMC, etc.

To compare the structural similarity of the compounds under study with ligands from the abovementioned databases, one can rely on molecular descriptors such as fingerprints [58, 28, 77], descriptors [43, 46], or molecular graphs [29]. There are also numerous software programs that can perform virtual screening based on the structures of small molecules. Here are some examples: ChemMapper [20], Ftrees [56], Topomer [12], etc. It is beyond the scope of this chapter to do a comprehensive review of them.

The FEMA GRAS Database This is maintained by the FEMA. It comprises a compilation of flavoring materials, whose safety has been reviewed by an expert panel of toxicologists and other specialists, and which are GRAS for human consumption within specified product categories and at specified usage levels. Materials on the GRAS list, together with certain FDA-approved food additives, are those that are legally permitted for use as flavorings (and for related purposes, such as taste modification) in the USA [22, 23].

New additions to the GRAS list (originally published approximately 50 years ago) appear in Food Technology every year or two. For example, GRAS 26 was published in August 2013 and included approximately 50 botanicals and discrete chemical entities. For each material, a FEMA #, principal name, and synonyms are listed, along with permitted food and beverage applications, including anticipated average usual and average maximum use levels (in ppm). To date, of the approximately 2800 GRAS materials, just more than 80% are discrete chemical entities. However, of these, in some cases, stereochemistry and even geometrical configuration are not fully specified.

The GRAS database is available online on FEMA's website (https://www.femaflavor.org), though exclusively for member companies. However, it is also available through third-party software, such as Flavor-Base 9 by John Leffingwell & Associates, or alternatively, it can be accessed in the public domain through web sites such as http://www.thegoodscentscompany.com.

4.2.2 *In Vitro* Models

4.2.2.1 Inflammation

The murine macrophage cell line RAW 264.7 is routinely used to assess anti-inflammatory activity and NF-κB signaling *in vitro*. Inflammation can be induced in RAW 264.7 macrophages with lipopolysaccharides (LPS), a component found on the outer membrane of Gram-negative bacteria. NO, cyclooxygenase (COX) 2, and prostaglandin E2 (PGE2) levels increase upon stimulation with LPS, as do the levels of proinflammatory cytokines tumor necrosis factor (TNF), and interleukin (IL) 1, and IL-6. Previously identified compounds isolated from plants, such as resveratrol, curcumin, and quercetin, have been shown to inhibit the proinflammatory effects of LPS treatment in RAW 264.7 macrophage cells. Initial experiments measuring nitrite concentration released into the RAW 264.7 culture medium were conducted to establish conditions that would be ideal for the efficient and consistent screening of selected GRAS list compounds. Nitrite, as stable intermediate of NO, is frequently used as a proxy for NO production using the Greiss reaction, an effective and inexpensive method for measuring NO activity.

RAW 264.7 macrophage cells were routinely cultured in Dulbecco's Modified Eagle's Medium–high glucose (DMEM), supplemented with 10% fetal bovine serum (FBS), penicillin (100 units/mL), and streptomycin (100 μg/mL) and maintained at 37 °C under 5% CO_2-humidified air. Cells were seeded in 96-well plates at 1 × 105 cells/100 μL and incubated for 24 h. After incubation, the culture medium was removed and replaced with 200 μL of fresh medium and several concentrations of LPS (0, 0.1, 1, 10, and 100 ng/mL) were added. Cells were incubated for an additional 24 h, then 100 μL of culture medium was removed from each well and mixed 1:1 with Greiss reagent (Sigma-Aldrich Co.) and read with a spectrophotometer at 550 nm after 15 min. Experiments were conducted under both serum and serum-free conditions to determine the appropriate concentration of LPS needed to stimulate nitrite production in RAW 264.7 cells. LPS concentration used in serum and serum-free conditions was established anticipating that some compounds may bind the serum component of the growth medium and become inactive.

Nitrite release in RAW 264.7 macrophage cells treated with LPS was found to be concentration dependent. In serum-containing conditions, nitrite was detected by 1 ng/mL LPS before leveling off at 10 ng/mL. In serum-free conditions, nitrite levels increased from 1 ng/mL LPS before reaching maximum levels at 100 ng/mL. Total RNA was extracted and purified from LPS-treated RAW 264.7 cells to establish the minimum amount of LPS needed to upregulate the gene expression of proinflammatory cytokines and other genes involved in the inflammatory process in both serum and serum-free conditions. Quantitative PCR was used to measure the levels of TNF-α, IL-1, IL-6, and COX-2 mRNA. Our results show that the minimum LPS concentration needed to induce NO activity and gene expression in RAW 264.7 cells is 10 ng/mL for serum conditions and 100 ng/mL for serum-free conditions. LPS treatment at 10 and 100 ng/mL LPS increased the mRNA expression of proinflammatory cytokines, such as TNF-α, IL-6, Cox-2, and IL-1, which are well-established markers for LPS stimulation in RAW 264.7 macrophages. Therefore, we proceeded to test compounds with both culture systems in the presence of 10% FBS and 10 ng/mL LPS. This was chosen because we did not want to increase the LPS concentration so high that it would overwhelm the cells and no protective effect would be observed.

4.2.2.2 Cytotoxicity by MTS Assay

RAW 264.7 macrophages were treated with LPS (50 ng/mL) in DMEM + 10% FBS or LPS (100 ng/ml) in DMEM (serum-free). Cells were incubated with LPS alone or in combination with the compounds at various concentrations (0.1–100 µM). After 24-h incubation, the media was removed and tested for nitric oxide activity. The remaining cells were treated with MTS to assess cell viability.

4.3 Results and Discussion

We have previously found interesting activities (e.g., inhibition of phosphodiesterases, cyclooxygenases, etc.) for several natural compounds employing RPG [3]. These illustrate the usefulness of RPG to identify potential applications for natural product molecules and the organisms that produce them. Below are two examples of studies we performed for two natural compounds which could be obtained in large quantities and which were devoid of toxicity.

4.3.1 Example of ε-Viniferin [16]

ε-Viniferin (EV) is a polyphenol and phytoalexin that can be extracted from leaves of the vine *Vitis vinifera* [35]. It is synthesized by plants in response to environment stress [35, 36]. EV consists of two fused resveratrol units. The naturally occurring stereoisomer is the E form. EV has numerous biological properties in oncology [2, 48], in CNS [9], as an antioxidant [54, 55], a hepatoprotector [52], and as an antibacterial [8]. EV was screened on a protein target database and phosphodiesterase 4 (PDE4) was found to be one of the most prominent targets. A binding assay confirmed the prediction with an IC50 = 4.6 µM. It was also shown that EV reduces the secretion of TNF-α and IL-8 in a dose-dependent manner [16]. So an extract of vine leaves may be useful for treating inflammatory conditions; likewise, any other sources that contain this molecule, provided there are no toxicity issues, etc. Table 4.1 lists the plants with the organ from which EV was purified.

Table 4.1 List of plants producing ε-viniferin (ND: Not Determined)

Family	Genus	Species	Botanist	Organ
Dipterocarpaceae	*Hopea*	*parviflora*	Bedd.	Stem bark
Dipterocarpaceae	*Shorea*	*seminis*	(De Vriese) Sloot.	Bark
Dipterocarpaceae	*Vateria*	*indica*	Linn	Stem bark
Dipterocarpaceae	*Vatica*	*affinis*	Thwaites	ND organ
Dipterocarpaceae	*Vatica*	*rassak*	(Korth.) Blume	Stem bark
Paeoniaceae	*Paeonia*	*suffruticosa*	Andrews	Seed
Vitaceae	*Vitis*	*coignetiae*	Pulliat ex Planch.	ND organ
Vitaceae	*Vitis*	*vinifera*	L.	Leaf

Table 4.2 List of plants producing meranzin

Family	Genus	Species	Botanist	Organ
Apiaceae	*Cnidium*	*monnieri*	(L.) Cusson ex Juss.	Fruit
Rutaceae	*Citrus*	*maxima*	(Burm. f.) Merr.	Peel
Rutaceae	*Citrus*	*paradisi*	Macfad. (pro sp.)	Pericarp
Rutaceae	*Limnocitrus*	*littoralis*	(Miq.) Swingle	Leaf
Rutaceae	*Murraya*	*gleinei*	Thwaites ex Oliv	Leaf

4.3.2 Example of Meranzin [17]

This molecule is a coumarin derivative characterized by an epoxide group. Meranzin may be found in the fruit of the traditional Chinese medicinal plant *Cnidium monnieri* (L.) Cusson [63]. Little is known about the biological properties of this molecule. We performed a study of meranzin by RPG and COX 1 and 2 were clearly identified by our *in silico* tool Selnergy as putative protein target partners for meranzin. Peroxisome proliferator-activated receptor (PPAR) δ was another interesting target for meranzin. *In vitro* validations were performed for the proteins. We could demonstrate that our product inhibits COX2 in a dose-dependent manner with %I=56% at 400 nM and that it activates PPARδ activity by 40% at 100 μM [17]. Taking these results together suggests that an extract of *Cnidium monnieri* with an appropriate amount of meranzin could be useful for treating inflammatory and metabolic conditions; likewise, any other sources that contain this molecule, provided there are no toxicity issues, etc. Table 4.2 lists the plants with the organ from which meranzin was purified.

4.3.3 Example of Studies on Selected FEMA GRAS Flavor Molecules

We now want to generalize the RPG approach to a group of compounds which are products of commerce and which are considered safe for human consumption. A list of food additives deemed GRAS is regularly updated by the US FDA. The definition of GRAS substances and the approach can be found at http://www.fda.gov/Food/IngredientsPackagingLabeling/GRAS:

"Under sections 201(s) and 409 of the Federal Food, Drug, and Cosmetic Act (the Act), any substance that is intentionally added to food is a food additive, that is subject to premarket review and approval by FDA, unless the substance is generally recognized, among qualified experts, as having been adequately shown to be safe under the conditions of its intended use, or unless the use of the substance is otherwise excluded from the definition of a food additive. Under sections 201(s) and 409 of the Act, and FDA's implementing regulations in 21 CFR 170.3 and 21 CFR 170.30, the use of a food substance may be GRAS either through scientific procedures or, for a substance used in food before 1958, through experience based on common use in food." The FEMA adopted the GRAS concept, and is responsible for the FEMA GRAS list of flavoring materials used in foods and beverages in the USA [22, 23]. The GRAS procedure is extremely well respected within the food, beverage, and associated industries.

A database of discrete chemical entities existing in the FEMA GRAS list was extracted, and the data comprised chemical name, structure, FEMA reference number, and CAS registry number. We selected a subset of 60 molecules to reposition them in cosmetics and/or food applications. We filtered them using appropriate rules [37, 70] to retain "lead-like" compounds, and used Unity fingerprints [43] and Optisim algorithm [10] from the Sybyl package to select the most chemically diverse structures. We screened the 60 compounds with Selnergy by either docking on protein 3D structures or comparing the chemical structures of the GRAS products to our known active ligand database.

We prioritized molecules that have putative anti-inflammatory properties, as inflammation is implicated in a wide range of ailments and anti-inflammatory products may have numerous applications in the health and wellness domain, including skin care.

Nine compounds were thus selected. Table 4.3 shows all the targets predicted for these GRAS molecules either by protein- or by ligand-based approaches. Some compounds were found to interact with numerous targets, e.g., β-naphthyl anthranilate and tolylaldehyde glyceryl acetyl. Others seem to be quite selective, e.g., phenoxaromate-681, vanillyl ethyl ether, and 2-methoxyphenyl acetate. We expect our putative modulators to be inhibitors of the listed enzymes (if indeed interaction is confirmed) as it is easier to block an enzyme or a receptor than to activate it. In the case of HST2—a homolog of sirtuin—an activator is sought.

In total, we have 24 different potential targets for the 9 GRAS molecules. For the sake of cost effectiveness and efficiency, we chose to employ a RAW 264.7 cell model—as described in the Materials and Methods section—to validate experimentally the putative anti-inflammatory effects of our compounds. This high-content assay allows one to measure several important inflammation-related parameters, such as NO, TNF-α, IL-1, IL-6, and PGE-2 activities. In our test assays, we also included compounds such as resveratrol as references, since it is known to have anti-inflammatory effects in this cell-based screening system as well as in other *in vitro* and *in vivo* models. Compounds were evaluated according to their maximum nontoxic concentration according to MTS assay (Table 4.4).

n-Propyl-2-furanacrylate is the only compound to exert a strong inhibition on NO synthesis, namely 65 % at 0.5 μM. We found that several compounds have very potent activities against PGE2, such as cinnamyl anthranilate, β-naphthyl anthranilate, and *n*-propyl-2-furanacrylate. No molecule shows activities on TNF-α. *n*-Propyl-2-furanacrylate has a strong effect on lowering IL-6 secretion (%I = 53 % at 0.5 μM). In the case of IL-1β, cinnamyl anthranilate and β-naphthyl anthranilate demonstrated strong inhibition at 1 μM. The activity of NF-κB was inhibited by cinnamyl anthranilate at 1 μM, and to a lesser extent by vanillyl ethyl ether and 2-methoxyphenyl acetate (at 25 μM). *n*-Propyl-2-furanacrylate seems to strongly inhibit

Table 4.3 Selected GRAS molecules with predicted protein partners and potential applications

Molecules	Putative protein partners	Potential applications related to predicted targets
Cinnamyl anthranilate	Fatty acid binding protein	Diabetes [19], obesity [45]
	Monoamine oxidase A and B	Antidepressant, anxiolytics [39]
	Phospholipase A2 (PLA2)	Inflammation [67]
	Retinol-binding protein	Skin protection [53]
β-Naphthyl anthranilate	Cyclooxygenase 1 (COX1)	Inflammation [65]
	Estrogen receptor alpha	Menopausal hot flash [5]
	Estrogen-related receptor alpha	Diabetes, obesity [74], osteoporosis [4]
	Fatty acid binding protein	Diabetes [19], obesity [45]
	Retinoic acid receptor gamma	Cancer, photoaging [59]
n-Propyl-2-furanacrylate	Aldose reductase (AR)	Diabetes complication [50]
	Neutrophil collagenase (NC)	Atopic dermatitis [24]
Tolylaldehyde glyceryl acetyl	N/A	Central nervous system stimulants, treat attention deficit hyperactivity disorder (Drugbank)
	Methionine aminopeptidase	Antibacterial [73]
	Phosphodiesterase 2A	Memory[72], anxiolytic [42]
	Phosphodiesterase 5B	Impotency, memory [72]
	Matrix metalloproteinase 3	Prophylaxis for diabetic nephropathy [71], skin protection [62]
	Adenosine deaminase	Cancer [60]
Phenoxaromate-681	Glycogen synthase kinase 3	Diabetes, inflammation, cancer, Alzheimer disease [57]
Vanillyl ethyl ether	Fatty acid binding protein	Diabetes [19], obesity [45]
2-Methoxyphenyl acetate	Cyclooxygenase 1 & 2	Inflammation [65]
Hesperetin	15-lipoxygenase (15-LOX)	Inflammation [27]
	Alpha-amylase	Diabetes [47]
	Aromatase (CYP19)	Male aging [11]
		Breast cancer [31]
	Phosphatidylinositol-3 kinase (PI3K)	Inflammation, cardioprotection
Phloretin	N/A	UV screen
	HST2 (homologue of sirtuin)	Aging (in case of activators)

GRAS generally recognized as safe; *N/A* not available; these predictions are exclusively based on structural similarity with known active ligands

the production of NO, PGE2, and IL-6. However, it also activates NF-κB. Cinnamyl anthranilate blocks three different markers of inflammation: PGE2, IL-1β, and NF-κB. We now compare the predictions of Selnergy with the experimental data.

Table 4.4 *In vitro* evaluation of selected GRAS compounds

Molecules	Inflammation markers	Test concentration (μM)	% Inhibition
Resveratrol	Nitrite (μM)	25	63
	PGE2	1	54
	TNF-α	50	43
	IL-6	25	63
	IL-1β	50	0
	NF-κB	50	29
Cinnamyl anthranilate	Nitrite (μM)	1	0
	PGE2	1	84
	TNF-α	1	8
	IL-6	1	0
	IL-1β	1	58
	NF-κB	1	62
β-Naphthyl anthranilate	Nitrite (μM)	1	0
	PGE2	1	93
	TNF-α	1	3
	IL-6	1	0
	IL-1β	1	61
	NF-κB	1	30
n-Propyl-2-furanacrylate	Nitrite (μM)	0.5	65
	PGE2	0.5	90
	TNF-α	1	0
	IL-6	0.5	53
	IL-1β	1	0
	NF-κB	1	−111
Tolylaldehyde glyceryl acetyl	Nitrite (μM)	49	12
	PGE2	25	96
	TNF-α	25	34
	IL-6	25	44
	IL-1β	49	0
	NF-κB	25	65
Phenoxaromate-681	Nitrite (μM)	25	73
	PGE2	1	99
	TNF-α	25	26
	IL-6	25	33
	IL-1β	52	0
	NF-κB	52	22
Vanillyl ethyl ether	Nitrite (μM)	53.4	29
	PGE2	25	90
	TNF-α	53.4	10
	IL-6	25	48
	IL-1β	53.4	0
	NF-κB	25	73

Table 4.4 (continued)

Molecules	Inflammation markers	Test concentration (μM)	% Inhibition
2-Methoxyphenyl acetate	Nitrite (μM)	58	32
	PGE2	25	92
	TNF-α	58	0
	IL-6	25	73
	IL-1β	58	0
	NF-κB	25	76
Hesperetin	Nitrite (μM)	50	0
	PGE2	50	89
	TNF-α	50	6
	IL-6	50	0
	IL-1β	50	71
	NF-κB	50	31
Phloretin	Nitrite (μM)	25	31
	PGE2	25	91
	TNF-α	25	6
	IL-6	25	23
	IL-1β	25	75
	NF-κB	25	45

GRAS generally recognized as safe, *PGE2* prostaglandin E2, TNF tumor necrosis factor, IL interleukin, *NF-κB* nuclear factor kappa-light-chain-enhancer of activated B cells

In Table 4.3, cinnamyl anthranilate was predicted to interact with fatty acid-binding protein, monoamine oxidase A and B, phospholipase A2 (PLA2; Fig. 4.1), and retinol-binding protein. Among these proteins, only PLA2 is clearly involved in the inflammation process. It was demonstrated by Huwiler et al. [30] that the inhibition of PLA2 led to a decrease in PGE2 synthesis by downregulation of IL-1β and inhibition of NF-κB. This seems to be consistent with our prediction of cinnamyl anthranilate as an inhibitor of PLA2.

Within the targets identified for β-naphthyl anthranilate, COX1 is implicated in inflammation. Choi et al. [9] demonstrated the contribution of COX1 in neuroinflammation induced by LPS. Using COX1 knockout mice or wild-type mice administered with SC-560, a nanomolar range COX1 selective inhibitor, they observed a significantly strong decrease in PGE2 ($P<0.01$), along with a decrease in IL-1β, IL-6, and TNF-α ($P<0.05$) via a reduction in the activation of NF-κB. We found that β-naphthyl anthranilate decreases PGE2, an indirect product of COX1 enzymatic activity, IL-1β, and the activity of NF-κB, though neither IL-6 nor TNF-α were decreased.

n-Propyl-2-furanacrylate may interact with aldose reductase (AR; Fig. 4.2) and neutrophil collagenase (NC). *In vitro*, we observed a lowering of nitrite, which relates to NO decrease, PGE2, IL-6, and increasing activity of NF-κB. The relationship between inhibition of AR and NO production seems to be dependent on the type of cells or tissues. In RAW264.7 cells [76] and vascular tissues [49], inhibiting AR results in a decrease of NO. The inverse effect is observed in neutrophil-endothelial

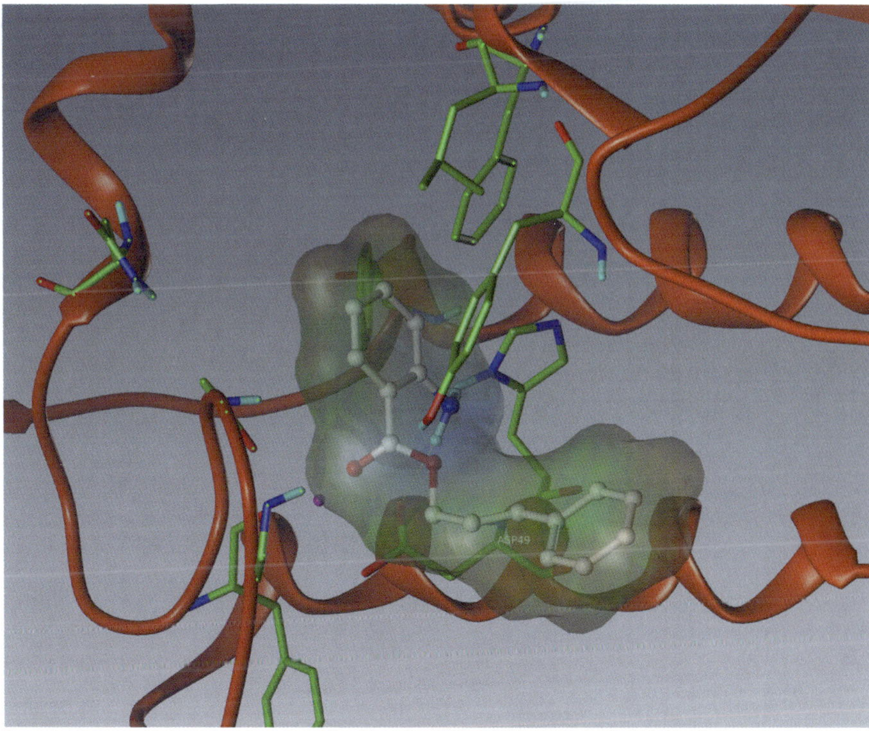

Fig. 4.1 Cinnamyl anthranilate, represented in *ball and stick* fashion, is docked into the active site of phospholipase A2. The *ribbon* represents the protein backbone. Protein residues are highlighted in *capped sticks*. The volume occupied by the ligand is delimited by the *transparent shape*. The carbonyl of the ligand forms a dative bond with a calcium cation, and the amine group forms a hydrogen bond with the ASP49 carboxylate

cells [51]. Shoeb et al. [64] demonstrated a link between the inhibition of AR and the decrease of PEG2. Fidarestat, an inhibitor of AR, provokes a significant lowering of IL-6 ($P<0.01$), IL-1β ($P<0.05$), and TNF-α ($P<0.05$) according to Takahashi et al. [68]. According to Wang et al. [76], AR inhibitors should also attenuate the activity of NF-κB, which is not the case here. To the best of our knowledge, there does not seem to have been any relationship between the inhibition of neutrophil collagenase and the listed markers according the scientific literature. Therefore, *n*-propyl-2-furanacrylate has a different profile compared to known AR inhibitors regarding its activation of NF-κB and its inactivity against IL-1β and TNF-α.

We could not relate *in vitro* observation of PGE2 level change with the inhibition of predicted targets for tolylaldehyde glyceryl acetyl. The attenuation of NF-κB activity may be linked to the inhibition of adenosine deaminase [14].

Only glycogen synthase kinase 3 (GSK3) was identified for phenoxaromate-681. There is some evidence that an inhibitor of GSK3 can exert a reduction of NO, PGE2, IL-1β, and TNF-α production [13]. We noticed that phenoxaromate-681 strongly attenuates PGE2 production, diminishes NO, and at a lesser level TNF-α,

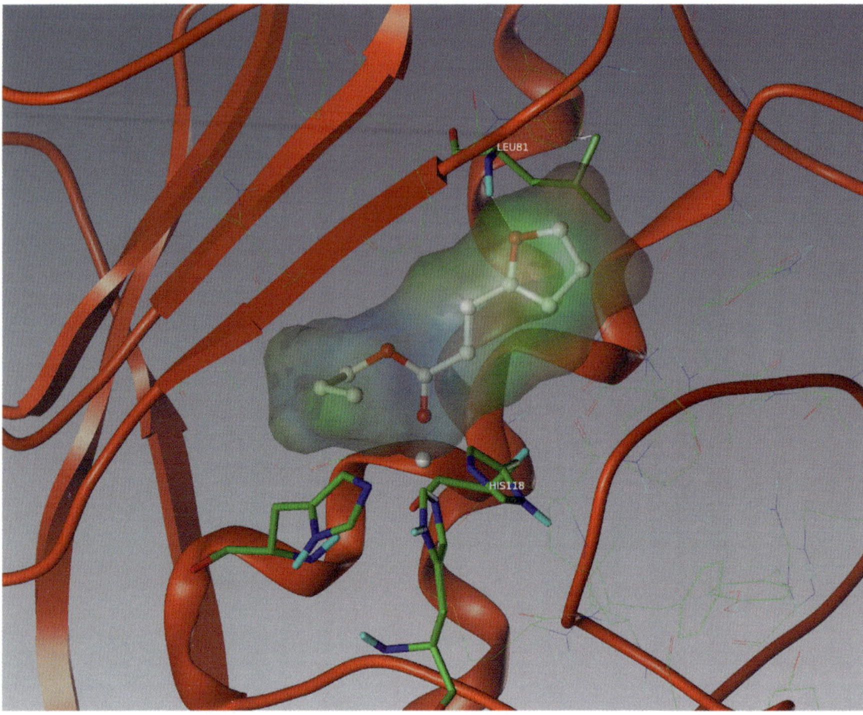

Fig. 4.2 n-Propyl-2-furanacrylate, represented in *ball and stick* fashion, is docked into the active site of neutrophile collagenase. The *ribbon* represents the protein backbone. Protein residues are highlighted in *capped sticks*. The volume occupied by the ligand is delimited by the *transparent shape*. The carbonyl of the ligand forms a dative bond with a zinc cation, and the oxygen of the furan forms a hydrogen bond with the nitrogen of the amidic group of LEU81

but observed no effect on IL-1β. However, we did observe a diminution in levels of IL-6 and NF-κB in the presence of phenoxaromate-681.

Vanillyl ethyl ether attenuates the activity of PGE2 and NF-κB. It was previously shown that blocking fatty acid-binding protein (FABP) can decrease the activation of NF-κB [40]. Nevertheless, we could not find any study in the scientific literature that reports the relationship between the inhibition of FABP and a diminution of PGE2 synthesis.

2-Methoxyphenyl acetate is a putative inhibitor of both COX1 and 2; *in vitro* results demonstrate that it modulates PGE2, IL-6, and NF-κB. Three proteins are listed as modulated by a COX1 inhibitor by Choi et al. [9].

There is no solid bibliographic evidence to support the inhibition of 15-LOX, alpha-amylase, or CYP19 with a change in the level of PGE2 or IL-1β. Phosphatidylinositol-3 kinase (PI3K) inhibitors, such as ZSTK474, were recently found to inhibit the production of PGE2 [25]. There is no clear evidence of a correlation of PI3K inhibition and the decrease of secretion of IL-1β through an experiment of LPS tolerance induction by Tanabe and Grenier [69] showed an attenuation of the increase of IL-1β but not TNF-α. Therefore, hesperetin seems to have a profile similar to a PI3K inhibitor.

Phloretin is known to inhibit PGE2, IL-1β, IL-6, TNF-α, and NF-κB [7]. Therefore, the *in vitro* values we found are consistent with the scientific literature—though its effect on TNF-α is not significant in our case. Probably the UV screen property identified for our molecule by *in silico* methods derives from this biological profile. Phloretin may interact with HST2, a yeast sirtuin. The activation of sirtuin 1 (SIRT1) is associated with antiaging, anticancer, and anti-inflammatory effects. We tested *in vitro* the activity of phloretin on human SIRT1. Unfortunately, our compound shows a dose-dependent inhibition towards this enzyme (data not shown). Though the biological effect is not of interest, this result suggests an interaction of phloretin with SIRT1, thus validating the prediction of Selnergy.

Overall, we could relate most of the Selnergy predictions with the values we found for the markers of inflammation. Of course, this is not a direct proof, and we cannot rule out the possibility that we might have the same profile of markers with other targets.

4.4 Conclusions and Perspectives

RPG has demonstrated its usefulness in the identification of new activity for (or repurposing of) natural compounds, which may then be extrapolated to plant extracts containing them. This approach also provides a hypothesis for substantiation of the ingredient based on the prediction of putative protein partners which may interact with the compound in question. Furthermore, a chemo-marker is provided for the development and production of the extract ingredient. Obviously, Selnergy, a key component of RPG, can also be applied to commercially sourced compounds, and we demonstrated this by studying nine compounds selected from the FEMA GRAS list of permitted flavoring materials. Though we could not validate all predicted small-molecule–protein interactions, we were able to find several cases of agreement between *in silico* predictions and *in vitro* results obtained, when focusing on targets related to inflammation. Cinnamyl anthranilate, β-naphthyl anthranilate, and tolylaldehyde glyceryl acetyl, better than being pursued as “actives” *per se*, may be good starting points (“hits”) for further development of a functional ingredient, employing the “hit-to-lead”-type approach. n-Propyl-2-furanacrylate needs further analysis to ascertain its effects related to activation of NF-κB.

Moreover, with targets identified by Selnergy for each molecule under study, we can explore combinations of compounds to inhibit complementary inflammation pathways and thus find potential synergies. For instance, n-propyl-2-furanacrylate and cinnamyl anthranilate may have putative synergistic effects on reducing inflammation.

One important, albeit obvious, limitation of RPG is the required presence of relevant data in the protein and known active ligand databases. Clearly, if a protein target, or a series of active ligands related to a target, is not in the database, we will not find the related biological activity. However, with the constant increase in database content in PDB and ChEMBL, DrugBank, etc., the impact of this limitation will gradually lessen over time.

The flavor industry has no intention of developing or promoting flavors for the purpose of treating, curing, preventing, or diagnosing disease, or even for the purpose of making health-related structure/function claims. Rather, there is curiosity in exploring flavors' secondary role as natural promoters of health and wellness by better understanding the occurrence of fortuitous relationships existing between some flavors and certain disease conditions (or parameters associated with them). In fact, numerous examples of this being the case are already present in the scientific literature. In any event, if there does indeed turn out to be a promising link between flavor molecule "A" and disease condition "B," then most likely the best practical results would be obtained by merely using identified flavor molecules as starting points for further development of functional ingredients. This work would most likely be carried out by companies actively involved in the development of bioactives.

Acknowledgments The authors wish to thank Robertet Flavors for permission to publish this work, and also Peter Lombardo for carefully reading through the manuscript and for making valuable suggestions.

References

1. Bernard P, Scior T, Didier B, Hibert M, Berthon JY (2001) Ethnopharmacology and bioinformatic combination for leads discovery: application to phospholipase A(2) inhibitors. Phytochemistry 58(6):865–874
2. Billard C, Izard JC, Roman V, Kern C, Mathiot C, Mentz F, Kolb JP (2002) Comparative antiproliferative and apoptotic effects of resveratrol, epsilon-viniferin and vine-shots derived polyphenols (vineatrols) on chronic B lymphocytic leukemia cells and normal human lymphocytes. Leuk Lymphoma 43(10):1991–2002
3. Blondeau S, Do QT, Scior T, Bernard P, Morin-Allory L (2010) Reverse pharmacognosy: another way to harness the generosity of nature. Curr Pharm Des 16(15):1682–1696
4. Bonnelye E, Aubin JE (2005) Estrogen receptor-related receptor alpha: a mediator of estrogen response in bone. J Clin Endocrinol Metab 90(5):3115–3121
5. Bowe J, Li XF, Kinsey-Jones J, Heyerick A, Brain S, Milligan S, O'Byrne K (2006) The hop phytoestrogen, 8-prenylnaringenin, reverses the ovariectomy-induced rise in skin temperature in an animal model of menopausal hot flushes. J Endocrinol 191(2):399–405
6. Bruneton J (1993) Pharmacognosie, Phytochimie, Plantes Médicinales. Lavoisier, Paris, p viii
7. Chang WT, Huang WC, Liou CJ (2012) Evaluation of the anti-inflammatory effects of phloretin and phlorizin in lipopolysaccharide-stimulated mouse macrophages. Food Chem 134(2):972–979
8. Cho HS, Lee JH, Ryu SY, Joo SW, Cho MH, Lee J (2013) Inhibition of *Pseudomonas aeruginosa* and *Escherichia coli* O157:H7 biofilm formation by plant metabolite ε-viniferin. J Agric Food Chem 61(29):7120–7126
9. Choi SH, Langenbach R, Bosetti F (2008) Genetic deletion or pharmacological inhibition of cyclooxygenase-1 attenuate lipopolysaccharide-induced inflammatory response and brain injury. FASEB J 22(5):1491–1501

10. Clark RD (1997) OptiSim: an extended dissimilarity selection method for finding diverse representative subsets. J Chem Inf Comput Sci 37(6):1181–1188
11. Cohen PG (2001) Aromatase, adiposity, aging and disease. The hypogonadal-metabolic- atherogenic-disease and aging connection. Med Hypotheses 56(6):702–708
12. Cramer RD, Jilek RJ, Guessregen S, Clark SJ, Wendt B, Clark RD (2004) Lead hopping. Validation of topomer similarity as a superior predictor of similar biological activities. J Med Chem 47(27):6777–6791
13. Cuzzocrea S, Crisafulli C, Mazzon E, Esposito E, Muia C, Abdelrahman M, Di Paola R, Thiemermann C (2006) Inhibition of glycogen synthase kinase-3β attenuates the development of carrageenan-induced lung injury in mice. Br J Pharmacol 149(6):687–702
14. de Araujo Junqueira AF, Dias AA, Vale ML, Spilborghs GM, Bossa AS, Lima BB, Carvalho AF, Guerrant RL, Ribeiro RA, Brito GA (2011) Adenosine deaminase inhibition prevents *Clostridium difficile* toxin A-induced enteritis in mice. Infect Immun 79(2):653–662
15. Do QT, Bernard P (2004) Pharmacognosy and reverse pharmacognosy: a new concept for accelerating natural drug discovery. IDrugs 7(11):1017–1027
16. Do QT, Renimel I, Andre P, Lugnier C, Muller CD, Bernard P (2005) Reverse pharmacognosy: application of selnergy, a new tool for lead discovery. The example of epsilon-viniferin. Curr Drug Discov Technol 2(3):161–167
17. Do QT, Lamy C, Renimel I, Sauvan N, André P, Himbert F, Morin-Allory L, Bernard P (2007) Reverse pharmacognosy: identifying biological properties for plants by means of their molecule constituents: application to meranzin. Planta Med 73(12):1235–1240
18. Fu J, Jin J, Cichewicz RH, Hageman SA, Ellis TK, Xiang L, Peng Q, Jiang M, Arbez N, Hotaling K, Ross CA, Duan W (2012) Trans-(-)-ε-viniferin increases mitochondrial sirtuin 3 (SIRT3), activates AMP-activated protein kinase (AMPK), and protects cells in models of Huntington disease. J Biol Chem 287(29):24460–24472
19. Furuhashi M, Tuncman G, Görgün CZ, Makowski L, Atsumi G, Vaillancourt E, Kono K, Babaev VR, Fazio S, Linton MF, Sulsky R, Robl JA, Parker RA, Hotamisligil GS (2007) Treatment of diabetes and atherosclerosis by inhibiting fatty-acid-binding protein aP2. Nature 447(7147):959–965
20. Gong J, Cai C, Liu X, Ku X, Jiang H, Gao D, Li H (2013) ChemMapper: a versatile web server for exploring pharmacology and chemical structure association based on molecular 3D similarity method. Bioinformatics 29(14):1827–1829
21. Gaulton A, Bellis LJ, Bento AP, Chambers J, Davies M, Hersey A, Light Y, McGlinchey S, Michalovich D, Al Lazikani B, Overington JP (2012) ChEMBL: a large-scale bioactivity database for drug discovery. Nucleic Acids Res 40(Database issue):D1100–D1107
22. Hallagan JB, Hall RL (1995) FEMA GRAS—a GRAS assessment program for flavor ingredients. Regul Toxicol Pharmacol 21:422–430
23. Hallagan JB, Hall RL (2009) Review: under the conditions of intended use—new developments in the FEMA GRAS program and the safety assessment of flavor ingredients. Food Chem Toxicol 47:267–278
24. Harper JI, Godwin H, Green A, Wilkes LE, Holden NJ, Moffatt M, Cookson WO, Layton G, Chandler S (2010) A study of matrix metalloproteinase expression and activity in atopic dermatitis using a novel skin wash sampling assay for functional biomarker analysis. Br J Dermatol 162(2):397–403
25. Haruna K, Mori S, Tamura N, Sasaki A, Nagamine M, Yaguchi S, Kamachi F, Enami J, Kobayashi S, Yamori T, Takasaki Y (2012) Inhibitory effects of ZSTK474, a phosphatidylinositol 3-kinase inhibitor, on adjuvant-induced arthritis in rats. Inflamm Res 61(6):551–562
26. Hasler CM (2002) Functional foods: benefits, concerns and challenges—a position paper from the american council on science and health. J Nutr 132(12):3772–3781
27. Herre S, Schadendorf T, Ivanov I, Herrberger C, Steinle W, Ruck-Braun K, Preissner R, Kuhn H (2006) Photoactivation of an inhibitor of the 12/15-lipoxygenase pathway. Chembiochem 7(7):1089–1095

28. Hert J, Willett P, Wilton DJ, Acklin P, Azzaoui K, Jacoby E, Schuffenhauer A (2004) Comparison of fingerprint-based methods for virtual screening using multiple bioactive reference structures. J Chem Inf Comput Sci 44(3):1177–1185
29. Hutter MC (2011) Graph-based similarity concepts in virtual screening. Future Med Chem 3(4):485–501
30. Huwiler A, Feuerherm AJ, Sakem B, Pastukhov O, Filipenko I, Nguyen T, Johansen B (2012) The?3-polyunsaturated fatty acid derivatives AVX001 and AVX002 directly inhibit cytosolic phospholipase A(2) and suppress PGE(2) formation in mesangial cells. Br J Pharmacol 167(8):1691–1701
31. John-Baptiste AA, Wu W, Rochon P, Anderson GM, Bell CM (2013) A systematic review and methodological evaluation of published cost-effectiveness analyses of aromatase inhibitors versus tamoxifen in early stage breast cancer. PLoS ONE 8(5):e62614
32. Kaiser J (2005) Science resources. Chemists want NIH to curtail database. Science 308(5723):774
33. Keiser MJ, Roth BL, Armbruster BN, Ernsberger P, Irwin JJ, Shoichet BK (2007) Relating protein pharmacology by ligand chemistry. Nat Biotechnol 25(2):197–206
34. Knox C, Law V, Jewison T, Liu P, Ly S, Frolkis A, Pon A, Banco K, Mak C, Neveu V, Djoumbou Y, Eisner R, Guo AC, Wishart DS (2011) DrugBank 3.0: a comprehensive resource for 'omics' research on drugs. Nucleic Acids Res 39(Database issue):D1035–D1041
35. Langcake P (1981) Disease resistance of Vitis spp. and the production of the stress metabolites resveratrol, ε-viniferin, α-viniferin and pterostilbene. Physiol Plant Pathol 18:213–226
36. Langcake P, Bryce RJ (1977) The production of resveratrol and viniferins, by grapevines in response to ultraviolet irradiation. Phytochemistry 16:1193–1196
37. Lipinski CA, Lombardo F, Dominy BW, Feeney PJ (2001) Experimental and computational approaches to estimate solubility and permeability in drug discovery and development settings. Adv Drug Deliv Rev 46(1–3):3–26
38. Liu T, Lin Y, Wen X, Jorissen RN, Gilson MK (2007) BindingDB: a web-accessible database of experimentally determined protein-ligand binding affinities. Nucleic Acids Res 35(Database issue):D198–D201
39. Mai A, Artico M, Esposito M, Ragno R, Sbardella G, Massa S (2003) Synthesis and biological evaluation of enantiomerically pure pyrrolyl-oxazolidinones as a new class of potent and selective monoamine oxidase type A inhibitors. Farmaco 58(3):231–241
40. Makowski L, Brittingham KC, Reynolds JM, Suttles J, Hotamisligil GS (2005) The fatty acid-binding protein, aP2, coordinates macrophage cholesterol trafficking and inflammatory activity. Macrophage expression of aP2 impacts peroxisome proliferator-activated receptor gamma and IkappaB kinase activities. J Biol Chem 280(13):12888–12895
41. Martinez-Mayorga K, Peppard TL, Lopez-Vallejo F, Yongye AB, Medina-Franco JL (2013) Systematic mining of generally recognized as safe (GRAS) flavor chemicals for bioactive compounds. J Agric Food Chem 61(31):7507–7514
42. Masood A, Huang Y, Hajjhussein H, Xiao L, Li H, Wang W, Hamza A, Zhan CG, O'Donnell JM (2009) Anxiolytic effects of phosphodiesterase-2 inhibitors associated with increased cGMP signaling. J Pharmacol Exp Ther 331(2):690–699
43. Matter H (1997) Selecting optimally diverse compounds from structure databases: a validation study of two-dimensional and three-dimensional molecular descriptors. J Med Chem 40(8):1219–1229
44. McClanahan C (2012) Functional foods. BioFiles 7(6):24–33
45. McDonnell PA, Constantine KL, Goldfarb V, Johnson SR, Sulsky R, Magnin DR, Robl JA, Caulfield TJ, Parker RA, Taylor DS, Adam LP, Metzler WJ, Mueller L, Farmer BT 2nd (2006) NMR structure of a potent small molecule inhibitor bound to human keratinocyte fatty acid-binding protein. J Med Chem 49(16):5013–5017
46. Medina-Franco JL, Martinez-Mayorga K, Peppard TL, Del Rio A (2012) Chemoinformatic analysis of GRAS (generally recognized as safe) flavor chemicals and natural products. PLoS One 7(11):e50798

47. Melzig MF, Funke I (2007) Inhibitors of alpha-amylase from plants—a possibility to treat diabetes mellitus type II by phytotherapy? Wien Med Wochenschr 157(13–14):320–324
48. Mishima S, Matsumoto K, Futamura Y, Araki Y, Ito T, Tanaka T, Iinuma M, Nozawa Y, Akao (2003) Antitumor effect of stilbenoids from *Vateria indica* against allografted sarcoma S-180 in animal model. J Exp Ther Oncol 3:283–288
49. Morales J, Dunbar JC, Ram JL (2002) Effect of aldose reductase inhibition on interleukin-1β-induced nitric oxide (NO) synthesis in vascular tissue. Int J Exp Diabetes Res 3(1):11–20
50. Muthenna P, Suryanarayana P, Gunda SK, Petrash JM, Reddy GB (2009) Inhibition of aldose reductase by dietary antioxidant curcumin: mechanism of inhibition, specificity and significance. FEBS Lett 583(22):3637–3642
51. Okayama N, Omi H, Okouchi M, Imaeda K, Kato T, Akao M, Imai S, Shimizu M, Fukutomi T, Itoh M (2002) Mechanisms of inhibitory activity of the aldose reductase inhibitor, epalrestat, on high glucose-mediated endothelial injury: neutrophil-endothelial cell adhesion and surface expression of endothelial adhesion molecules. J Diabetes Complications 16(5):321–326
52. Oshima Y, Namao K, Kamijou A, Matsuoka S, Nakano M, Terao K, Ohizumi Y (1995) Powerful hepatoprotective and hepatotoxic plant oligostilbenes, isolated from the oriental medicinal plant *Vitis coignetiae* (vitaceae). Experientia 51:63–66
53. Pavicic T, Steckmeier S, Kerscher M, Korting HC (2009) Evidence-based cosmetics: concepts and applications in photoaging of the skin and xerosis. Wien Klin Wochenschr 121(13–14):431–439
54. Piver B, Berthou F, Dreano Y, Lucas D (2003) Differential inhibition of human cytochrome P450 enzymes by epsilon-viniferin, the dimer of resveratrol: comparison with resveratrol and polyphenols from alcoholized beverages. Life Sci 73:1199–1213
55. Privat C, Telo JP, Bernardes-Genisson V, Vieira A, Souchard JP, Nepveu F (2002) Antioxidant properties of trans-epsilon-viniferin as compared to stilbene derivatives in aqueous and nonaqueous media. J Agric Food Chem 50:1213–1217
56. Rarey M, Stahl M (2001) Similarity searching in large combinatorial chemistry spaces. J Comput Aided Mol Des 15(6):497–520
57. Rayasam GV, Tulasi VK, Sodhi R, Davis JA, Ray A (2009) Glycogen synthase kinase 3: more than a namesake. Br J Pharmacol 156(6):885–898
58. Riniker S, Landrum GA (2013) Open-source platform to benchmark fingerprints for ligand-based virtual screening. J Cheminform 5(1):26
59. Sakuta T, Kanayama T (2006) Marked improvement induced in photoaged skin of hairless mouse by ER36009, a novel RARgamma-specific retinoid, but not by ER35794, an RXR-selective agonist. Int J Dermatol 45(11):1288–1295
60. Sauter C, Lamanna N, Weiss MA (2008) Pentostatin in chronic lymphocytic leukemia. Expert Opin Drug Metab Toxicol 4(9):1217–1222
61. Schreyer AM, Blundell TL (2013) CREDO: a structural interactomics database for drug discovery. Database (Oxford) 2013:bat049
62. Senni K, Gueniche F, Foucault-Bertaud A, Igondjo-Tchen S, Fioretti F, Colliec-Jouault S, Durand P, Guezennec J, Godeau G, Letourneur D (2006) Fucoidan a sulfated polysaccharide from brown algae is a potent modulator of connective tissue proteolysis. Arch Biochem Biophys 445(1):56–64
63. Shin E, Lee C, Sung SH, Kim YC, Hwang BY, Lee MK (2011) Antifibrotic activity of coumarins from *Cnidium monnieri* fruits in HSC-T6 hepatic stellate cells. J Nat Med 65(2):370–374
64. Shoeb M, Yadav UC, Srivastava SK, Ramana KV (2011) Inhibition of aldose reductase prevents endotoxin-induced inflammation by regulating the arachidonic acid pathway in murine macrophages. Free Radic Biol Med 51(9):1686–1696
65. Smith CJ, Zhang Y, Koboldt CM, Muhammad J, Zweifel BS, Shaffer A, Talley JJ, Masferrer JL, Seibert K, Isakson PC (1998) Pharmacological analysis of cyclooxygenase-1 in inflammation. Proc Natl Acad Sci U S A 95(22):13313–13318

66. Sprous DG, Salemme FR (2007) A comparison of the chemical properties of drugs and FEMA/FDA notified GRAS chemical compounds used in the food industry. Food Chem Toxicol 45(8):1419–1427
67. Suckling KE (2009) Phospholipase A2 inhibitors in the treatment of atherosclerosis: a new approach moves forward in the clinic. Expert Opin Investig Drugs 18(10):1425–1430
68. Takahashi K, Mizukami H, Kamata K, Inaba W, Kato N, Hibi C, Yagihashi S (2012) Amelioration of acute kidney injury in lipopolysaccharide-induced systemic inflammatory response syndrome by an aldose reductase inhibitor, fidarestat. PLoS One 7(1):e30134
69. Tanabe SI, Grenier D (2008) Macrophage tolerance response to *Aggregatibacter actinomycetemcomitans* lipopolysaccharide induces differential regulation of tumor necrosis factor-alpha, interleukin-1 β and matrix metalloproteinase 9 secretion. J Periodontal Res 43(3):372–377
70. Teague SJ, Davis AM, Leeson PD, Oprea T (1999) The Design of Leadlike Combinatorial Libraries. Angew Chem Int Ed Engl 38(24):3743–3748
71. Thrailkill KM, Clay Bunn R, Fowlkes JL (2009) Matrix metalloproteinases: their potential role in the pathogenesis of diabetic nephropathy. Endocrine 35(1):1–10
72. van Donkelaar EL, Rutten K, Blokland A, Akkerman S, Steinbusch HW, Prickaerts J (2008) Phosphodiesterase 2 and 5 inhibition attenuates the object memory deficit induced by acute tryptophan depletion. Eur J Pharmacol 600(1–3):98–104
73. Vaughan MD, Sampson PB, Honek JF (2002) Methionine in and out of proteins: targets for drug design. Curr Med Chem 9(3):385–409
74. Villena JA, Kralli A (2008) ERRalpha: a metabolic function for the oldest orphan. Trends Endocrinol Metab 19(8):269–276
75. Wallach I, Lilien R (2009) The protein-small-molecule database, a non-redundant structural resource for the analysis of protein-ligand binding. Bioinformatics 25(5):615–620
76. Wang ZL, Deng CY, Zheng H, Xie CF, Wang XH, Luo YF, Chen ZZ, Cheng P, Chen LJ (2012) (Z)2-(5-(4-methoxybenzylidene)-2, 4-dioxothiazolidin-3-yl) acetic acid protects rats from CCl(4)-induced liver injury. J Gastroenterol Hepatol 27(5):966–973
77. Xue L, Godden JW, Bajorath J (2003) Mini-fingerprints for virtual screening: design principles and generation of novel prototypes based on information theory. SAR QSAR Environ Res 14(1):27–40

Chapter 5
Molecular Approaches to Explore Natural and Food-Compound Modulators in Cancer Epigenetics and Metabolism

Alberto Del Rio and Fernando B. Da Costa

5.1 Introduction

Let food be thy medicine and medicine be thy food (Hippocrates)

The biological activity of chemical constituents from natural sources and food is crucial in many cellular processes. Several clinical, physiopathological, and epidemiological studies highlight the detrimental or beneficial role of natural/food factors in conjunction with epigenetic and metabolic alterations. Chemical constituents isolated from various sources can interfere with many different biological targets and have been considered as possible starting points for therapeutic purposes. These agents include, for example, curcumin (turmeric), genistein (soybean), polyphenols (green tea, berries, and cocoa), resveratrol (grapes), and sulforaphane (cruciferous vegetables). Moreover, a wide variety of compounds from medicinal plants, spices, bees, or fish can also be mentioned as examples in this category. Among pathways and functions of cells that are notably modulated by these natural constituents, metabolism and epigenetics have emerged in the context of cancer prevention and therapy. Interestingly, epigenetic changes are tightly linked to metabolism, thus adding a higher level of complexity to elucidate the biological role of these compounds. A deeper understanding on how metabolism and epigenetics are influenced by compounds from natural sources and food can be achieved at molecular level by using a variety of chemoinformatic and computer-aided techniques. These include data mining, molecular databasing, and molecular design techniques such as

A. Del Rio (✉)
Institute of Organic Synthesis and Photoreactivity (ISOF), National Research Council (CNR), Via P. Gobetti 101, 40129 Bologna, Italy
e-mail: alberto.delrio@gmail.com, alberto.delrio@isof.cnr.it

Department of Experimental, Diagnostic and Specialty Medicine (DIMES), Alma Mater Studiorum, University of Bologna, Via S.Giacomo 14, 40126 Bologna, Italy

F. B. Da Costa
School of Pharmaceutical Sciences of Ribeirão Preto, University of São Paulo, Avenida do Café s/n, Ribeirão Preto, SP 14040-903, Brazil

K. Martinez-Mayorga, J. L. Medina-Franco (eds.), *Foodinformatics*,
DOI 10.1007/978-3-319-10226-9_5

pharmacophore-based methods or molecular docking. An overview of these techniques will be described in this chapter in the view of using them as valuable tools to elucidate molecular determinants, mechanism of actions, and polypharmacological role of chemical constituents of food and natural sources.

5.2 Bioactivity of Natural and Food Compounds

The idea that nature is a rich source of bioactive constituents is a 4000-year-old concept. Indians, Egyptians, and Chinese have used natural sources as medicines in early periods of the human civilization. Hippocrates often described diet as a valuable way to treat diseases such as diabetes. Dioscorides, in his five-volume encyclopedia, described the medical uses of herbs, animals, and minerals, and this fantastic work remained alive for more than 15 centuries. Today, lifestyle modifications based on healthy diet, thus on the intake of food and natural compounds, is called lifestyle medicine. The perception that bioactivity of nutraceuticals may have causal relations with the cure or treatment of diseases and, therefore, influence the biological balance of our organism, was spurred starting from the early 1900s. A valuable example of this concept is the treatment of goiter, a disease caused, for over the 90 % of cases, by an iodine deficiency, successfully carried out by the administration of iodine-rich foods or potassium iodine. Yet, the beneficial role of natural compounds has been progressively associated to specific food intake. For instance, it has been observed that consuming fish could contribute to keep in good health heart of healthy people as well as positively influence people who are affected by cardiovascular diseases and are exposed to correlated risks. Thanks to the progress in the analytical techniques of food chemicals, fish was identified to be a good source of omega-3 fatty acids (Fig. 5.1a). Indeed, this class of compounds has the capacity to decrease the risks of arrhythmia, triglycerides level, the rate of atherosclerotic plaque, and to lower blood pressure. Consequently, the beneficial effects of fish have been linked to omega-3 fatty acids.

The awareness that natural compounds and food have beneficial or detrimental effects on our life has been also fuelled by the growing epidemiological evidences that have been made possible by the effective exchange of scientific data, the growing availability of specific natural sources, and the effective number of scientists dedicated to the study of phytochemicals, e.g., in the field of pharmacognosy. This kind of research has also assumed in the past decades the "multidisciplinary" dimension involving not only pharmacists, chemists, and pharmacologists but also biochemists, cellular and molecular biologists, toxicologists, and clinicians, among others. Despite the growing information about natural and food components that would suggest their usage as valuable chemicals to prevent and/or treat diseases, contributing to people well-being, there are still several hurdles to clear in this field pervaded by misinformation, not only in the scientific literature but also in the common knowledge. For instance, a common misconception is the assumption that "natural

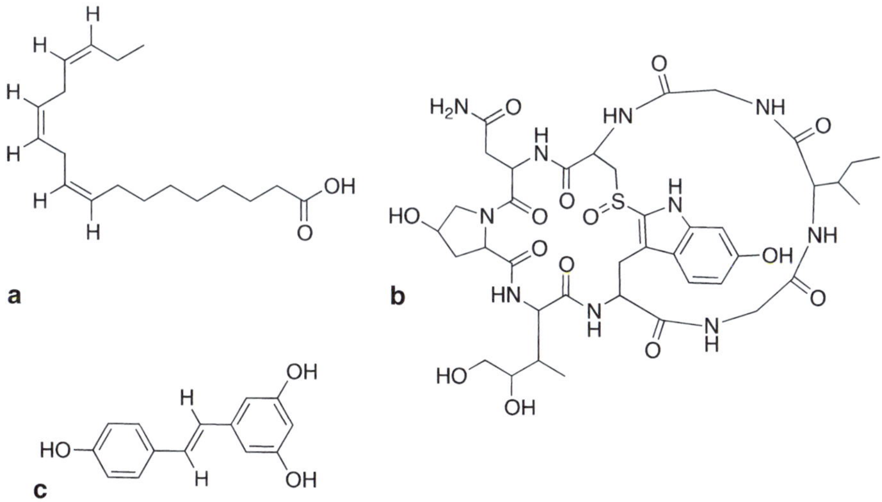

Fig. 5.1 Chemical structures reporting examples of natural compounds with different biological effects. **a** α-Linolenic acid, an essential omega-3 fatty acid. Omega-3 is known to decrease the risk of cardiovascular diseases; **b** α-amantine is a deadly natural compound found in the *Amanita phalloides* mushrooms; **c** resveratrol, a polyphenolic stilbenoid produced in plants with several reported pharmacological actions

is always good," which is an easily falsifiable statement. Indeed, a large number of phytochemicals are known to be harmful for health and, in several cases, also lethal. For example, α-amantine (Fig. 5.1b) is a natural cyclic octapeptide contained in the *Amanita phalloides* fungus, which is widely distributed across Europe and resembles several species of edible mushrooms. α-Amantine is an example of highly poisonous and deadly natural compound which was proved to bind to the bridge helix in RNA polymerase II, interfering with the translocation of RNA and DNA, leading to a drastically reduced rate of synthesis of the RNA molecule [1]. There are numerous classical examples of natural constituents from plants or food which are dangerous to health, such as strychnine from *Strychnos* species, cyanogenic glucosides from cassava (*Manihot* species) or myristicin from nutmeg (*Myristica fragrans*).

To clarify the role of natural and dietary compounds, an elucidation of the interaction mechanisms of these molecules with the human biological network is required, especially at a molecular level. This includes the uncover of the biophysical mechanisms by which these compounds bind to receptors or enzymes (i.e., allosteric regulation and inhibition/activation profile) and their kinetics (i.e., reversible/irreversible, substrate and cofactors competition/non-competition) that could underlie to specific pharmacological actions. These studies are far to be accomplished because, in many cases, it is experimentally difficult to isolate large amounts of compounds from the natural source and, even when this is possible, it is complicated to dissect their intrinsically polypharmacological roles, rendering this area of research extremely challenging. An exemplifying case of polypharmacology is resveratrol (Fig. 5.1c), a polyphenolic stilbenoid produced in plants and found in wine which possesses several reported pharmacological actions, including anti-inflammatory, anticarcinogenic, antimutagenic, antiaging, antioxidant, and anticoagulant. Many examples reporting bioactivity of resveratrol in different molecular pathways can be found in the literature [2–6].

5.2.1 Pharmaceutical Development of Natural Compounds

It was only after the advent of advanced technologies for isolation, purification, and structure elucidation of organic compounds that scientists could realize how natural sources were able to deliver an important amount of diverse chemical entities. Nowadays, it is well known that the natural product landscape constitute a very varied supply of building blocks and intermediates useful for the drug discovery process, which, in many cases, represent the starting point for generating lead compounds. The latter can be further synthetically modified in order to create and develop specific therapeutically relevant pharmaceuticals [7]. The impressive chemical diversity along with the structural complexity of natural compounds represents a source of inspiration for the generation of chemical libraries belonging, in most cases, to an unexplored and "intellectual property free" chemical space, allowing pharmaceutical companies to protect composition of matter together with medical uses [8]. In this sense, we assist to a conceptual shift, passing from the classical era of combinatorial chemistry, during which pharmaceutical companies essentially disregarded the development of natural products as potential drug candidates, to the development of targeted or focused compound libraries inspired by natural sources [9]. The accumulating evidence that the natural selection process represents a unique way to diversify the chemistry of natural compounds and the way in which the latter evolved in biological organisms has favored this process. For these reasons, the interactions of natural compounds with other biological macromolecules reflect, in different cases, high specificity and potency profiles. Since natural products can be considered the richest source of novel chemical scaffolds for biological studies, technologies and strategies to extract them from different sources have evolved rapidly in the past years [10]. A number of advanced separation and structure elucidation techniques are now available for chemists/pharmacists that can now have access to an increasing number of purified natural compounds [11]. Among the separation procedures, high-performance liquid chromatography (HPLC) is the technique of choice because it allows isolation of compounds from the analytical to the preparative scale level. In addition, HPLC can also be coupled to ultraviolet (UV), mass spectrometry (MS), or nuclear magnetic resonance (NMR), comprising the so-called hyphenated or tandem techniques (LC-MS or LC-NMR), which greatly increase the efficiency of compound identification [11].

However, despite the advance in purification techniques, natural products resources are still largely unexplored, mostly due to the technical obstacles to collect samples, especially from the most concealed places on earth, e.g., deep sea level, arid or extremely cold regions. Historically, the most widely used natural compounds have been isolated from plants and animals by means of classical chromatographic techniques such as column or thin-layer chromatography. Subsequently, cultured soil microorganisms, or the direct access to the genome of soil organisms clonable into culturable organisms, provided a rich source of natural products [12]. In the last decade, compounds recovered from the marine environment have come into focus: Indeed, oceans harbor one of the widest variety of ecosystems on earth, a fact reasonably reflected by an unprecedented discovery of new chemical entities of marine origin.

Food compounds, most of them plant secondary metabolites, can be seen as a particular class of natural compounds since they have to be considered as materials designated as "generally recognized as safe" (GRAS) [13]. There is currently a great deal of interest in exploring benefits of bioactive food components and relate them to health and wellness. However, despite the efforts made by researchers to identify food-compounds, few studies report the systematic extraction and purification of a specific bioactive component from different food sources, with the notable exceptions of fruits, vegetables, beverages, and essential oils [14, 15].

5.2.2 *Anticancer Compounds from Natural and Food Sources*

Natural and dietary compounds present molecular scaffolds that are particularly attracting as sources of lead compounds for cancer therapy. Indeed, more than 60% of the anticancer drugs have natural origin or are the result of chemical optimizations of natural scaffolds. Accordingly, it is not surprising that the interest in natural products have gained momentum in the past years, as their application as lead compounds is source of novel chemical entities (NCEs) in different areas of anticancer drug design [16–18]. With their unique chemical diversity, the usage of natural compounds in cancer therapies is even more justified if considered the wide range of variability in terms of biochemical and biological pathways that are present in cancer pathologies. The result of the drift toward natural compounds and their derivatives is reflected by the wide range of chemical compounds from very different sources already associated to bioactivities of oncogenic targets.

Historically, this discovery resulted mainly in the development of anticancer agents from plants (e.g., vinca alkaloids like vincristine and vinblastine; *Podophyllum* lignans like podophyllotoxin; taxanes like paclitaxel and docetaxel; and quinoline alkaloid like camptothecin, topotecan, and irinotecan), marine organisms (i.e., toxins like latrunculins; didemnins like aplidine and trabectedin; and strongylophorines) and microorganisms (e.g., anthracyclines like doxorubicin, daunorubicin, mitoxantrone and idarubicin; chromomycins like dactinomycin and plicamycin; and miscellaneous antibiotics like mitomycin and bleomycin). More recently, different types of terpenoids have been demonstrated to inhibit the NF-kB signaling, to suppress inflammation processes and to reduce cancer progression

[19] while α-methylene-γ-lactones, in particular sesquiterpene lactones (especially found in Asteraceae species), have proven to be promising candidates for treatment of various types of cancer [20–22]. Salinosporamides, a class of marine natural compounds, present in *Salinispora tropica* bacterium, were identified to be potent inhibitors of proteasome [23].

Among natural sources, several food-component agents have also been identified as beneficial for anticancer therapy. Dietary sources including fruits, vegetables, and spices have drawn a great deal of attention from the scientific community due to their demonstrated ability to interfere with cancer mechanisms; nevertheless, speculations by the general public has fomented the idea that fabricated supplements can be a panacea [24]. Scientific literature provided evidence that the regular consumption of fruits, vegetables and spices lowers the incidence of cancers (i.e., stomach, esophagus, lung, oral, endometrium, pancreas, and colon) [25]. These agents include curcumin (turmeric), resveratrol (red grapes, peanuts and berries), genistein (soybean), diallylsulfide (allium), *S*-allyl cysteine (allium), allicin (garlic), lycopene (tomato), capsaicin (red chili), diosgenin (fenugreek), 6-gingerol (ginger), ellagic acid (pomegranate), ursolic acid (apple, pears, prunes), silymarin (milk thistle), anethol (anise, camphor, and fennel), catechins (green and white tea, berries and cocoa), eugenol (cloves), indole-3-carbinol (cruciferous vegetables), limonene (citrus fruits), beta-carotene (carrots), and several dietary fibers.

Many other examples of natural and dietary compounds that have a role in cancer-related diseases underline the importance of this topic in oncological research. In the following paragraphs, we provide an overview of these compounds that specifically modulate cell pathways and functions connected to epigenetic and metabolic changes in cancer diseases.

5.3 Epigenetic and Metabolic Pathophysiology of Cancer

Cancer is a complex set of diseases. Genetic aberrations, epigenetic alterations, and inflammations constitute some of the known mechanisms by which normal cells develop and progress towards neoplastic pathologies. While last decades marked a major understating in cancer genetics, it is now evident that the dissection of the mechanisms of this multifaceted set of diseases requires a deeper look in other paradigms of cancer biology in order to conceive new prevention or therapeutic approaches. This larger framework has evolved in the recent years on novel lines of research, for instance, toward the understanding of the immune system regulation [26, 27] and the epigenetic modifications, but also on the reinterpretation of old studies by means of new scientific awareness that marked a return to cancer metabolism [28–31]. In the next paragraphs, we will discuss cancer metabolism and epigenetics, focusing on the possibilities to interfere with the mechanism of pathogenesis and progression of cancer diseases by means of small molecules of natural and food origin.

5.3.1 Natural Compounds Modulating Epigenetic and Metabolic Mechanisms

Epigenetics is a general term that refers to modifications of genes expression through alteration of chromatin structure and/or DNA methylation occurring without changes in the DNA sequence, from which the term *epi*-(from greek: over, outside of, around)*genetics*. Global modifications of chromatin packaging and its influence in the transcription of associated genes fuelled the research on cancer epigenetics in the past years. The ensemble of known epigenetic mechanisms can be categorized into three classes: i) histone posttranslational modifications (PTMs) that represent one of the major way to arrange the different states of chromatin; ii) DNA methylation, i.e., the methylation of DNA cytosines to 5-methylcytosines; and iii) regulation of gene expression by non-coding RNA (ncRNA). The elucidation of epigenetic phenomena, representing nowadays an important topic of research, is necessary to understand the basis of several biological processes and is progressively translating into the development of new therapeutic *epi-compounds* or *epi-drugs* [32–34]. Different studies have highlighted how alterations in the epigenetic code contribute to the onset and growth of a variety of cancers [35–48]. Consequently, epigenetic modifications are constituting attractive therapeutic targets for the development of new cancer therapies [33, 49–52]. An increasing number of reports describe, in particular, new types of histone post-translational modifications (PTMs) associated with the characterization of the enzymes that are in charge of operating these chemical reactions [53]. Yet, other studies point on the validation of these PTMs in the context of chromatin remodeling and regulation, as well as their clinicopathological relevance in human diseases [54]. It is important to point out that the increasing evidences linking epigenetic targets and cancer pathologies have been boosted by the surge of structural data describing these proteins, thus creating the basis to develop specific probe compounds and start new drug discovery campaigns [54, 55]. However, although the ensemble of these data promises to shed light on cancer epigenetics, the way in which epigenetic modifications relate to cancer and, consequently, their therapeutic relevance in cancer diseases, is still largely unknown. Most of these targets, despite being linked to cancer pathologies, may not have causal role in specific malignant transformations. Some notable exceptions [56, 57] are the recent success stories documenting the potential to interfere with these mechanisms by means of small organic molecules [34]. In particular, the first clinical results have been obtained with histone deacetylases (class I, II and IV HDACs) inhibitors [58], DNA methyltransferases inhibitors (DNMTi) [59] and histone methyltransferases inhibitors [60]. Other classes of epigenetic enzymes are rapidly reaching the potential to become pharmaceutically validated biological targets. Among them are sirtuins, which are NAD^+ dependent histone deacetylases also known as class III HDACs [6], and histone demethylases [61]. Apart from histones PTMs and DNA methylation, growing evidences indicate that modulating microRNAs expression might be useful to interfere with epigenetic mechanisms and develop novel RNA-based drugs for a wide range of diseases [62–65]. Indeed, the deregulation of microRNAs expression and activity is frequently observed in a variety of human pathologies including cancer [66]. Therefore, in addition to the general strategy of increasing or decreasing miRNA abundance and activity by using oligonucleotides or plasmid- and virus-based constructs, a novel paradigm aims to target miRNA expression by means of specific compounds targeting miRNA transcription and processing. Clearly, the potential success of small molecules can be ascribed to their capacity to circumvent the issue of delivery into most tissues making them very attractive as a therapeutic tool.

Metabolic changes have been rediscovered in the context of cancer diseases after the initial observations of Otto Warburg in the early 1920s [30, 31, 67]. Warburg noticed that proliferating cancer cells consume glucose at a high rate, releasing lactate and not carbon dioxide. Indeed, one of the primary metabolic changes in cancer transformation is constituted by an increased catabolic glucose metabolism characterized by high rates of anaerobic glycolysis, regardless of oxygen concentration. While the underlying mechanisms that alter metabolic programs of cancer cells are still to be fully elucidated, it is known that several genetic alterations in cell pathways responsible for the regulation of cells metabolism contribute to cancer growth and progression. For instance, the conversion of glucose to glucose-6-phosphate (G6P) is critical to different cancer phenotypes, a process catalyzed by the enzyme hexokinase-II. Thus, intermediates of glycolysis like G6P can therefore accumulate, creating a highly advantageous environment for cancer survival and growth. On these bases, the pharmacological modulation of specific metabolic enzymes is currently under investigation by various research groups as a viable strategy to block cancer cell proliferation [68–72].

Several natural and dietary components have been already identified as capable to interfere with different epigenetic and metabolic mechanisms [29, 73, 74] (Fig. 5.2). Dietary components like phenolics from green tea, genistein from soybean, isothiocyanates from plant foods (e.g., from Brassicaceae species), diallylsulfide from garlic, curcumin from turmeric, resveratrol from grapes, and sulforaphane from cruciferous vegetables have been studied for their ability to target the epigenome, in relation, for instance, to breast cancer [73, 75–79]. While in most of the cases the mechanisms of action of natural compounds are still poorly understood, some of them have been identified. For instance, luteolin (Fig. 5.2), a common flavonoid found in parsley and celery has been demonstrated to inhibit DNMTs and sirtuins (SIRTs), while retinoic acid, found in carrots, spinach and eggs, and used nowadays to treat leukemias, is an HDAC inhibitor. Among polyphenols, epigallocatechin-3-gallate, the major compound found in green tea, was reported to have a complex polypharmacology, as inhibitor of histone acetyltransferases (HATs), HDACs, SIRTs, DNMTs, retinoic acid receptor (RARβ), proteasome, 78 kDa glucose regulated protein (Grp78) and heat shock protein 90 (Hsp90). Similarly, curcumin (Fig. 5.2) and curcuminoids have also been widely studied for their anti-inflammatory, antiangiogenic, antioxidant, wound healing, and anticancer effects. Importantly, curcumin analogs, like dihydrocoumarins, have been demonstrated to inhibit sirtuins. Since the isoform SIRT1 has been shown to have a role in deacetylating p53, a master regulator of metabolic function in the cell, the inhibition of enzymes like SIRT1 likely contributes to the

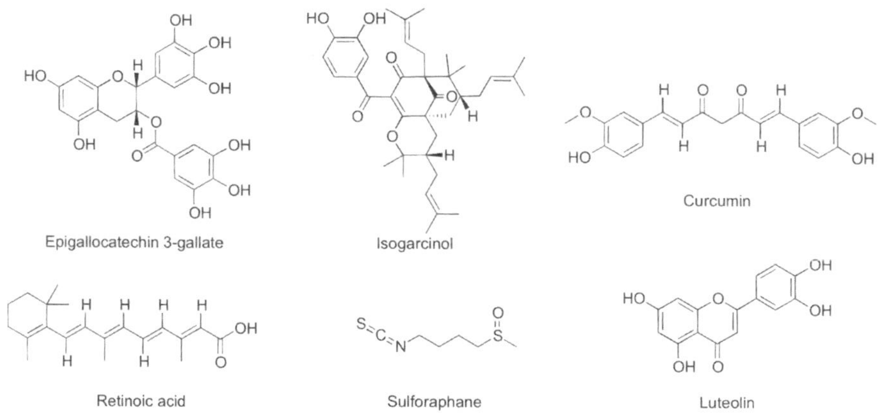

Fig. 5.2 Examples of the chemical diversity of natural compounds with a role in epigenetics and metabolic pathways

regulation of both epigenetic mechanisms and metabolic pathways like glycolysis. Other classes of natural compounds, such as anacardic acid and related compounds from cashew nut, alkaloids such as sanguinarine, quinone derivatives, peptides and peptide conjugates, and polyisoprenylated benzophenone derivatives (PBDs), have been demonstrated to have activities against HATs [80]. As previously pointed out, the discovery of natural scaffolds is allowing the development of focused libraries of compounds that are able to act on epigenetic enzymes with more potent and specific profiles. An example of this strategy is given by Kundu and co-workers, who could generate garcinol derivatives starting from isogarcinol (Fig. 5.2), in order to develop inhibitors for p300 and PCAF HATs [81]. Because of the tight connection with epigenetic and metabolic changes, it is known that specific cancer conditions are strongly influenced by lifestyle and environmental factors, including the intake of food and nutrients [82]. For instance, the absorption of compounds like flavonoids and folates through diet has been shown to alter DNA methylation and modify the risk of human colon cancer and cardiovascular diseases, even though their mechanisms of action have to be ascertained, yet [83–85]. Additional researches on the effects that nutraceuticals have on epigenetic and metabolic changes promise to be relevant for devising new preventive and therapeutic interventions.

5.3.2 *Linking Metabolism and Epigenetic Mechanism*

Growing evidences show how epigenetic changes are linked to cancer metabolism in different cancer pathologies [29]. It is meaningful to stress on how many enzymes, substrates, and co-factors are common in metabolic and epigenetic pathways/targets, as shown in Fig. 5.3. For example, sirtuins deacetylate histone proteins and have also a primary role in metabolic regulation which is dependent on the pool of intracellular NAD^+, whose biosynthesis and signaling became an emerging area in

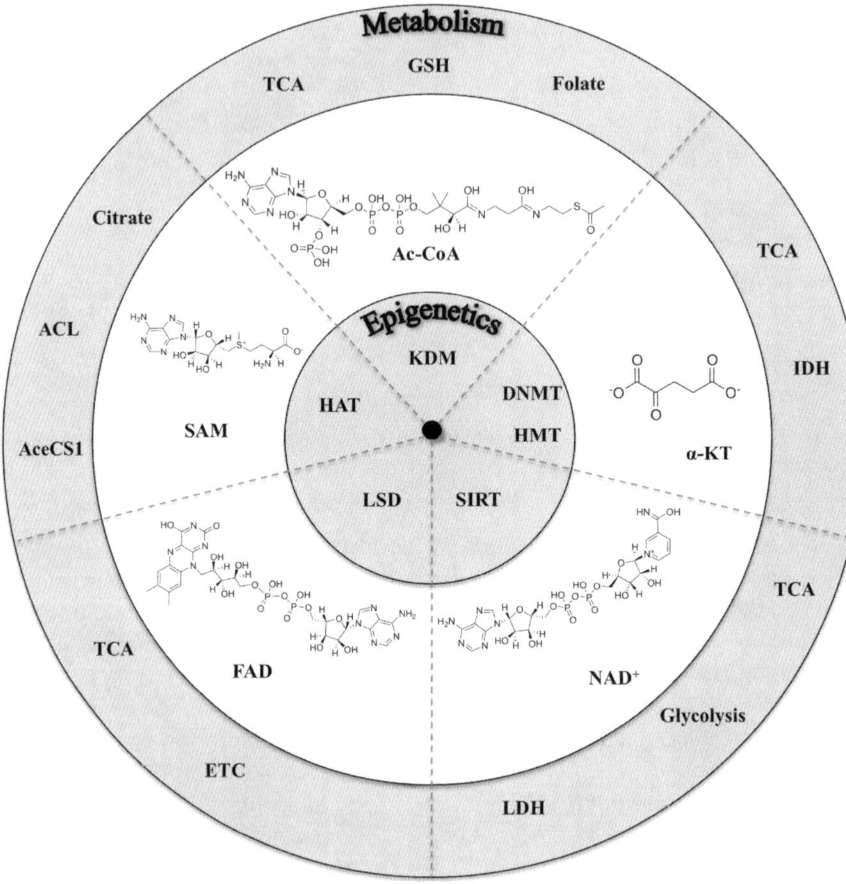

Fig. 5.3 Examples of connections between epigenetics and metabolic pathways. (Abbreviations: *α-KT* α-ketoglutarate; *AcCoA* acetyl coenzyme A; *AcsCS1* acetyl-CoA synthase 1; *ACL* ATP-citrate lyase; *ETC* electron-transport chain; *FAD* flavin adenine dinucleotide; *GSH* glutathione; *IDH* isocitrate dehydrogenase; *LDH* lactate dehydrogenase; *NAD*$^+$ nicotinamide adenine dinucleotide; *SAM* *S*-adenosyl methionine; *TCA* tricarboxylic acid cycle)

medicinal chemistry [86]. Many cancer cells rely on glycolysis to satisfy their energy requirements, a process that leads to the production of lactate and not of acetyl-CoA (AcCoA), like for healthy cells. Since AcCoA is also a substrate of epigenetic enzymes, such as histone acetyltransferases (HATs), depletion of the AcCoA in cancer cells might contribute to epigenetic alterations. A similar consideration can be drawn for other metabolic co-substrate and co-factors like *S*-adenosylmethionine (SAM), flavin adenine dinucleotide (FAD), and α-ketoglutarate (Fig. 5.3), which are all involved in the epigenetic regulation through various enzymatic mechanisms [87]. Moreover, compounds of natural and food origin can be converted by cell metabolites into chemical intermediates implicated in epigenetic and metabolic alterations [25, 29, 44, 75, 82, 88]. So, it is evident that a molecular-level knowledge of the connections between metabolism and epigenetic mechanisms is required in order to define the polypharmacological role of small molecules. It should be noted that the biological effect of many chemical scaffolds, especially of natural origin, is in most cases ascribable to a promiscuous activity towards biological targets that uses common substrates and cofactors like NAD+/NADH, FAD, SAM, AcCoA, α-ketoglutarate, and ATP. Therefore, in the framework of developing compound libraries from natural and food origin, it is essential to assess compounds against their impact on epigenome and metabolism by looking at their polypharmacological behavior. In particular, the screening of biological activities acquires importance if considered that the detrimental or beneficial effects of natural compounds for the treatment of a specific disease, is dependent on the physiopathological context [89].

5.4 Computer-Aided Molecular Design Approaches

Computer-aided molecular techniques are heavily used in academia and industrial settings to assist the selection of new compounds with predefined biological activity. Several examples testify their successful applications in the development of new chemical entities [90–92] and a wide range of disciplines nowadays revolve around computer-aided drug discovery (CADD), including chemoinformatics, computational chemistry, structural biology, biophysics, medicinal chemistry, organic chemistry, and pharmacology. Among the various computational techniques available, virtual screening is certainly the most popular to screen rapidly and cost-effectively new chemicals from large libraries of compounds [93–95]. In principle, this technique can be divided in ligand- and structure-based drug design techniques (LBDD and SBDD): the first category usually takes advantage of information from known bioactive compounds (ligand), while the second usually exploit three-dimensional structure of the biological target (protein) in order to identify putative modulators of the protein activity. In the past years, the growing availability of protein structures, resolved by structural biologists, progressively raised the possibility to deploy SBDD. Nevertheless, ligand-based techniques are still essential tools, for example when structural information of a biological target is missing or when the molecular design is not directed towards a target-centric approach, but point to modulate cellular pathways or phenotypic traits without a precise knowledge of the mechanism of action. In addition, it should be noted that, despite the apparent advantage and the success of the target-centric approach, which consist in the design of small molecules having high-selectivity profiles against a specific target, it has failed in many other cases [96].

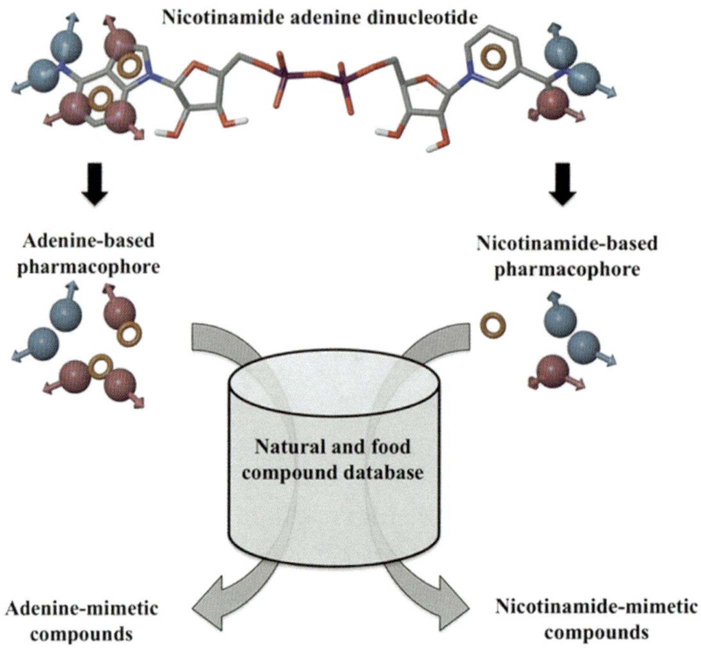

Fig. 5.4 Example of pharmacophore-based *in silico* approach to discover adenine and nicotinamide mimetic compound of NAD^+ originating from natural or food sources

5.4.1 CADD Approaches on Epigenetic and Metabolic Targets

As seen in previous sections, the research aiming at developing new therapeutic anticancer strategies against epigenetic and metabolic targets has flourished in the past years. Several reports describe rationales, targets, new drugs, approaches, novel compounds, and methodologies [34, 35, 37, 41–48, 52, 53, 56, 97–107]. Computational techniques are being actively used in this field and several reviews and articles have been published recently on this topic [56, 57, 59, 105, 108, 109]. A valuable example is the extensive usage of computer-aided techniques for epigenetic enzymes like sirtuins [6, 108, 110–114]. A variety of computational tools like molecular docking and pharmacophore mapping have been used to identify novel modulating compounds while trying to explain the mechanism of actions of these small molecules. Equally, theoretical tools have also been applied to identify and elucidate pharmacological mechanisms of metabolic enzymes like lactate dehydrogenase and hexokinase-II [69, 114, 115]. Of note, many of these targets use NAD^+ as a cofactor and several computational strategies were directed to find competitive compounds of either the adenine or the nicotinamide pocket, or both. As an example, Fig. 5.4 depicts a typical *in silico* screening workflow that uses pharmacophore techniques to identify NAD^+ competitive inhibitors with natural or dietary-derived scaffolds mimicking the adenine or the nicotinamide moieties. In fact, three-dimensional pharmacophore modeling techniques revealed to be useful for virtual screening and computational purposes to analyze diverse compound databases in order to define pharmaceutical values of new compounds [116, 117]. Interestingly, the use of less-sophisticated techniques based on topological-structural descriptors and subsequent statistical treatment, i.e., discriminant analysis, have also been demonstrated as very efficient methodologies for the selection of new natural compounds. Even in this case, the validated model could be readily applied for searching new chemicals of natural origin in large databases [118, 119]. It is expected that the usage and combination of various *in silico* approaches and the availability of compound databases of natural and dietary sources (see below) could constitute an effective step toward the identification, development, and pharmacological definition of natural and dietary-derived components in metabolic and epigenetic mechanism of cancer.

5.4.2 Chemical Space of Natural and Food Compounds

Since natural products and dietary components are known to represent a vast chemical diversity with underlying scaffold complexities and architectures, exploring the chemical space of these compounds it currently a major field of research for different groups [13, 120–125]. Most of the natural products are assorted by chemical groups reflecting novel molecular properties/features as compared to synthetic compounds and available drugs. Several chemoinformatic analyzes, in fact, highlight this behavior and, at the same time, recognize the adherence to drug- and lead-like rules purporting the idea that several classes can be considered as pharmaceutically relevant entities [13, 124]. In addition, despite this diversity, natural products insure the presence of privileged scaffolds that could offer the advantage to address the coverage of poorly explored chemical space [121, 126]. As previously indicated, this feature is particularly appealing for industrial settings to insure the appropriate intellectual property protection requested for the pharmaceutical development. In this direction, it should be noted also that natural products are providing line principles for novel library design in combinatorial chemistry and targeted compound libraries inspired by nature [126–128].

From the chemical point of view, the analysis of natural products databases available in the public domain shows a low-molecular overlap of compounds and highlight as the most representative molecular fragment benzene, acyclic compounds, flavones, coumarins, and flavanones [121, 122, 125]. A particular class of natural compounds that can be considered as dietary component are flavoring substances like menthol, camphor, and anethol, that are discrete chemical entities that usually are considered "generally recognized as safe" (GRAS) compounds. Interestingly, the comparison of collections of compounds including GRAS, natural products, approved drugs, and dataset from commercial molecules by means of chemoinformatic analysis demonstrated that GRAS products are an important source of bioactive compounds that possess all the characteristics for drug discovery and nutraceutical purposes [13].

Among computational approaches that can help driving the discovery of new bioactive compounds, a prominent workflow is the screening of large database of readily available molecules. It is with surprise that the scientific community has not developed yet a freely available and fully chemically annotated database of food components [8, 9]. Despite this lack, some examples are starting to appear in the literature and on the Internet. Among them, we can list the INFOODS of FAO [129], the USDA national nutrient database [130] and the FooDB that has been recently released [131]. In the direction of the creation of a comprehensive and freely available collection of food chemicals, it should be noted also the necessity to include the possible procurement from commercial sources of purified samples of food components that, ideally, should complement the major efforts that have been done in the past years for other natural sources [123].

5.5 Conclusions

Many anticancer drugs have natural origin or are the result of chemical optimizations of natural scaffolds. Because the natural product landscape constitutes a varied supply of building blocks and intermediates, they can represent the starting point for generating lead compounds with bioactive relevance. A thriving topic in cancer research deals with metabolism and epigenetics mechanisms that lead to malignant transformation and the way to interfere pharmacologically with the pathogenesis and progression of cancer diseases by means of small molecules. Natural and food-derived compounds able to modulate epigenetic and metabolic mechanisms are of great interest because they promise to provide new therapeutic interventions, as they are capable to exert anti-inflammatory, antiangiogenic and antioxidant effects that could also be beneficial for anticancer purposes. In this framework, it is expected that advances in computational approaches, with emphasis on pharmacophore and docking-based techniques, together with the systematic cataloguing of natural and dietary-related components, would greatly help to track molecular mechanisms involved in nutriepigenomics and nutrimetabolomics, and therefore constitute a launching platform for new drug-discovery pipelines.

Acknowledgments Authors thank Greta Varchi and Federico Andreoli for useful discussions and proof reading. A. Del Rio would like to thank the Italian Association for Cancer Research (Start Up grant N.6266) and F. B. Da Costa FAPESP and CNPq for financial support.

References

1. Bushnell DA, Cramer P, Kornberg RD (2002) Structural basis of transcription: alpha-amanitin-RNA polymerase II cocrystal at 2.8 A resolution. P Natl Acad Sci U S A 99:1218–1222
2. Beher D, Wu J, Cumine S, Kim KW, Lu S-C, Atangan L, Wang M (2009) Resveratrol is not a direct activator of SIRT1 enzyme activity. Chem Biol Drug Des 74:619–624

3. Denu JM (2012) Fortifying the link between SIRT1, resveratrol, and mitochondrial function. Cell Metab 15:566–567
4. Price NL, Gomes AP, Ling AJY et al (2012) SIRT1 is required for AMPK activation and the beneficial effects of resveratrol on mitochondrial function. Cell Metab 15:675–690
5. Moniot S, Weyand M, Steegborn C (2012) Structures, substrates, and regulators of mammalian sirtuins—opportunities and challenges for drug development. Front Pharmacol 3:16
6. Bruzzone S, Parenti MD, Grozio A, Ballestrero A, Bauer I, Del Rio A, Nencioni A (2013) Rejuvenating sirtuins: the rise of a new family of cancer drug targets. Curr Pharm Des 19:614–623
7. Newman DJ, Cragg GM (2012) Natural products as sources of new drugs over the 30 years from 1981 to 2010. J Nat Prod 75:311–335
8. Hong J (2011) Role of natural product diversity in chemical biology. Curr Opin Chem Biol 15:350–354
9. Sheridan C (2012) Recasting natural product research. Nat Biotechnol 30:385–387
10. Paterson I, Anderson EA (2005) Chemistry. The renaissance of natural products as drug candidates. Science 310:451–453
11. Pauli GF, Chen S-N, Friesen JB, McAlpine JB, Jaki BU (2012) Analysis and purification of bioactive natural products: the AnaPurNa study. J Nat Prod 75:1243–1255
12. Handelsman J, Rondon MR, Brady SF, Clardy J, Goodman RM (1998) Molecular biological access to the chemistry of unknown soil microbes: a new frontier for natural products. Chem Biol 5:R245–R249
13. Medina-Franco JL, Martínez-Mayorga K, Peppard TL, Del Rio A (2012) Chemoinformatic analysis of GRAS (Generally Recognized as Safe) flavor chemicals and natural products. PLoS One 7:e50798
14. Barbosa-Pereira L, Pocheville A, Angulo I, Paseiro-Losada P, Cruz JM (2013) Fractionation and purification of bioactive compounds obtained from a brewery waste stream. Biomed Res Int 2013(2013):408491
15. Angela A, Meireles M (eds) (2008) Extracting bioactive compounds for food products. CRC, Boca Raton. doi:10.1201/9781420062397
16. Kinghorn AD, Chin Y-W, Swanson SM (2009) Discovery of natural product anticancer agents from biodiverse organisms. Curr Opin Drug Discov Dev 12:189–196
17. Newman DJ (2008) Natural products as leads to potential drugs: an old process or the new hope for drug discovery? J Med Chem 51:2589–2599
18. Gordaliza M (2008) Natural products as leads to anticancer drugs. Clin Transl Oncol 9:767–776
19. Salminen A, Lehtonen M, Suuronen T, Kaarniranta K, Huuskonen J (2008) Terpenoids: natural inhibitors of NF-kappaB signaling with anti-inflammatory and anticancer potential. Cell Mol Life Sci 65:2979–2999
20. Merfort I (2011) Perspectives on sesquiterpene lactones in inflammation and cancer. Curr Drug Targets 12:1560–1573
21. Ghantous A, Gali-Muhtasib H, Vuorela H, Saliba NA, Darwiche N (2010) What made sesquiterpene lactones reach cancer clinical trials? Drug Discov Today 15:668–678
22. Janecka A, Wyrębska A, Gach K, Fichna J, Janecki T (2012) Natural and synthetic α-methylenelactones and α-methylenelactams with anticancer potential. Drug Discov Today 17:561–572
23. Gulder TAM, Moore BS (2010) Salinosporamide natural products: Potent 20 S proteasome inhibitors as promising cancer chemotherapeutics. Angew Chem Int Edit 49:9346–9367
24. vel Szic KS, Palagani A, Hassannia B (2011) Phytochemicals and cancer chemoprevention: epigenetic friends or foe? In: Rasooli I (ed) Phytochemicals—Bioactivities and Impact on Health. InTech, Croatia, pp 159–198. doi:10.5772/28499
25. vel Szic KS, Ndlovu MN, Haegeman G, Vanden Berghe W (2010) Nature or nurture: let food be your epigenetic medicine in chronic inflammatory disorders. Biochem Pharmacol 80:1816–1832
26. Galluzzi L, Senovilla L, Zitvogel L, Kroemer G (2012) The secret ally: immunostimulation by anticancer drugs. Nat Rev Drug Discov 11:215–233

27. Cavallo F, De Giovanni C, Nanni P, Forni G, Lollini P-L (2011) 2011: the immune hallmarks of cancer. Cell 60:319–326
28. Cairns RA, Harris IS, Mak TW (2011) Regulation of cancer cell metabolism. Nat Rev Cancer 11:85–95
29. Gerhäuser C (2012) Cancer cell metabolism, epigenetics and the potential influence of dietary components—a perspective. Biomed Res 23:1–21
30. Semenza GL (2011) A return to cancer metabolism. J Mol Med 89:203–204
31. Koppenol WH, Bounds PL, Dang CV (2011) Otto Warburg's contributions to current concepts of cancer metabolism. Nat Rev Cancer 11:325–337
32. Best JD, Carey N (2010) Epigenetic therapies for non-oncology indications. Drug Discov Today 15:1008–1014
33. Best JD, Carey N (2010) Epigenetic opportunities and challenges in cancer. Drug Discov Today 15:65–70
34. Dhanak D (2012) Cracking the code: the promise of epigenetics. ACS Med Chem Lett 3(7):521–523. doi:10.1021/ml300141h
35. Ellis L, Atadja PW, Johnstone RW (2009) Epigenetics in cancer: targeting chromatin modifications. Mol Cancer Ther 8:1409–1420
36. Altucci L, Minucci S (2009) Epigenetic therapies in haematological malignancies: searching for true targets. Eur J Cancer 45:1137–1145
37. Chi P, Allis CD, Wang GG (2010) Covalent histone modifications—miswritten, misinterpreted and mis-erased in human cancers. Nat Rev Cancer 10:457–469
38. Herranz M, Esteller M (2006) New therapeutic targets in cancer: the epigenetic connection. Clin Transl Oncol 8:242–249
39. Graham JS, Kaye SB, Brown R (2009) The promises and pitfalls of epigenetic therapies in solid tumours. Eur J Cancer 45:1129–1136
40. Santos-Rosa H, Caldas C (2005) Chromatin modifier enzymes, the histone code and cancer. Eur J Cancer 41:2381–2402
41. Rodríguez-Paredes M, Esteller M (2011) Cancer epigenetics reaches mainstream oncology. Nat Med 17:330–339
42. Rius M, Lyko F (2012) Epigenetic cancer therapy: rationales, targets and drugs. Oncogene 31:4257–4265
43. Kulis M, Esteller M (2010) DNA methylation and cancer. Adv Genet 70:27–56
44. Meeran SM, Ahmed A, Tollefsbol TO (2010) Epigenetic targets of bioactive dietary components for cancer prevention and therapy. Clin Epigenetics 1:101–116
45. Ljungman M (2009) Targeting the DNA damage response in cancer. Chem Rev 109:2929–2950
46. Claes B, Buysschaert I, Lambrechts D (2010) Pharmaco-epigenomics: discovering therapeutic approaches and biomarkers for cancer therapy. Heredity 105:152–160
47. Pollock RM, Richon VM (2009) Epigenetic approaches to cancer therapy. Drug Discov Today Ther Strategy 6:71–79
48. Spannhoff A, Sippl W, Jung M (2009) Cancer treatment of the future: inhibitors of histone methyltransferases. Int J Biochem Cell Biol 41:4–11
49. Sala A, Corona DFV (2008) Epigenetics: More than genetics. Fly 2:165–168
50. Baylin SB (2008) Epigenetics and cancer. Mol Basis Cancer. 2:57–65
51. Lohrum M, Stunnenberg HG, Logie C (2007) The new frontier in cancer research: deciphering cancer epigenetics. Int J Biochem Cell Biol 39:1450–1461
52. Inche AG, La Thangue NB (2006) Chromatin control and cancer-drug discovery: realizing the promise. Drug Discov Today 11:97–109
53. Bannister AJ, Kouzarides T (2011) Regulation of chromatin by histone modifications. Cell Res 21:381–395
54. Andreoli F, Barbosa AJM, Parenti MD, Del Rio A (2013) Modulation of epigenetic targets for anticancer therapy: clinicopathological relevance, structural data and drug discovery perspectives. Curr Pharm Des 19:578–613
55. Hou H, Yu H (2010) Structural insights into histone lysine demethylation. Curr Opin Struct Biol 20:739–748

56. Sippl W, Jung M (2009) Epigenetic drug discovery special issue. Bioorg Med Chem 19:3603–3604
57. Sippl W, Jung M (2009) Epigenetic targets in drug discovery. Wiley, Weinheim
58. Lombardi PM, Cole KE, Dowling DP, Christianson DW (2011) Structure, mechanism, and inhibition of histone deacetylases and related metalloenzymes. Curr Opin Struct Biol 21:735–743
59. Yoo J, Medina-Franco JL (2012) Inhibitors of DNA methyltransferases: insights from computational studies. Curr Med Chem 19(21):3475–3487
60. Copeland RA, Solomon ME, Richon VM (2009) Protein methyltransferases as a target class for drug discovery. Nat Rev Drug Discov 8:724–732
61. Lohse B, Kristensen JL, Kristensen LH, Agger K, Helin K, Gajhede M, Clausen RP (2011) Inhibitors of histone demethylases. Bioorg Med Chem 19:3625–3636
62. Cho WC (2012) Exploiting the therapeutic potential of microRNAs in human cancer. Expert Opin Ther Targets 16:345–350
63. Esau CC, Monia BP (2007) Therapeutic potential for microRNAs. Adv Drug Deliv Rev 59:101–114
64. Bratkovič T, Glavan G, Strukelj B, Zivin M, Rogelj B (2012) Exploiting microRNAs for cell engineering and therapy. Biotechnol Adv 30:753–765
65. Ling H, Fabbri M, Calin GA (2013) MicroRNAs and other non-coding RNAs as targets for anticancer drug development. Nat Rev Drug Discov 12:847–865
66. De Santa F, Iosue I, Del Rio A, Fazi F (2013) microRNA biogenesis pathway as a therapeutic target for human disease and cancer. Curr Pharm Des 19:745–764
67. Bayley J-P, Devilee P (2012) The Warburg effect in 2012. Curr Opin Oncol 24:62–67
68. Le A, Cooper CR, Gouw AM, Dinavahi R, Maitra A, Deck LM, Royer RE, Vander Jagt DL, Semenza GL, Dang C V (2010) Inhibition of lactate dehydrogenase A induces oxidative stress and inhibits tumor progression. Proc Natl Acad Sci U S A 107:2037–2042
69. Salani B, Marini C, Del Rio A et al (2013) Metformin impairs glucose consumption and survival in Calu-1 cells by direct inhibition of hexokinase-II. Sci Rep 3:2070
70. Zhao Y, Butler EB, Tan M (2013) Targeting cellular metabolism to improve cancer therapeutics. Cell Death Dis 4:e532
71. Birsoy K, Sabatini DM, Possemato R (2012) Untuning the tumor metabolic machine: targeting cancer metabolism: a bedside lesson. Nat Med 18:1022–1023
72. Vander Heiden MG (2011) Targeting cancer metabolism: a therapeutic window opens. Nat Rev Drug Discov 10:671–684
73. Stefanska B, Karlic H, Varga F, Fabianowska-Majewska K, Haslberger A (2012) Epigenetic mechanisms in anti-cancer actions of bioactive food components—the implications in cancer prevention. Br J Pharmacol 167:279–297
74. Kirk H, Cefalu WT, Ribnicky D, Liu Z, Eilertsen KJ (2008) Botanicals as epigenetic modulators for mechanisms contributing to development of metabolic syndrome. Metabolism 57:16–23
75. Khan SI, Aumsuwan P, Khan IA, Walker LA, Dasmahapatra AK (2012) Epigenetic events associated with breast cancer and their prevention by dietary components targeting the epigenome. Chem Res Toxicol 25:61–73
76. Su Y, Shankar K, Rahal O, Simmen RCM (2011) Bidirectional signalling of mammary epithelium and stroma: implications for breast cancer—preventive actions of dietary factors. J Nutr Biochem 22:605–611
77. Lustberg MB, Ramaswamy B (2010) Epigenetic therapy in breast cancer. Curr Breast Cancer Rep 3:34–43
78. Ramaswamy B, Sparano JA (2010) Targeting epigenetic modifications for the treatment and prevention of breast cancer. Curr Breast Cancer Rep 2:198–207
79. Thornburg KL, Shannon J, Thuillier P, Turker MS (2010) In utero life and epigenetic predisposition for disease. Adv Genet 71:57–78
80. Piaz FD, Vassallo A, Rubio OC, Castellano S, Sbardella G, De Tommasi N (2011) Chemical biology of histone acetyltransferase natural compounds modulators. Mol Divers 15:401–416

81. Mantelingu K, Reddy BAA, Swaminathan V et al (2007) Specific inhibition of p300-HAT alters global gene expression and represses HIV replication. Chem Biol 14:645–657
82. Nyström M (2009) Diet and epigenetics in colon cancer. World J Gastroenterol 15:257
83. Duthie SJ (2011) Epigenetic modifications and human pathologies: cancer and CVD. P Nutr Soc 70:47–56
84. Van Engeland M, Herman JG (2010) Viewing the epigenetics of colorectal cancer through the window of folic acid effects. Cancer Prev Res 3:1509–1512
85. Garagnani P, Pirazzini C, Franceschi C (2013) Colorectal cancer microenvironment: among nutrition, gut microbiota, inflammation and epigenetics. Curr Pharm Des 19:765–778
86. Nencioni A, Bruzzone S, Del Rio A (2013) Editorial: NAD+ biosynthesis and signaling as an emerging area in medicinal chemistry. Curr Top Med Chem 13:2905–2906
87. Teperino R, Schoonjans K, Auwerx J (2010) Histone methyl transferases and demethylases; can they link metabolism and transcription? Cell Metab 12:321–327
88. Rajendran P, Williams DE, Ho E, Dashwood RH (2011) Metabolism as a key to histone deacetylase inhibition. Crit Rev Biochem Mol Biol 46:181–199
89. Petrelli A, Giordano S (2008) From single- to multi-target drugs in cancer therapy: when aspecificity becomes an advantage. Curr Med Chem 15:422–432
90. Talele TT, Khedkar SA, Rigby AC (2010) Successful applications of computer aided drug discovery: moving drugs from concept to the clinic. Curr Top Med Chem 10:127–141
91. Van Drie JH (2007) Computer-aided drug design: the next 20 years. J Comput Aided Mol Des 21:591–601
92. Borhani DW, Shaw DE (2012) The future of molecular dynamics simulations in drug discovery. J Comput Aided Mol Des 26:15–26
93. Moura Barbosa AJ, Del Rio A (2012) Freely accessible databases of commercial compounds for high- throughput virtual screenings. Curr Top Med Chem 12:866–877
94. Del Rio A, Barbosa AJM, Caporuscio F, Mangiatordi GF (2010) CoCoCo: a free suite of multiconformational chemical databases for high-throughput virtual screening purposes. Mol Biosyst 6:2122–2128
95. Del Rio A, Barbosa AJM, Caporuscio F (2011) Use of large multiconformational databases with structure-based pharmacophore models for fast screening of commercial compound collections. J Cheminform 3:P27
96. Bottegoni G, Favia AD, Recanatini M, Cavalli A (2012) The role of fragment-based and computational methods in polypharmacology. Drug Discov Today 17:23–34
97. Copeland RA, Olhava EJ, Scott MP (2010) Targeting epigenetic enzymes for drug discovery. Curr Opin Chem Biol 14:505–510
98. De Koning L, Corpet A, Haber JE, Almouzni G (2007) Histone chaperones: an escort network regulating histone traffic. Nat Struct Mol Biol 14:997–1007
99. Golbabapour S, Abdulla MA, Hajrezaei M (2011) A concise review on epigenetic regulation: insight into molecular mechanisms. Int J Mol Sci 12:8661–8694
100. Mai A, Cheng D, Bedford MT et al (2008) Epigenetic multiple ligands: mixed histone/protein methyltransferase, acetyltransferase, and class III deacetylase (sirtuin) inhibitors. J Med Chem 51:2279–2290
101. Suganuma T, Workman JL (2008) Crosstalk among histone modifications. Cell 135:604–607
102. Jones P (2012) Development of second generation epigenetic agents. Med Chem Comm 3:135
103. Mani S, Herceg Z (2010) DNA demethylating agents and epigenetic therapy of cancer. Adv Genet 70:327–340
104. Karberg S (2009) Switching on epigenetic therapy. Cell 139:1029–1031
105. Medina-Franco JL, Caulfield T (2011) Advances in the computational development of DNA methyltransferase inhibitors. Drug Discov Today 16:418–425
106. Kristensen LS, Nielsen HM, Hansen LL (2009) Epigenetics and cancer treatment. Eur J Pharmacol 625:131–142
107. Hamm CA, Costa FF (2011) The impact of epigenomics on future drug design and new therapies. Drug Discov Today 16:626–635

108. Heinke R, Carlino L, Kannan S, Jung M, Sippl W (2011) Computer- and structure-based lead design for epigenetic targets. Bioorg Med Chem 19:3605–3615
109. Yoo J, Medina-Franco JL (2011) Discovery and optimization of inhibitors of DNA methyltransferase as novel drugs for cancer therapy. In: Rundefeldt C (ed) Drug development—a case study based insight into modern strategies. InTech, Croatia
110. Neugebauer RC, Uchiechowska U, Meier R, Hruby H, Valkov V, Verdin E, Sippl W, Jung M (2008) Structure-activity studies on splitomicin derivatives as sirtuin inhibitors and computational prediction of binding mode. J Med Chem 51:1203–1213
111. Costantini S, Sharma A, Raucci R, Costantini M, Autiero I, Colonna G (2013) Genealogy of an ancient protein family: the Sirtuins, a family of disordered members. BMC Evol Biol 13:60
112. Sakkiah S, Arooj M, Kumar MR, Eom SH, Lee KW (2013) Identification of inhibitor binding site in human sirtuin 2 using molecular docking and dynamics simulations. PLoS One 8:e51429
113. Chen L (2011) Medicinal chemistry of sirtuin inhibitors. Curr Med Chem 18:1936–1946
114. Mak L, Liggi S, Tan L, Kusonmano K, Rollinger JM, Glen RC, Kirchmair J, Koutsoukas A (2012) Anti-cancer drug development: computational strategies to identify and target proteins involved in cancer metabolism. Curr Pharm Des 19(4):532–577
115. Manerba M, Vettraino M, Fiume L, Di Stefano G, Sartini A, Giacomini E, Buonfiglio R, Roberti M, Recanatini M (2012) Galloflavin (CAS 568–80-9): a novel inhibitor of lactate dehydrogenase. Chem Med Chem 7:311–317
116. Sanders MPA, Barbosa AJM, Zarzycka B, Nicolaes GAF, Klomp JPG, de Vlieg J, Del Rio A (2012) Comparative analysis of pharmacophore screening tools. J Chem Inf Model 52:1607–1620
117. Schuster D, Wolber G (2010) Identification of bioactive natural products by pharmacophore-based virtual screening. Curr Pharm Des 16:1666–1681
118. Galvez-Llompart M, Zanni R, García-Domenech R (2011) Modeling natural anti-inflammatory compounds by molecular topology. Int J Mol Sci 12:9481–9503
119. Gálvez-Llompart M, Recio MC, García-Domenech R (2011) Topological virtual screening: a way to find new compounds active in ulcerative colitis by inhibiting NF-κB. Mol Divers 15:917–926
120. Harvey AL (2008) Natural products in drug discovery. Drug Discov Today 13:894–901
121. Rosén J, Gottfries J, Muresan S, Backlund A, Oprea TI (2009) Novel chemical space exploration via natural products. J Med Chem 52:1953–1962
122. Singh N, Guha R, Giulianotti MA, Pinilla C, Houghten RA, Medina-Franco JL (2009) Chemoinformatic analysis of combinatorial libraries, drugs, natural products, and molecular libraries small molecule repository. J Chem Inf Model 49:1010–1024
123. Füllbeck M, Michalsky E, Dunkel M, Preissner R (2006) Natural products: sources and databases. Nat Prod Rep 23:347–356
124. Lachance H, Wetzel S, Kumar K, Waldmann H (2012) Charting, navigating, and populating natural product chemical space for drug discovery. J Med Chem 55:5989–6001
125. Yongye AB, Waddell J, Medina-Franco JL (2012) Molecular scaffold analysis of natural products databases in the public domain. Chem Biol Drug Des 80:717–724
126. Bauer RA, Wurst JM, Tan DS (2010) Expanding the range of "druggable" targets with natural product-based libraries: an academic perspective. Curr Opin Chem Biol 14:308–314
127. Kumar K, Waldmann H (2009) Synthesis of natural product inspired compound collections. Angew Chem Int Ed Engl 48:3224–3242
128. Over B, Wetzel S, Grütter C, Nakai Y, Renner S, Rauh D, Waldmann H (2012) Natural-product-derived fragments for fragment-based ligand discovery. Nat Chem 5:21–28
129. International network of food data systems. http://www.fao.org/infoods/infoods/en/ . Accessed 7 Sept 2014
130. USDA National Nutrient Database. http://ndb.nal.usda.gov/ . Accessed 7 Sept 2014
131. FooDB. http://www.foodb.ca/. Accessed 7 Sept 2014

Chapter 6
Discovery of Natural Products that Modulate the Activity of PPARgamma: A Source for New Antidiabetics

Santiago Garcia-Vallve, Laura Guasch and Miquel Mulero

6.1 Introduction

Both the prevalence and incidence of diabetes are increasing worldwide, particularly in developing countries. The sixth edition of the International Diabetes Federation (IDF) Diabetes Atlas estimated that 382 million people, or 8.3 % of the worldwide adult population, had diabetes in 2013 and that the number of people with the disease will rise to 592 million by 2035, an increase of the 55 % [1]. Diabetes caused approximately 5.1 million deaths in 2013 in people aged between 20 and 79 years, an equivalent of one death every 6 s [1]. People with diabetes have an increased risk of developing a number of serious health problems. Over time, diabetes can damage the heart, blood vessels, eyes, kidneys and nerves, causing an increased risk of cardiovascular disease, blindness, kidney failure and lower limb amputation. The overall risk of dying among people with diabetes is at least double the risk of their peers without diabetes [2]. In financial terms, the burden of diabetes is enormous, costing US$ 548 billion in health spending in 2013 [1]. This accounts for 10.8 % of total health expenditure worldwide [1]. By 2035, this number is projected to exceed US$ 627 billion [1].

Type 2 diabetes (T2D), also called noninsulin-dependent diabetes mellitus or adult-onset diabetes, is the most common type of diabetes. At least 90 % of people

S. Garcia-Vallve (✉) M. Mulero
Cheminformatics and Nutrition Group, Departament de Bioquímica i Biotecnologia, Universitat Rovira i Virgili (URV), Tarragona, Catalonia, Spain
e-mail: santi.garcia-vallve@urv.cat

S. Garcia-Vallve
Centre Tecnològic de Nutrició i Salut (CTNS), TECNIO, Reus, Catalonia, Spain

L. Guasch
Computer-Aided Drug Design Group, Chemical Biology Laboratory, Center for Cancer Research, National Cancer Institute, National Institutes of Health, Frederick, MD, USA

K. Martinez-Mayorga, J. L. Medina-Franco (eds.), *Foodinformatics*,
DOI 10.1007/978-3-319-10226-9_6

around the world with diabetes have T2D [3]. In T2D, the body is able to produce insulin but this either is insufficient or the body is unable to respond to the effects of insulin (also known as insulin resistance), leading to a build-up of glucose in the blood. People with T2D can remain undiagnosed for many years, unaware of the long-term damage being caused by the disease. T2D is often, but not always, associated with being overweight or obese, which itself can cause insulin resistance and lead to high blood glucose levels. Many people with T2D are able to manage their condition through a healthy diet and increased physical activity. People whose blood glucose levels are high but not as high as those in people with diabetes are said to have impaired glucose tolerance (commonly referred to as IGT) or impaired fasting glucose (IFG). IGT is defined as high blood glucose levels after eating, whereas IFG is defined as high blood glucose after a period of fasting. People with IFG and IGT are at increased risk of developing diabetes, although this is reversible. The global prevalence of IGT was estimated to be 6.9% in 2013 and will rise to 8.0% in 2035 [1]. Adding the global prevalence of diabetes and IGT results, 15.2% of the worldwide adult population, or almost 700 million people, had diabetes or were at a high risk of developing diabetes in 2013. If these trends continue, by 2035 more than 1 billion people will suffer from diabetes or be at high risk of developing it.

Once diabetes is established, it is difficult to delay the complications associated with the disease even if a tight glycemic control is established [4]. Thus, the key is to prevent progression of glucose dysregulation and ideally, correct and reverse any disorder of glucose homeostasis at the earliest possible stage [4]. Although blood glucose levels return to normality over a period of several years in more than one third of IGT cases [5], the best evidence for preventing T2D comes from studies involving people with IGT. A healthy diet, regular physical activity, maintaining a normal body weight and avoiding tobacco use can prevent the progression of diabetes. Functional foods could add a new mode for the prevention and management of T2D [6–8]. Increasing insulin secretion, enhancing glucose uptake by adipose and skeletal muscle tissues, inhibiting intestinal glucose absorption and inhibiting hepatic glucose production are potential strategies by which functional foods could reduce blood glucose levels [8]. It is therefore evident that functional foods have a broad potential in terms of cost-effective public health policies [8].

Thiazolidinediones (TZDs) are a class of antidiabetic drugs developed in the late 1990s that have been widely used for the treatment of type II diabetes. TZDs work as insulin sensitisers that lower serum glucose without increasing pancreatic insulin secretion by binding to the peroxisome proliferator-activated receptor gamma (PPARγ), inducing the transactivation activity of this nuclear receptor. PPARγ-binding compounds are an active area of investigation for the prevention and treatment of T2D. In this chapter, we will review the following:

- The evidence needed to demonstrate the beneficial effects in vitro and in vivo, as well as the absence of adverse effects of the PPARγ-targeted compounds.
- The natural products and plant extracts that have been described to bind PPARγ.
- The way that these compounds can be discovered through VS procedures.

6.2 PPARγ-Targeted Antidiabetic Compounds

PPARγ is a member of the nuclear receptor family [9] that plays a central role in adipogenesis, acting as a cellular sensor that activates transcription in response to the binding of endogenous ligands, i.e. free fatty acids and eicosanoids. Activation of PPARγ induces the differentiation of preadipocytes into adipocytes and favours lipid storage pathways. Synthetic ligands of PPARγ, such as rosiglitazone and pioglitazone from the TZD family, have been widely used as a novel class of insulin sensitisers to treat T2D. These compounds act as PPARγ full agonists by binding to PPARγ and making the cells more responsive to insulin, thus decreasing the insulin resistance that is prevalent in T2D. Several trials have shown that TZDs can reduce the risk of developing diabetes [10–13]. In addition, it has been suggested that herbal and traditional natural medicines may provide an alternative mode of preventing or delaying the progression of diabetic retinopathy through the activity of PPARγ [14]. Despite the therapeutic benefits of rosiglitazone, its use has been highly restricted in the USA and withdrawn in Europe because an elevated risk of cardiovascular events, such as heart attack and stroke, was observed in patients treated with this drug [15]. Pioglitazone has recently been associated with a possible increased risk of bladder cancer [16] and has been withdrawn in some countries. In addition, TZDs present serious side effects such as weight gain, increased risk of bone fractures, fluid retention leading to oedema and heart failure [17–19]. For these reasons, the drug expenditures of TZDs in ambulatory visits for treatment of T2D in the USA have declined from 41% in 2005 to 16% in 2012 [20]. To overcome the adverse effects of TZDs, a new class of compounds called PPARγ partial agonists or selective modulators of PPARγ, were developed [21]. These compounds showed enhanced therapeutic efficacy as insulin sensitisers but had reduced adverse effects. Full and partial agonists bind differently to the ligand-binding domain (LBD) of PPARγ [22–25] (see Fig. 6.1 and Fig. 6.2). However, the binding differences between full and partial agonists do not explain the antidiabetic properties of both types of compounds.

In 2010, Choi and coworkers [26] suggested a new mechanism by which PPARγ agonists act to improve insulin sensitivity. This mechanism is independent of the classical receptor activity of PPARγ and consists of blocking the cyclin-dependent kinase 5 (CDK5)-mediated phosphorylation of PPARγ at Ser273 [26]. Inflammatory signals such as cytokines are commonly observed in obesity. These signals activate CDK5, which phosphorylates PPARγ at Ser273 [26]. TZDs and other PPARγ agonists inhibit the CDK5-mediated phosphorylation of PPARγ at Ser273, preventing the transcription of some genes that include adipsin (a fat-cell-selective gene, the expression of which is altered in obesity) and adiponectin (an insulin-sensitising adipokine) [26]. Interestingly, the CDK5-mediated phosphorylation of PPARγ is completely independent of classical receptor transcriptional agonism [26]. This new mechanism explain how partial agonists can exhibit similar or higher antidiabetic effects than full agonists and how the two types of agonists can have differing side effect profiles. It seems likely that partial and full agonists achieve comparable efficacy in insulin sensitisation through a similar inhibitory effect on

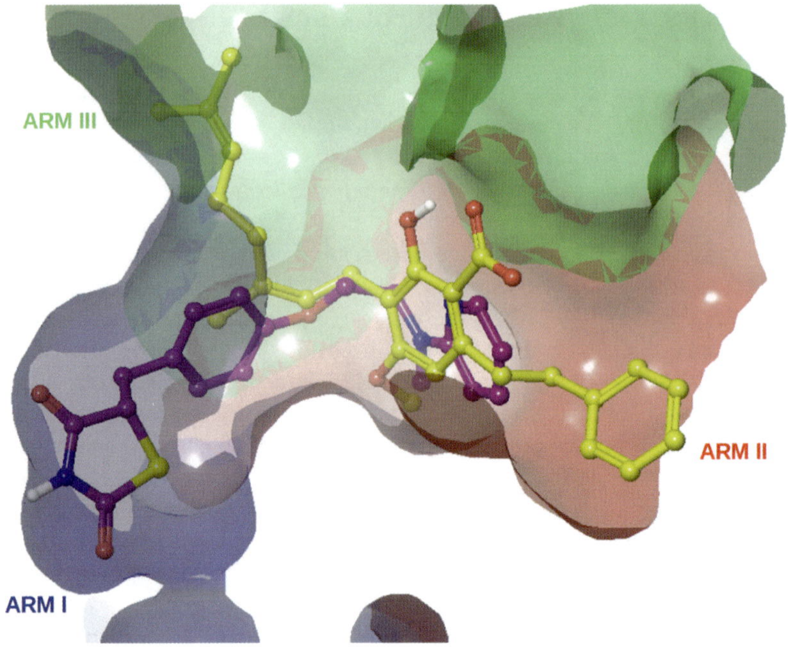

Fig. 6.1 Binding differences between proliferator-activated receptor (PPARγ) partial and full agonists. The ligand-binding domain (LBD) of PPARγ forming a complex with amorfrutin B (a partial agonist, in *yellow*) from the protein data bank (PDB) entry 4A4 is superimposed with the structure of rosiglitazone (a full agonist, in *purple*) from the PDB entry 1FM6. Amorfrutin B is a natural product with high binding affinity to PPARγ, but it only shows a 20 % PPARγ transactivation activity with respect to the maximum activation of rosiglitazone. The partial agonist occupies mainly *arm II* and *arm III* of the LDB of PPARγ, but the full agonist occupies mainly *arm I* and *arm II*. Both structures were validated by VHELIBS software and then were aligned by Maestro (Schrodinger)

PPARγ phosphorylation, whereas the differences in their agonistic potency could explain the differences in their side effects [27]. With this new knowledge, effective and safe PPARγ agonists designed as antidiabetic compounds must maximise the inhibition of PPARγ phosphorylation at Ser273 while reducing the PPARγ transactivation activity.

6.3 Experimental Evidences Needed to Demonstrate the Action of a PPARγ Mediated Antidiabetic Compounds

The identification of novel antidiabetic PPARγ agonists in vitro has been usually performed by evaluating their binding affinity to PPARγ and studying their activity in functional assays that assess transactivation and lipogenesis activities [28]. However, recent evidence suggests that the classical transactivation activity of

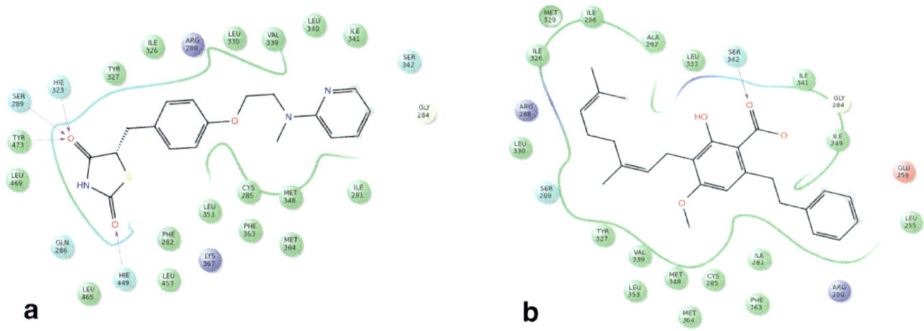

Fig. 6.2 Key interactions between proliferator-activated receptor (PPARγ) full and partial agonists with the ligand-binding domain of PPARγ. Schematic diagrams of atomic interactions between **a** rosiglitazone (pdb:2PRG) and **b** amorfrutin B (pdb:4A4W) bound to the ligand-binding domain (LBD) of PPARγ. The diagrams were obtained with Maestro (Schrodinger) using the ligand interaction diagram. Residues coloured in *green* are hydrophobic while residues coloured in cyan are polar. *Red* residues are negatively charged and could act as acceptors, whereas *purple* residues are positively charged and could act as donors. Ligand exposure to the solvent is coloured in *light grey*. Hydrogen bonds to the protein backbone are shown by *solid pink lines* and hydrogen bonds to the protein side chains are shown by *dotted pink* lines

PPARγ could be responsible for the adverse effects of PPARγ agonists and that the inhibition of CDK5-mediated phosphorylation of PPARγ at Ser273 is a key determinant of their antidiabetic effects. For these reasons, a PPARγ-mediated antidiabetic compound must have a high glucose-lowering activity, while lacking adipogenic activity. To characterise such compounds and demonstrate their beneficial effects on glucose metabolism, compounds must bind to PPARγ with a good affinity without promoting the transactivation activity of PPARγ (or promoting less than PPARγ full agonists). In addition, these compounds must not stimulate adipocyte differentiation while blocking the phosphorylation of PPARγ at Ser273, which would increase the insulin-induced glucose uptake in adipocytes. Below, we summarise the techniques available for these analyses.

- **Calculation of the binding affinity (IC50) to PPARγ.** Fluorescence polarisation (FP) is a homogeneous method that allows the rapid, quantitative analysis of diverse molecular interactions and enzyme activities. FP detection is based on the excitation of a fluorophore in a manner similar to standard fluorescence intensity. An easy and reliable calculation of the binding affinity of a test compound for the PPARγ nuclear receptor could be done with the PolarScreen™ PPARγ Competitor Assay Kit from Life Technologies, which is based on FP. When the nuclear receptor binds to the Fluormone™ ligand, the resulting complex yields a high polarisation value. If the test compound displaces the Fluormone™ ligand from the complex, the polarisation value is lowered. Because this occurs only in the presence of a test compound, the shift in polarisation value enables the accurate and convenient determination of the relative affinity of a test compound for the nuclear receptor. The concentration of the test compound that resulted in a half-maximal shift in polarisation value is defined as IC_{50}. This value is a measure of the relative affinity of the test compound for the PPARγ LBD.

- **In vitro transactivation activity on PPARγ and adipogenic activity assay.** Reporter gene assays are the most common and widespread in vitro test systems for quantifying the transactivation activity of a nuclear receptor in the presence of its ligand [29]. In these assays, cells such as HeLa cells are transfected with a plasmid expressing the full-length PPARγ and a second vector containing a reporter gene, e.g., firefly luciferase, under the control of the PPARγ response element. This enables the quantification of the transcriptional activity of PPARγ after treatment with PPARγ ligands [28]. In general, reporter gene assays can be used to characterise agonists and antagonists. For agonist testing, the transfected cells are incubated with varying concentrations of the test compound. From the resulting sigmoidal curve an EC_{50} value can be estimated as well as the maximum activation activity compared to a known PPARγ full agonist (which is set as 100 %). Figure 6.3 compares the reporter gene activity between a PPARγ full agonist and a PPARγ partial agonist. The maximum activation activities of the PPARγ partial agonists are less than the values for full agonists. For the characterisation of an antagonist, the transfected cells can be incubated with varying concentrations of the test compounds and a constant concentration of a known agonist, in a competitive assay. Comparison of the relative transcriptional activity of ligands in cells transfected with each of the three different PPAR subtypes (α, δ and γ) allows for the study of the selectivity of these compounds. PPARγ plays an important role in the regulation of adipocyte differentiation. In the absence of PPARγ or a PPARγ agonist, adipocytes fail to develop. The lipogenic activity of a compound can be therefore assessed in vitro by analysing during their development the triglyceride (TG) accumulation of preadipocytes such as murine 3T3-L1 cells. Antidiabetic compounds must not have an adipogenic activity to avoid weight gain and other adverse effects showed by PPARγ full agonists.

- **Analyses to show the inhibition of phosphorylation at Ser273.** A specific antibody against PPARγ phosphorylated at Ser273 is required for the development

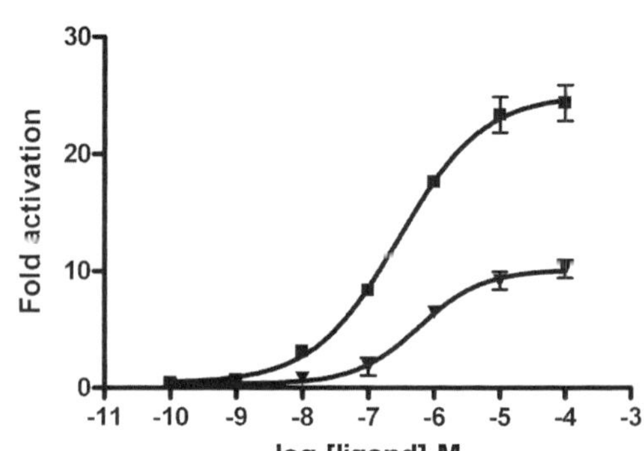

Fig. 6.3 Comparison of the in vitro proliferator-activated receptor (PPARγ) transactivation activity, measured with a luciferase reporter assay, between a full agonist (represented by *squares*) and a partial agonist (represented by *triangles*)

of an in vitro assay to study the inhibitory capacity of the natural products on PPARγ phosphorylation at Ser273. The assay could be developed as follows: Purified PPAR LBD is incubated with active CDK5 p35 (Sigma) in the presence of ATP and a full agonist, partial agonist or the test compounds at several concentrations. Proteins are resolved by SDS-PAGE, and PPARγ phosphorylation is assessed by immunoblotting with the anti Ser273 antibody. The concentration-dependent reduction of the Ser273 phosphorylation band is a reflex of the specific phosphorylation inhibitory capacity of the tested compounds. In order to normalise the signal, the total content of non-phosphorylated PPARγ must also be assessed by using one of the commercially available PPARγ antibodies.

- **Effects on insulin-induced glucose uptake in adipocytes and in vivo analyses.** For an in vitro measurement of the glucose uptake induced by the test compounds, the radioactive glucose (2-deoxy-d-[^{3}H]glucose) assay in differentiated adipocytes, such as the murine cell line 3T3-L1, could be used [30]. This assay measures the incorporation of the radioactive signal inside the cell, which is induced by the test compound. Administration of PPARγ agonists to several insulin-resistant animal models has been used to evaluate the agonists' ability to reduce plasma glucose levels and lower insulin in vivo [28]. There are several genetic animal models, such as ob/ob mice, db/db mice, obese Zucker (fa/fa) rats, Zucker fatty diabetic (ZDF) rats and diabetic KKAy mice that present this insulin-resistance state. Alternatively, several non-genetic approaches, such as streptozotocin-treated mice and high-fat-diet-induced obese C57BL/6J mice, could also be developed to induce insulin resistance. Independently of the animal model used, the reduction of plasma glucose and insulin levels demonstrates the antidiabetic effectiveness of the tested compound. The weight of the animals can also be checked to assess if the administration of the test compounds produces weight gain, an adverse effect of PPARγ full agonists.

6.4 Natural Products that Modulate the Action of PPARγ

Natural products, especially plants extracts, have been traditionally used for the treatment of T2D [31]. More than 111 plant families, including *Leguminoseae, Lamiaceae, Liliaceae, Cucurbitaceae, Asteraceae, Moraceae, Rosaceae, Euphorbiaceae* and *Araliaceae*, have been identified to have antidiabetic properties [31, 32]. However, there are few studies that demonstrate the mechanisms of action of the bioactive compounds responsible for the antidiabetic properties of natural extracts.

Natural products offer a privileged starting point in the search for highly specific and potent modulators of biomolecular function as well as novel drugs [33]. Several plant and fungi extracts have been shown to modulate the activity of PPARγ [7, 34–46]. Mueller and Jungbauer [38] analysed the influence of 70 plants, herbs and spices on PPARγ activation or antagonism. Approximately, 50 out of the 70 plant extracts, such as pomegranate, apple, clove, cinnamon, thyme, green coffee, bilberry and bay leaves, were found to bind PPARγ in a competitive ligand-binding assay [38]. Only five spices, nutmeg, licorice, black pepper, holy basil and sage, were found to transactivate PPARγ [38]. Interestingly, nearly all plant extracts antagonised rosiglitazone-mediated coactivator recruitment [38], suggesting that there are many candidate plant extracts that may have antidiabetic properties through the modulation of PPARγ, without the adverse effects presented by TZDs and other full agonists. This opens the possibility of using these extracts for the development of new functional foods with antidiabetic action. One of the main problems of using plant extracts for experimental research is that in some cases, the active compounds that exert the biological action are not yet completely identified [44]. In some cases, the molecule responsible for the PPARγ-mediated activity of a natural extract has been suggested (see Table 6.1). What is lacking, however, are deeper studies of the metabolic effects of PPARγ modulation. In most cases, by similarity with TZDs, potential antidiabetic compounds and natural extracts are suggested by their capacity of promoting the transactivation activity of PPARγ, identifying PPARγ full agonists as suitable candidates for the treatment of T2D or metabolic syndrome. With the new antidiabetic mechanism proposed for TZDs [26, 27], deeper analyses are needed to demonstrate the antidiabetic action of a compound or extract. In addition, the adverse effects caused by TZDs and other PPARγ full agonists must be considered when a new (PPARγ-mediated) antidiabetic natural compound or extract is suggested. Some of the natural compounds that bind to PPARγ seem to be weak transactivators of PPARγ or do not stimulate adipocyte differentiation [39, 42, 47]. Some of them, such as amorfrutin 1 and pseudoginsenoside F11, have been shown to block the CDK5-mediated phosphorylation of PPARγ at Ser273 [48, 49]. These compounds are the interesting ones. Some PPARγ antagonists, i.e. compounds that inhibit the PPARγ-induced adipocyte differentiation, such as ginsenosides Rh2 and Rg3 and tanshinone IIA, are able to improve glucose tolerance in vivo [50–53]. These PPARγ antagonists could be compounds that do not promote the transactivation activity of PPARγ, but still have antidiabetic properties through the inhibition of CDK5-mediated phosphorylation of PPARγ at Ser273. In addition, if these compounds antagonise the transactivation activity of PPARγ and adipocyte differentiation, they could also possess antiobesity effects.

Glycyrrhiza uralensis or Glycyrrhiza Radix is one of the herbs used in traditional Chinese medicine for the treatment of diabetes [54]. Glycyrin is a component found in the roots of *G. uralensis* that has a high transactivation activity on PPARγ that is similar to troglitazone, a member of the TZDs, and significantly decreases the blood glucose levels of genetically diabetic mice (Table 6.1) [55]. A fraction of flavonoid oil from the roots of *Glycyrrhiza glabra*, or licorice, has been shown to suppress weight gain and the increase of blood glucose levels in genetically diabetic mice fed with a high-fat diet [56]. An ethanolic extract from licorice stimulates human adipocyte differentiation in vitro [56], suggesting that its hypoglycemic effects are possibly mediated via the activation of PPARγ [56]. Several phenolics compounds isolated from *G. glabra* exhibit significant PPARγ ligand-binding activity and their transactivation activities on PPARγ are similar or higher than troglitazone [57]. Other natural products identified as full agonists of PPARγ (Table 6.1) are psi-baptigenin, hesperidin and chrysin [58]. However, their effect as antidiabetic

Table 6.1 Natural products described as PPARγ agonists or antagonists

Compound	Natural source	Type of PPARγ agonist	Binding affinity IC_{50} µM	Transactivation activity (% of max. activation relative to rosiglitazone)	Effect on glucose metabolism	Reference
Saufuran A Saufuran B	Roots of *Saururus chinensis*	Full Partial		High (comparable to ciglitazone) weak		[102]
Dehydrotrametenolic acid	*Poria cocos* Wolf (Polyporaceae)				It reduces hyperglycemia and act as an insulin sensitiser in mouse models	[68, 69]
Glycyrin	Roots from *Glycyrrhiza uralensis*	Full		High (similar to troglitazone)	Significantly decreases the blood glucose levels of genetically diabetic mice	[55]
Daidzein	*Pueraria thomsonii*	Dual[a]		Moderate (25 % relative to pioglitazone)		[74]
Genistein	Plants such lupin, fava beans, soybeans, kudzu and psorale	Dual[a]	Ki = 5.7 µM	Moderate (35 % relative to pioglitazone)[b]		[74, 107]
Formononetin	*Astragalus membranaceus*	Dual[a]		Moderate (17 % relative to pioglitazone)		[74]
Biochanin A	Legumes such as red clover, soy, alfalfa sprouts, peanuts, chickpea, oregano	Dual[a]	23.7	Moderate (26 %)		[71, 74]
Ginsenoside Rh2 Ginsenoside Rg3	Ginseng (*Panax ginseng*)	Antagonist		Significantly inhibits the rosiglitazone-induced transcriptional activity	Significantly enhances glucose uptake in the insulin-resistant muscle cells	[50, 53]
Psi-baptigenin	Plants such as Red clover (*Trifolium pratense*), Hen's eye (*Ardisia crenata* Sims) and the bark of Brazilian Tulipwood (*Dalbergia frutescens*)	Full		High (similar to rosiglitazone)		[58]

Table 6.1 (continued)

Compound	Natural source	Type of PPARγ agonist	Binding affinity IC_{50} μM	Transactivation activity (% of max. activation relative to rosiglitazone)	Effect on glucose metabolism	Reference
Hesperidin	Citrus fruits	Full		High		[58]
Chrysin	Passion flowers *Passiflora caerule* and *Passiflora incarnata*, *Oroxylum indicum*, chamomile, the mushroom *Pleurotus ostreatus* and in honeycomb	Full		High		[58]
Apigenin	Plants such as parsley, celery and chamomile tea	Partial	80	Moderate (16 %)		[38, 58]
Tanshinone IIA	*Salvia miltiorrhiza*	Antagonist	3.90		Improves glucose tolerance in a high-fat-diet-induced obese animal model	[52]
7-Chloroarctinone-b	Roots of *Rhaponticum uniflorum*	Antagonist	KD = 2.63 μM			[101]
Quercitin	Plants such as dill, bay leaves, oregano	Antagonist	3.0	None		[71]
Rosmarinic acid	Marjoram, oregano, sage, thyme, rosemary	Antagonist/ PPARα agonist	32.4	None		[71]
Diosmetin	Oregano	Antagonist	13	None		[71]
Naringenin	Grapefruit, oranges, oregano	Partial	81	Moderate (16 %)		[71]
Several flavone and isoflavones derivatives	Roots from *Glycyrrhiza glabra*	Full		High (similar to troglitazone)	Suppresses the increase of blood glucose levels in genetically diabetic mice	[56, 57]
2′-hydroxy chalcone	Cinnamon	Partial	3.8	High (48 %)		[38]

Table 6.1 (continued)

Compound	Natural source	Type of PPARγ agonist	Binding affinity IC_{50} µM	Transactivation activity (% of max. activation relative to rosiglitazone)	Effect on glucose metabolism	Reference
Coumestrol	Alfalfa	Partial	11	Moderate (25 %)		[38]
Resveratrol	Bilberry	Partial	62	Moderate (39 %)		[38]
Oleanonic acid	Oleoresin of *Pistacia lentiscus* var. Chia (chios mastic gum)	Partial		Moderate (20 %)		[99]
Dieugenol	Dried flower buds of *Syzygium aromaticum* (clove)	Partial	Ki = 0.24 µM	Moderate[b]		[98]
Tetrahydrodieugenol	Dried flower buds of *Syzygium aromaticum* (clove)	Partial	Ki = 0.32 µM	Moderate[b]		[98]
Magnolol	Bark of *Magnolia officinalis* Rehc. and Wils	Partial	Ki = 2.04 µM	Moderate[b]		[98]
Artepillin C	*Baccharis dracunculifolia*		Weaker affinity than rosiglitzane	[b]	In mature 3T3-L1 adipocytes, it significantly enhanced the basal and insulin-stimulated glucose uptake	[103]
Luteolin	Marjoram, sage, rosemary, tarragon thyme, parsley and alfalfa	Partial	0.50	Moderate (35 %)	Luteolin-5-O-b-rutinoside reduces glycemia and increases pancreatic insulin in diabetic rats	[47, 59]
Decanoic acid	Coconut and palm kernel oil, milk of mammals	Partial	Ki = 41.7 µM	At 10 and 50 µM increased the reporter expression by 3.3- and 4.3-fold	Its triglyceride form decreases the fasted blood glucose levels in diabetic mice	[104]
Tirotundin Tagitinin A	*Tithonia diversifolia*	Dual[a]	27 55	Moderate–high		[73]

Table 6.1 (continued)

Compound	Natural source	Type of PPARγ agonist	Binding affinity IC_{50} μM	Transactivation activity (% of max. activation relative to rosiglitazone)	Effect on glucose metabolism	Reference
Amorfrutins	Roots of *Glycyrrhiza foetida* (licorice) and fruits of *Amorpha fruticosa*	Partial	0.24–0.34	Moderate (15–39 %)	Amorfrutin 1 reduces plasma insulin and glucose concentrations in leptin receptor-deficient db/db mice	[48, 60]
Honokiol	Bark of *Magnolia officinalis*	Partial	Ki = 22.86 μM	Moderate (17 % relative to pioglitazone)	Enhances the glucose uptake in adipocytes Significantly improves the glucose tolerance and insulin levels of diabetic mice	[61]
Falcarindiol	Rhizomes and roots of *Notopterygium incisum*	Partial	Ki = 3.07 μM	Moderate (35 % relative to pioglitazone)[b]		[105]
Pseudoginsenoside F11	Roots and leaves of *Panax quinquefolium L.* (American ginseng)	Partial		Moderate (30 %)[b]		[49]
Isosilybin A	Milk thistle (*Silybum marianum*)	Partial		Moderate		[106]

PPAR proliferator-activated receptor
[a] Dual: PPARα-γ dual agonist
[b] It promotes the adipocyte differentiation of pre-adipocytes

compounds has not been analysed. Glycyrin and other PPARγ full agonists are potential antidiabetic natural products, although their high transactivation activity and putative adverse effects must be taken into account.

Luteolin, amorfrutins and honokiol are natural products with a low or moderate transactivation activity on PPARγ that show beneficial effects on glucose metabolism (see Table 6.1). Luteolin is found in marjoram, sage, rosemary, tarragon, thyme, parsley and alfalfa [38]. It has a moderate transactivation activity on PPARγ [47] and a luteolin derivative (luteolin-5-O-b-rutinoside) reduces glycemia while increasing pancreatic insulin in diabetic rats [59]. Amorfrutins are PPARγ partial agonists found in the roots of *Glycyrrhiza foetida* that have a moderate capacity of promoting the transactivation activity of PPARγ, showing a transactivation activity of 15–39 % relative to full PPARγ activation by rosiglitazone [48, 60]. Amorfrutin 1 reduces plasma insulin and glucose concentrations in leptin receptor-deficient db/db mice and blocks the CDK5-mediated phosphorylation of PPARγ at Ser273 [48]. Honokiol is found in the bark of *Magnolia officinalis* and has a moderate transactivation activity (maximal activity of 17 % relative to pioglitazone) [61]. Honokiol enhances the glucose uptake in adipocytes and significantly improves the glucose tolerance and insulin levels of diabetic mice [61]. 2′-Hydroxy chalcone is a natural product found in cinnamon and has a moderately high capacity to promote the transactivation activity of PPARγ [38]. Traditional Native American treatments of diabetes now use cinnamon [62], as cinnamon-derived active compounds have been shown to exert beneficial effects on glucose metabolism and insulin sensitivity [8]. However, a recent review has concluded that there is insufficient evidence to support the use of cinnamon for type 1 or type 2 diabetes mellitus [63]. Further randomised clinical trials are required to establish the therapeutic safety and efficacy of cinnamon. Other natural products that act as PPARγ partial agonists are saufuran B, apigenin, naringenin, coumestrol, resveratrol, oleanonic acid, diugenol, tetrahydrodieugenol, magnolol and falcarindiol (Table 6.1). However, deeper studies on the effects of these compounds on the glucose metabolism and their potential capacity to block the CDK5-mediated phosphorylation of PPARγ are needed.

Curcumin from turmeric (*Curcuma longa*, a spice used in Indian cuisine and in curry) ameliorates diabetes in high-fat-diet-induced obese and leptin-deficient ob/ob male C57BL/6J mice as determined by glucose and insulin-tolerance testing and hemoglobin A1c percentages [64]. The beneficial effects of curcumin are significantly abolished by pretreatment with PPARγ antagonists, suggesting that the beneficial effects are mediated through the activation of PPARγ [36, 65, 66]. However, curcumin has not been suggested to be a PPARγ ligand because it does not induce the differentiation of preadipocytes, does not increase the relative transcriptional activity of PPARγ and does not displace [^{3}H]-rosiglitazone from the PPARγ-LBD [67]. Similarly, dehydrotrametenolic acid from *Poria cocos* Wolf, a mushroom used in traditional Chinese medicine to treat diabetes, was suggested to act as an insulin sensitiser through the action of PPARγ [68]. However, because it does not activate the PPARγ pathway, the enhanced insulin sensitivity induced by dehydrotrametenolic acid has been suggested to be irrespective of PPARγ [69]. It is important to remark that the lack of adipogenic activity and/or transcriptional activity of PPARγ in the presence of a compound must not be considered an evidence that their antidiabetic activity is not mediated by PPARγ. Further investigations into the potential ability of a compound to block the CDK5-mediated phosphorylation of PPARγ are needed to characterise its antidiabetic mechanisms. The lack of transcriptional activity on PPARγ makes curcumin and dehydrotrametenolic acid potential effective antidiabetic compounds that might not have the adverse effects present in TZDs and other full agonists.

Ginseng has been used in traditional medicine for more than 2000 years. Several reports have described that several ginsenosides from *Panax ginseng* (Asian ginseng) and *Panax quinquefolius* (American ginseng) show antidiabetic properties [8, 31]. However, further studies that take into account the chemical differences between the types of ginseng are needed to shed light on its therapeutic potential [8]. Ginsenoside Rh2 and ginsenoside Rg3 have been suggested as candidates for preventing metabolic disorders such as obesity through their capacity to inhibit adipocyte differentiation via PPARγ inhibition [50, 51]. In addition, both compounds also significantly enhance glucose uptake in insulin-resistant muscle cells [53]. These two compounds could be, at least in part, responsible for the antidiabetic effect of ginseng, with the additional benefit as anti-obesity compounds. Tanshinone IIA from *Salvia miltiorrhiza* is another PPARγ antagonist that improves glucose tolerance in a high-fat-diet-induced obese animal model [52]. *S. miltiorrhiza* has been used traditionally to treat diabetes [31]. The molecules deoxyneocryptotanshinone and miltionone I from *S. miltiorrhiza* are extremely similar to tanshinone IIA, and have been predicted to be PPARγ partial agonists [70]. A possible mechanism for the antidiabetic activity of PPARγ antagonists is that they might block the CDK5-mediated phosphorylation of PPARγ at Ser273. Other PPARγ antagonists are diosmetin and quercitin [71], although there are no studies on the effect of these compounds on glucose metabolism (see Table 6.1).

Dual PPARα/γ agonists are compounds that are used to treat dyslipidemia and diabetes; combining the therapeutic effects of both PPAR-γ and PPAR-α selective agonists [72]. Several natural products have been suggested to be dual PPARα/γ agonists [45]. Tirotundin and tagitinin A are sesquiterpene lactones derived from *Tithonia diversifolia* (a traditional Chinese medicine used for treating diabetes), which have been suggested to be dual PPARα/γ agonists [73], along with rosmarinic acid [71], daidzein, genistein and formononetin [74] (Table 6.1). However, more evidence is required to demonstrate their antidiabetic effects and their molecular mechanism. In addition, their transactivation activity must be low in order to avoid the adverse effects of PPARγ full agonists. The failed development of several dual PPARα/γ agonists represents the increased awareness of potential toxicities with this class of compounds [72].

A food can be regarded as 'functional' if it satisfactorily demonstrates beneficial effects (beyond adequate nutrition) on one or more target functions in the body in a way that is relevant to either an improved state of health and wellbeing and/or the reduced risk of disease [75]. To develop new functional foods for diabetes prevention mediated by PPARγ, nutraceuticals and natural compounds that modulate PPARγ activity should be identified. However, only rigorous analyses could establish the pharmacological and toxicological profiles of these compounds and their potential in influencing human health [76]. In this sense, results should be validated through large-scale population trials [8]. Although there are different PPARγ-targeted molecules that have shown promising results as antidiabetic compounds, the new antidiabetic mechanism suggested for PPARγ modulators makes the acquisition of more evidence necessary in order to demonstrate their beneficial effects and the absence of adverse effects.

6.5 Cheminformatic Tools for the Discovery of PPARγ-Mediated Antidiabetic Compounds

Computer-aided drug design methods have had a huge impact on drug discovery. A preliminary application of these methods optimises time and cost in introducing a drug to the market. One of the most widely used techniques is virtual screening (VS) [77]. VS is a computational technique to search libraries of small molecules in order to identify those structures which are most likely to bind to a target and become potential drugs. Figure 6.4 shows an example of a hypothetical VS workflow based on five usual in silico techniques, which are summarised below, for finding novel active compounds.

- **Prediction of absorption, distribution, metabolism and excretion/toxicity (ADMET) properties.** To develop its pharmacological activity, a drug candidate has to penetrate various physiological barriers, move to its effector site, be modified by specialised enzymes and finally be removed from the body. In other words, it requires some particular properties of absorption, distribution, metabolism and excretion without being toxic. ADMET properties have been identified

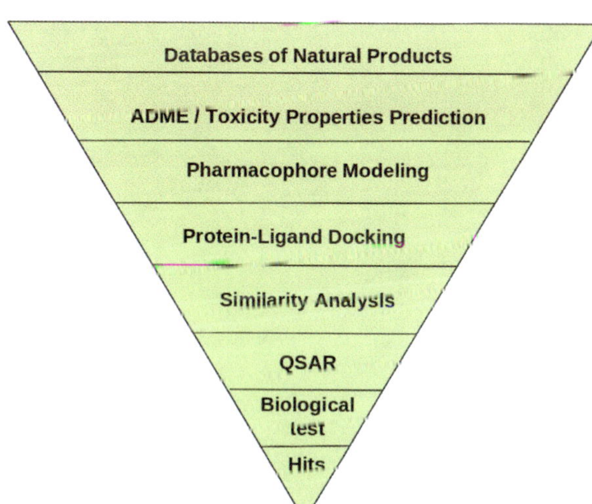

Fig. 6.4 Hypothetical virtual screening workflow. Schematic overview of a virtual screening workflow for identifying lead compounds from large and chemically diverse databases. This workflow consists of applying several computer-aided drug design methods with one usually used after another in a filter-like process in order to select potential hits

as defining characteristics for the success or failure of drug development. Thus, it is important to assess and predict the pharmacokinetic properties of bioactive compounds in the early stages of drug discovery projects [78]. Several software programs and databases can be used for predicting ADMET properties in silico [79].

- **Pharmacophore modelling.** Pharmacophore modelling has become a popular tool for VS to discover novel scaffolds. A pharmacophore is a specific 3D arrangement of steric and electronic features that are essential to a compound's biological activity [80]. Typical pharmacophore features include hydrogen bond acceptors or donors, hydrophobic centroids, aromatic rings, cations and anions. A pharmacophore can be established based on the knowledge of which active ligands bind to the same receptor (a ligand-based pharmacophore model) or based on the 3D structure of the target protein to generate a topological description of the ligand–receptor interactions (a structure-based pharmacophore model) [81]. A variety of pharmacophore-modelling approaches has been implemented by packages such as Catalyst/Discovery Studio, Phase [82], MOE and LigandScout [83]. Figure 6.5 shows a structure-based common pharmacophore derived from the alignment of several PPARγ partial agonists [30]. The pharmacophore is formed by one hydrogen bond acceptor (AP1) coloured in pink and three hydrophobic sites (HP1, HP2 and HP3) coloured in green. Amorfrutin B, a recently described PPARγ partial agonist (from the protein data bank (PDB) entry 4A4W) [60], perfectly matches this pharmacophore (Fig. 6.5).

- **Protein-ligand docking.** Protein–ligand docking is a widely used structure-based drug discovery approach that predicts the binding orientation of small

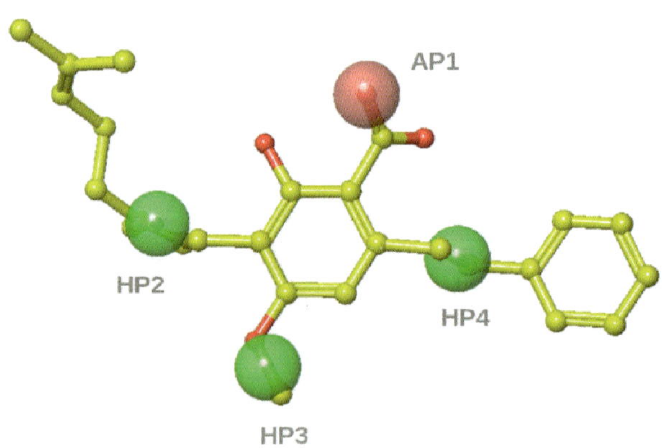

Fig. 6.5 Proliferator-activated receptor (PPARγ) partial agonist pharmacophore. Structure-based common pharmacophore derived from the alignment of PPARγ partial agonists. The pharmacophore is formed by one hydrogen bond acceptor (*AP1*) coloured in *pink* and three hydrophobic sites (*HP1*, *HP2* and *HP3*) coloured in *green*. The ligand amorfrutin B (from the protein data bank (PDB) entry 4A4W) is also represented as a spatial reference. The pharmacophore was generated by Phase (Schrodinger)

molecule drug candidates to their protein targets in order to predict the affinity and activity of the small molecule [84]. Docking protocols can be described as a combination of search algorithms and scoring functions to rank and evaluate the orientation and conformation of a ligand [85]. Most docking programs account for ligand flexibility. Efficient handling of the flexibility of the protein receptor and the scoring function are considered to be the main challenges in the field of docking. Several protein–ligand docking software applications, such as Glide, AutoDock, GOLD and eHiTS, are available [84].
- **Similarity analysis.** Molecular similarity, clustering and diversity analysis has played a significant role in ligand-based drug discovery [86, 87]. Similarity search algorithms use 2D fingerprints descriptors (fingerprint similarity analysis) or 3D shape descriptors (electrostatic/shape similarity analysis) to compare a biologically active query molecule to a database molecule. Along with other metrics, the Tanimoto coefficient is used to quantify the similarity. OpenEye suite has similarity algorithms for comparison of shape (ROCS) and electrostatic (EON) properties (OpenEye Scientific Software, Inc., Santa Fe, New Mexico, USA; http://www.eyesopen.com).
- **QSAR: quantitative structure activity relationship.** QSAR models have been applied in the development of relationships between physicochemical properties of molecules and their biological activities to obtain a reliable statistical model for predicting the activities of new drug candidates [88]. This method is only fruitful if the dataset contains compounds that are structurally related to the molecules used to construct the model. Therefore, in contrast to lead discovery techniques, such as similarity analysis and pharmacophore modelling, QSARs are frequently used in the optimisation phases of drug design [89]. Many different 1D, 2D, 3D and multidimensional QSAR approaches have been developed during the past several decades [88]. The major differences in these methods include the chemical descriptors and mathematical approaches that are used to establish the correlation between the target properties and the descriptors. QSAR models are typically created using a training set of ligands, and the models are then tested against the test set of ligands. From an application point of view, numerous software programs and websites exist for predicting a wide range of properties in either a qualitative or quantitative way.

VS has emerged as an important tool in identifying bioactive compounds by employing knowledge about the protein target or known bioactive ligands [90]. For VS to be successful, it is essential to ensure the reliability and accuracy of the data used. Taking into account that crystal structures are models, it is important to validate the experimental PDB complexes before using them in structure-based drug discovery approaches [91]. Different validation tools are available for evaluating the binding site and ligand against the electron density [92]. The number of complexes classified as good, dubious or bad after applying the VHELIBS tool [92] to 173 ligand/PPARγ binding site complexes is shown in Table 6.2. Only 5 of the 173 complexes are defined as good, i.e. the electron density map perfectly matches the coordinates of the PDB model, simultaneously for the ligand and binding site. Most of the complexes on Table 6.2 are classified as a dubious. This does not mean that

Table 6.2 Number of complexes classified as good, dubious or bad after applying VHELIBS to 173 ligand/PPARγ binding site complexes using the PDB profile with default values

		Binding site			
		Good	Dubious	Bad	
Ligand	Good	5	29	5	39
	Dubious	9	89	20	118
	Bad	2	11	3	16
		16	129	28	173

PPAR proliferator-activated receptor, *PDB* protein data bank

these models are wrong, but a visual inspection to check if the coordinates fit well with the electron density is necessary prior to using these models in any structure-based approach.

Successful VS relies on the ability to discriminate between active and inactive compounds in order to provide a set of compounds for experimental screening that is highly enriched in active molecules [93]. Sets of known active and inactive compounds are needed for the assessment of VS approaches. Decoys are molecules that are presumed to be inactive against a target, which can be used when too few inactive compounds are available for such testing [94]. Many metrics are currently used to quantify the effectiveness of a VS [95]. The enrichment factor (EF) represents one of the most prominent metrics in VS. EF measures how many more active compounds are found within a defined 'early recognition' fraction of the ordered list relative to a random distribution. Sensitivity and specificity are also descriptors that assess the enrichment of active molecules from a database. Sensitivity (Se, or true positive rate) describes the ratio of the number of active molecules found by the VS method to the number of all active compounds in the database. Specificity (Sp, or true negative rate) represents the ratio of the number of inactive compounds that were not selected by the VS protocol to the number of all inactive molecules included in the database [93].

There are successful examples of the application of drug design methods in the discovery of new PPARγ-mediated antidiabetic compounds. Table 6.3 shows a selection of VS examples that used natural products or derivatives as a starting database for the screening. While the first studies did not specify between a search for full and partial PPARγ agonists, the profiles of the hit compounds follow the full PPARγ agonist features. Most of the studies apply protein–ligand docking after the VS workflow in order to get a deeper mechanistic understanding of the binding of compounds to the PPARγ ligand-binding pocket. Salam and coworkers [58] used a docking approach against a natural product library of 200 compounds to reveal 29 potential PPARγ full agonists. Of these 29 potential hits, 6 flavonoids that included apigenin, chrysin, hesperidin and psi-baptigenin were shown to stimulate PPARγ transcriptional activity in vitro. Tanrikulu and coworkers [96] used a structure-based pharmacophore to search 15,590 compounds from the AnalytiCon Discovery collection of natural-product-derived combinatorial database. Of the eight compounds tested, two were derived from the natural compound α-santonin and were able to promote the PPARγ transactivation activity in a cell-based reported

Table 6.3 Several successful examples of VS procedures used for identifying PPARγ agonists among natural products or derivatives

Methods used	Databases used	Type of PPARγ agonist	VS hits	Hits with proved activity towards PPARγ	Reference
Docking	200 natural products from the Herbal Medicines Research and Education Center subset (Univ. of Sydney)	Full	29	6	[58]
Pharmacophore	15,590 compounds from the AnalytiCon Discovery collection of natural-product-derived combinatorial compounds (v01/2007)	Full	8	2	[96]
Machine learning	360,000 compounds from Asinex Gold and Platinum collections (v11/2007)	Full	15	4	[97]
Pharmacophore	9676 compounds from the DIOS database of natural products found in ancient herbal medicines described in *De materia medica*; and 10,216 compounds from the Chinese herbal medicine (CHM) database	Partial	4	3	[98]
Pharmacophore Lipinski rule of five	57,346 compounds from the Chinese natural product database (CNPD, v2004.1)	Partial	1	1	[99]
ADME/toxicity prediction, anti-pharmacophore, pharmacophore, electrostatic similarity analysis, fingerprint diversity analysis	89,165 compounds from the natural product subset from ZINC database	Partial	10	5	[30]
ADME/toxicity prediction, anti-pharmacophore, pharmacophore, electrostatic similarity analysis, fingerprint diversity analysis	29,779 compounds from an in-house dataset of natural compounds and natural sources	Partial	65		[70]

ADME absorption, distribution metabolism and excretion; *PPAR* proliferator-activated receptor; *VS* virtual screening gene assay, with values of 31 and 8% for maximal PPARγ activation relative to pioglitazone [96]. Rupp and coworkers [97] combined several machinelearning methods to virtually screen a database of 360,000 compounds. They tested 15 compounds in a cellular reporter assay [97]. Eight compounds exhibit agonistic activity towards PPARα, PPARγ or both. The most potent PPARγ-selective hit was a derivative of the natural product truxillic acid [97]. Using a pharmacophore-based VS of 19,892 natural products, Fakhrudin and coworkers [98], identified several neolignans, such as dieugenol, tetrahydrodieugenol and magnolol, as PPARγ partial agonists. However, these three compounds induce 3T3-L1 preadipocyte differentiation [98], suggesting that they could have some adverse effects when used as antidiabetic compounds. Using a 4-point pharmacophore based on 13 PPARγ partial agonists, Petersen and coworkers [99] scanned a database of 57,346 compounds from the Chinese natural product database and identified methyl oleanonate as a PPARγ partial agonist [99]. The in vitro analysis of several subfractions of Chios mastic gum, where methyl oleanonate is found, confirmed their biological activity towards PPARγ [99]. Guasch and coworkers [30] developed a VS procedure using structure-based pharmacophore, protein–ligand docking and electrostatic/shape similarity to discover novel scaffolds of PPARγ partial agonists. Interestingly, the VS procedure of Guasch and coworkers [30] is the only approach that includes a structure-based anti-pharmacophore to exclude possible PPARγ full agonists. This VS procedure was used to identify 135 compounds as potential PPARγ partial agonists [30] from an initial set of 89,165 natural products and natural product derivatives from the ZINC database [100]. Five out of the eight tested compounds were confirmed to be PPARγ partial agonists as they bind to PPARγ, do not or only moderately stimulate the transactivation activity of PPARγ, do not induce adipogenesis of preadipocyte cells and stimulate insulin-induced glucose uptake by adipocytes [30]. Using a slightly modified version of their VS workflow, Guasch and coworkers [70] predicted, as potential PPARγ partial agonists, 12 molecules from 11 natural extracts known to have antidiabetic activity. In addition, they also identified 10 molecules from 16 plants with undescribed antidiabetic activity but that are related to plants with known antidiabetic properties [70].

6.6 Conclusions

Although several natural compounds and plant extracts have been shown to modulate the activity of PPARγ, deeper analyses of the active compounds, their molecular mechanisms and their metabolic effects are needed. The new antidiabetic mechanism of blocking the CDK5-mediated phosphorylation of PPARγ at Ser273 suggests that new classes of PPARγ-mediated antidiabetic compounds must be based on preventing this specific phosphorylation. The classical transactivation activity of PPARγ is not enough to prove the antidiabetic properties of a compound or extract, and this activity must be absent in order to avoid the adverse effects of TZDs and other PPARγ agonists. VS procedures and other cheminformatics tools may be useful for finding PPARγ-mediated antidiabetic compounds with the desired properties. More research into the molecular mechanisms and the efficacy of PPARγ-mediated antidiabetic compounds is needed prior to developing PPARγ-based functional foods for the prevention of diabetes.

Acknowledgments This manuscript was edited for English-language fluency by American Journal Experts. This study was supported by grant AGL2011-25831/ALI from the Spanish Government and ACC1Ó program [TECCT11-1-0012] as well as grant XRQTC from 'Generalitat de Catalunya'.

References

1. International Diabetes Federation (2013) IDF Diabetes Atlas, 6th edn. http://www.idf.org/diabetesatlas
2. Roglic G, Unwin N, Bennett PH et al (2005) The burden of mortality attributable to diabetes: realistic estimates for the year 2000. Diabetes Care 28:2130–2135
3. World Health Organization (2013) Diabetes. Fact sheet N. 312. http://www.who.int/mediacentre/factsheets/fs312/en/
4. Khavandi K, Amer H, Ibrahim B, Brownrigg J (2013) Strategies for preventing type 2 diabetes: an update for clinicians. Ther Adv Chronic Dis 4:242–261. doi:10.1177/2040622313494986
5. Shaw JE, Zimmet PZ, de Courten M et al (1999) Impaired fasting glucose or impaired glucose tolerance. What best predicts future diabetes in Mauritius? Diabetes Care 22:399–402
6. Rudkowska I (2009) Functional foods for health: focus on diabetes. Maturitas 62:263–269. doi:10.1016/j.maturitas.2009.01.011
7. Perera P, Li Y (2011) Mushrooms as a functional food mediator in preventing and ameliorating diabetes. Funct Foods Health Dis 4:161–171
8. Ballali S, Lanciai F (2012) Functional food and diabetes: a natural way in diabetes prevention? Int J Food Sci Nutr 63 Suppl 1:51–61. doi:10.3109/09637486.2011.637487
9. Garcia-Vallvé S, Palau J (1998) Nuclear receptors, nuclear-receptor factors, and nuclear-receptor-like orphans form a large paralog cluster in Homo sapiens. Mol Biol Evol 15:665–682
10. Buchanan TA, Xiang AH, Peters RK et al (2002) Preservation of pancreatic beta-cell function and prevention of type 2 diabetes by pharmacological treatment of insulin resistance in high-risk hispanic women. Diabetes 51:2796–2803
11. Knowler WC, Hamman RF, Edelstein SL et al (2005) Prevention of type 2 diabetes with troglitazone in the Diabetes Prevention Program. Diabetes 54:1150–1156
12. Gerstein HC, Yusuf S, Bosch J et al (2006) Effect of rosiglitazone on the frequency of diabetes in patients with impaired glucose tolerance or impaired fasting glucose: a randomised controlled trial. Lancet 368:1096–1105. doi:10.1016/S0140-6736(06)69420-8
13. DeFronzo RA, Abdul-Ghani MA (2011) Preservation of β-cell function: the key to diabetes prevention. J Clin Endocrinol Metab 96:2354–2366. doi:10.1210/jc.2011-0246
14. Song MK, Roufogalis BD, Huang THW (2012) Modulation of diabetic retinopathy pathophysiology by natural medicines through PPAR-γ-related pharmacology. Br J Pharmacol 165:4–19. doi:10.1111/j.1476-5381.2011.01411.x
15. Nissen SE, Wolski K (2007) Effect of rosiglitazone on the risk of myocardial infarction and death from cardiovascular causes. N Engl J Med 356:2457–2471. doi:10.1056/NEJMoa072761
16. Lewis JD, Ferrara A, Peng T et al (2011) Risk of bladder cancer among diabetic patients treated with pioglitazone: interim report of a longitudinal cohort study. Diabetes Care 34:916–922. doi:10.2337/dc10-1068
17. Feldman PL, Lambert MH, Henke BR (2008) PPAR modulators and PPAR pan agonists for metabolic diseases: the next generation of drugs targeting peroxisome proliferator-activated receptors? Curr Top Med Chem 8:728–749

18. Amato AA, Rocha Neves F de A (2012) Idealized PPARγ-based therapies: lessons from bench and bedside. PPAR Res 2012:978687. doi:10.1155/2012/978687
19. Bortolini M, Wright MB, Bopst M, Balas B (2013) Examining the safety of PPAR agonists—current trends and future prospects. Expert Opin Drug Saf 12:65–79. doi:10.1517/14740338.2013.741585
20. Turner LW, Nartey D, Stafford RS et al (2014) Ambulatory treatment of Type 2 diabetes mellitus in the United States, 1997–2012. Diabetes Care 37:985–992. doi:10.2337/dc13-2097
21. Gelman L, Feige JN, Desvergne B (2007) Molecular basis of selective PPARgamma modulation for the treatment of Type 2 diabetes. Biochim Biophys Acta 1771:1094–1107. doi:10.1016/j.bbalip.2007.03.004
22. Bruning JB, Chalmers MJ, Prasad S et al (2007) Partial agonists activate PPARgamma using a helix 12 independent mechanism. Structure 15:1258–1271. doi:10.1016/j.str.2007.07.014
23. Pochetti G, Godio C, Mitro N et al (2007) Insights into the mechanism of partial agonism: crystal structures of the peroxisome proliferator-activated receptor gamma ligand-binding domain in the complex with two enantiomeric ligands. J Biol Chem 282:17314–17324. doi:10.1074/jbc.M702316200
24. Farce A, Renault N, Chavatte P (2009) Structural insight into PPARgamma ligands binding. Curr Med Chem 16:1768–1789
25. Guasch L, Sala E, Valls C et al (2011) Structural insights for the design of new PPARgamma partial agonists with high binding affinity and low transactivation activity. J Comput Aided Mol Des 2011:717–728. doi:10.1007/s10822-011-9446-9
26. Choi JH, Banks AS, Estall JL et al (2010) Anti-diabetic drugs inhibit obesity-linked phosphorylation of PPARgamma by Cdk5. Nature 466:451–456. doi:10.1038/nature09291
27. Choi JH, Banks AS, Kamenecka TM et al (2011) Antidiabetic actions of a non-agonist PPARγ ligand blocking Cdk5-mediated phosphorylation. Nature 477:477–481. doi:10.1038/nature10383
28. Vázquez M, Silvestre JS, Prous JR (2002) Experimental approaches to study PPAR gamma agonists as antidiabetic drugs. Methods Find Exp Clin Pharmacol 24:515–523
29. Merk D, Steinhilber D, Schubert-Zsilavecz M (2014) Characterizing ligands for farnesoid X receptor-available in vitro test systems for farnesoid X receptor modulator development. Expert Opin Drug Discov 9:27–37
30. Guasch L, Sala E, Castell-Auví A et al (2012) Identification of PPARgamma partial agonists of natural origin (I): development of a virtual screening procedure and in vitro validation. PLoS One 7:e50816. doi:10.1371/journal.pone.0050816
31. Simmonds MSJ, Howes M-JR (2005) Plants used in the treatment of diabetes. In: Soumyanath A (ed) Traditional medicines for Modern Times. Antidiabetic plants. Taylor & Francis Group, Abingdon, pp 19–82
32. Bnouham M, Ziyyat A, Mekhfi H et al (2006) Medicinal plants with potential antidiabetic activity-A review of ten years of herbal medicine research (1990–2000). Int J Diabetes Metab 14:1–25
33. Hong J (2011) Role of natural product diversity in chemical biology. Curr Opin Chem Biol 15:350–354. doi:10.1016/j.cbpa.2011.03.004
34. Huang TH-W, Kota BP, Razmovski V, Roufogalis BD (2005) Herbal or natural medicines as modulators of peroxisome proliferator-activated receptors and related nuclear receptors for therapy of metabolic syndrome. Basic Clin Pharmacol Toxicol 96:3–14. doi:10.1111/j.1742-7843.2005.pto960102.x
35. Rau O, Wurglics M, Dingermann T et al (2006) Screening of herbal extracts for activation of the human peroxisome proliferator-activated receptor. Pharmazie 61:952–956
36. Jacob A, Wu R, Zhou M, Wang P (2007) Mechanism of the anti-inflammatory effect of curcumin: PPAR-gamma activation. PPAR Res 2007:89369. doi:10.1155/2007/89369
37. Huang TH-W, Teoh AW, Lin B-L et al (2009) The role of herbal PPAR modulators in the treatment of cardiometabolic syndrome. Pharmacol Res 60:195–206. doi:10.1016/j.phrs.2009.03.020
38. Mueller M, Jungbauer A (2009) Culinary plants, herbs and spices—a rich source of PPARγ ligands. Food Chem 117:660–667. doi:10.1016/j.foodchem.2009.04.063

39. Christensen KB, Minet A, Svenstrup H et al (2009) Identification of plant extracts with potential antidiabetic properties: effect on human peroxisome proliferator-activated receptor (PPAR), adipocyte differentiation and insulin-stimulated glucose uptake. Phytother Res 23:1316–1325. doi:10.1002/ptr.2782
40. Christensen KB, Jørgensen M, Kotowska D et al (2010) Activation of the nuclear receptor PPARγ by metabolites isolated from sage (*Salvia officinalis L.*). J Ethnopharmacol 132:127–133. doi:10.1016/j.jep.2010.07.054
41. Christensen KB, Petersen RK, Kristiansen K, Christensen LP (2010) Identification of bioactive compounds from flowers of black elder (*Sambucus nigra L.*) that activate the human peroxisome proliferator-activated receptor (PPAR) gamma. Phytother Res 24 (Suppl 2):S129–132. doi:10.1002/ptr.3005
42. Jungbauer A, Medjakovic S (2012) Anti-inflammatory properties of culinary herbs and spices that ameliorate the effects of metabolic syndrome. Maturitas 71:227–239. doi:10.1016/j.maturitas.2011.12.009
43. Rozema E, Atanasov AG, Fakhrudin N et al (2012) Selected extracts of Chinese herbal medicines: their effect on NF-κB, PPARα and PPARγ and the respective bioactive compounds. Evid Based Complement Alternat Med 2012:983023. doi:10.1155/2012/983023
44. Ortuño Sahagún D, Márquez-Aguirre AL, Quintero-Fabián S et al (2012) Modulation of PPAR-γ by Nutraceutics as complementary treatment for obesity-related disorders and inflammatory diseases. PPAR Res 2012:318613. doi:10.1155/2012/318613
45. Yang MH, Avula B, Smillie T et al (2013) Screening of medicinal plants for PPARα and PPARγ activation and evaluation of their effects on glucose uptake and 3T3-L1 adipogenesis. Planta Med 79:1084–1095. doi:10.1055/s-0033-1350620
46. Weidner C, Wowro SJ, Rousseau M et al (2013) Antidiabetic effects of chamomile flowers extract in obese mice through transcriptional stimulation of nutrient sensors of the peroxisome proliferator-activated receptor (PPAR) Family. PLoS One 8:e80335. doi:10.1371/journal.pone.0080335
47. Puhl AC, Bernardes A, Silveira RL et al (2012) Mode of peroxisome proliferator-activated receptor γ activation by luteolin. Mol Pharmacol 81:788–799. doi:10.1124/mol.111.076216
48. Weidner C, de Groot JC, Prasad A et al (2012) Amorfrutins are potent antidiabetic dietary natural products. Proc Natl Acad Sci U S A 109:7257–7262. doi:10.1073/pnas.1116971109
49. Wu G, Yi J, Liu L et al (2013) Pseudoginsenoside F11, a novel partial PPAR γ agonist, promotes adiponectin oligomerization and secretion in 3T3-L1 adipocytes. PPAR Res 2013:701017. doi:10.1155/2013/701017
50. Hwang J-T, Kim S-H, Lee M-S et al (2007) Anti-obesity effects of ginsenoside Rh2 are associated with the activation of AMPK signaling pathway in 3T3-L1 adipocyte. Biochem Biophys Res Commun 364:1002–1008. doi:10.1016/j.bbrc.2007.10.125
51. Hwang J-T, Lee M-S, Kim H-J et al (2009) Antiobesity effect of ginsenoside Rg3 involves the AMPK and PPAR-gamma signal pathways. Phytother Res 23:262–266. doi:10.1002/ptr.2606
52. Gong Z, Huang C, Sheng X et al (2009) The role of tanshinone IIA in the treatment of obesity through peroxisome proliferator-activated receptor gamma antagonism. Endocrinology 150:104–113. doi:10.1210/en.2008-0322
53. Lee H-M, Lee O-H, Lee B-Y (2010) Effect of Ginsenoside Rg3 and Rh2 on Glucose uptake in insulin-resistant muscle cells. J Korean Soc Appl Biol Chem 53:106–109. doi:10.3839/jksabc.2010.018
54. Yoshikawa M, Matsuda H (2005) Traditional Chinese and Kampo medicines. In: Soumyanath A (ed) Traditional medicines for Modern Times. Antidiabetic plants. Taylor & Francis Group, Abingdon, pp 135–149
55. Kuroda M, Mimaki Y, Sashida Y et al (2003) Phenolics with PPAR-gamma ligand-binding activity obtained from licorice (*Glycyrrhiza uralensis* roots) and ameliorative effects of glycyrin on genetically diabetic KK-A(y) mice. Bioorg Med Chem Lett 13:4267–4272
56. Nakagawa K, Kishida H, Arai N et al (2004) Licorice flavonoids suppress abdominal fat accumulation and increase in blood glucose level in obese diabetic KK-A(y) mice. Biol Pharm Bull 27:1775–1778

57. Kuroda M, Mimaki Y, Honda S et al (2010) Phenolics from *Glycyrrhiza glabra* roots and their PPAR-gamma ligand-binding activity. Bioorg Med Chem 18:962–970. doi:10.1016/j.bmc.2009.11.027
58. Salam NK, Huang TH-W, Kota BP et al (2008) Novel PPAR-gamma agonists identified from a natural product library: a virtual screening, induced-fit docking and biological assay study. Chem Biol Drug Des 71:57–70. doi:10.1111/j.1747-0285.2007.00606.x
59. Zarzuelo A, Jiménez I, Gámez MJ et al (1996) Effects of luteolin 5-O-beta-rutinoside in streptozotocin-induced diabetic rats. Life Sci 58:2311–2316
60. De Groot JC, Weidner C, Krausze J et al (2013) Structural characterization of amorfrutins bound to the peroxisome proliferator-activated receptor γ. J Med Chem 56:1535–1543. doi:10.1021/jm3013272
61. Atanasov AG, Wang JN, Gu SP et al (2013) Honokiol: a non-adipogenic PPARγ agonist from nature. Biochim Biophys Acta 1830:4813–4819. doi:10.1016/j.bbagen.2013.06.021
62. Cichewicz RH, Clifford LJ (2005) Native American medicine. In: Soumyanath A (ed) Traditional medicines for Modern Times. Antidiabetic plants. Taylor & Francis Group, Abingdon, pp 169–177
63. Leach MJ, Kumar S (2012) Cinnamon for diabetes mellitus. Cochrane database Syst Rev 9:CD007170. doi:10.1002/14651858.CD007170.pub2
64. Weisberg SP, Leibel R, Tortoriello D V (2008) Dietary curcumin significantly improves obesity-associated inflammation and diabetes in mouse models of diabesity. Endocrinology 149:3549–3558. doi:10.1210/en.2008-0262
65. Rinwa P, Kaur B, Jaggi AS, Singh N (2010) Involvement of PPAR-gamma in curcumin-mediated beneficial effects in experimental dementia. Naunyn Schmiedebergs Arch Pharmacol 381:529–539. doi:10.1007/s00210-010-0511-z
66. Wang H-M, Zhao Y-X, Zhang S et al (2010) PPARgamma agonist curcumin reduces the amyloid-beta-stimulated inflammatory responses in primary astrocytes. J Alzheimers Dis 20:1189–1199. doi:10.3233/JAD-2010-091336
67. Narala VR, Smith MR, Adapala RK et al (2009) Curcumin is not a ligand for peroxisome proliferator-activated receptor-γ. Gene Ther Mol Biol 13:20–25
68. Sato M, Tai T, Nunoura Y et al (2002) Dehydrotrametenolic acid induces preadipocyte differentiation and sensitizes animal models of noninsulin-dependent diabetes mellitus to insulin. Biol Pharm Bull 25:81–86
69. Li T-H, Hou C-C, Chang CL-T, Yang W-C (2011) Anti-hyperglycemic properties of crude extract and triterpenes from Poria cocos. Evid Based Complement Alternat Med 2011:128402. doi:10.1155/2011/128402
70. Guasch L, Sala E, Mulero M et al (2013) Identification of PPARgamma partial agonists of natural origin (II): in silico prediction in natural extracts with known antidiabetic activity. PLoS One 8:e55889. doi:10.1371/journal.pone.0055889
71. Mueller M, Lukas B, Novak J et al (2008) Oregano: a source for peroxisome proliferator-activated receptor gamma antagonists. J Agric Food Chem 56:11621–11630. doi:10.1021/jf802298w
72. Fiévet C, Fruchart J-C, Staels B (2006) PPARalpha and PPARgamma dual agonists for the treatment of type 2 diabetes and the metabolic syndrome. Curr Opin Pharmacol 6:606–614. doi:10.1016/j.coph.2006.06.009
73. Lin H-R (2012) Sesquiterpene lactones from *Tithonia diversifolia* act as peroxisome proliferator-activated receptor agonists. Bioorg Med Chem Lett 22:2954–2958. doi:10.1016/j.bmcl.2012.02.043
74. Shen P, Liu MH, Ng TY et al (2006) Differential effects of isoflavones, from *Astragalus membranaceus* and *Pueraria thomsonii*, on the activation of PPARalpha, PPARgamma, and adipocyte differentiation in vitro. J Nutr 136:899–905
75. Agget P, Alexander J, Alles M et al (1999) Scientific concepts of functional foods in Europe. Consensus document. Br J Nutr 81(Suppl 1):S1–27
76. Penumetcha M, Santanam N (2012) Nutraceuticals as ligands of PPARγ. PPAR Res 2012:858352. doi:10.1155/2012/858352

77. Song CM, Lim SJ, Tong JC (2009) Recent advances in computer-aided drug design. Brief Bioinform 10:579–591. doi:10.1093/bib/bbp023
78. Di L, Kerns EH, Carter GT (2009) Drug-like property concepts in pharmaceutical design. Curr Pharm Des 15:2184–2194
79. Peach ML, Zakharov AV, Liu R et al (2012) Computational tools and resources for metabolism-related property predictions. 1. Overview of publicly available (free and commercial) databases and software. Future Med Chem 4:1907–1932. doi:10.4155/fmc.12.150
80. Wermuth CG, Ganellin CR, Lindberg P, Mitscher LA (1998) Glossary of terms used in medicinal chemistry (IUPAC Recommendations 1998). Pure Appl Chem 70:1129–1143. doi:10.1351/pac199870051129
81. Caporuscio F, Tafi A (2011) Pharmacophore modelling: a forty year old approach and its modern synergies. Curr Med Chem 18:2543–2553
82. Dixon SL, Smondyrev AM, Knoll EH et al (2006) PHASE: a new engine for pharmacophore perception, 3D QSAR model development, and 3D database screening: 1. Methodology and preliminary results. J Comput Aided Mol Des 20:647–671. doi:10.1007/s10822-006-9087-6
83. Wolber G, Langer T (2005) LigandScout: 3-D Pharmacophores derived from protein-bound ligands and their use as virtual screening filters. J Chem Inf Model 45:160–169. doi:10.1021/ci049885e
84. Sousa SF, Ribeiro AJM, Coimbra JTS et al (2013) Protein-ligand docking in the new millennium—a retrospective of 10 years in the field. Curr Med Chem 20:2296–2314. doi:10.2174/0929867311320180002
85. Sousa SF, Fernandes PA, Ramos MJ (2006) Protein–ligand docking: current status and future challenges. Proteins Struct Funct Bioinforma 65:15–26. doi:10.1002/prot.21082
86. Kitchen DB, Stahura FL, Bajorath J (2004) Computational techniques for diversity analysis and compound classification. Mini Rev Med Chem 4:1029–1039
87. Maggiora GM, Vogt M, Stumpfe D, Bajorath J (2013) Molecular similarity in medicinal chemistry. J Med Chem. doi:10.1021/jm401411z
88. Verma J, Khedkar VM, Coutinho EC (2010) 3D-QSAR in drug design—a review. Curr Top Med Chem 10:95–115
89. Fischer PM (2008) Computational chemistry approaches to drug discovery in signal transduction. Biotechnol J 3:452–470. doi:10.1002/biot.200700259
90. Kar S, Roy K (2013) How far can virtual screening take us in drug discovery? Expert Opin Drug Discov 8:245–261. doi:10.1517/17460441.2013.761204
91. Hawkins PCD, Warren GL, Skillman AG, Nicholls A (2008) How to do an evaluation: pitfalls and traps. J Comput Aided Mol Des 22:179–190. doi:10.1007/s10822-007-9166-3
92. Cereto-Massagué A, Ojeda MJ, Joosten RP et al (2013) The good, the bad and the dubious: VHELIBS, a validation helper for ligands and binding sites. J Cheminform 5:36. doi:10.1186/1758-2946-5-36
93. Kirchmair J, Markt P, Distinto S et al (2008) Evaluation of the performance of 3D virtual screening protocols: RMSD comparisons, enrichment assessments, and decoy selection–what can we learn from earlier mistakes? J Comput Aided Mol Des 22:213–228. doi:10.1007/s10822-007-9163-6
94. Cereto-Massagué A, Guasch L, Valls C et al (2012) DecoyFinder: an easy-to-use python GUI application for building target-specific decoy sets. Bioinformatics 28:1661–1662. doi:10.1093/bioinformatics/bts249
95. Truchon J-F, Bayly CI (2007) Evaluating virtual screening methods: good and bad metrics for the "Early Recognition" problem. J Chem Inf Model 47:488–508. doi:10.1021/ci600426e
96. Tanrikulu Y, Rau O, Schwarz O et al (2009) Structure-based pharmacophore screening for natural-product-derived PPARgamma agonists. Chembiochem 10:75–78. doi:10.1002/cbic.200800520
97. Rupp M, Schroeter T, Steri R et al (2010) From machine learning to natural product derivatives that selectively activate transcription factor {PPAR} gamma. ChemMedChem 5:191–194. doi:10.1002/cmdc.200900469

98. Fakhrudin N, Ladurner A, Atanasov AG et al (2010) Computer-aided discovery, validation, and mechanistic characterization of novel neolignan activators of peroxisome proliferator-activated receptor gamma. Mol Pharmacol 77:559–566. doi:10.1124/mol.109.062141
99. Petersen RK, Christensen KB, Assimopoulou AN et al (2011) Pharmacophore-driven identification of PPARγ agonists from natural sources. J Comput Aided Mol Des 25:107–116. doi:10.1007/s10822-010-9398-5
100. Irwin JJ, Sterling T, Mysinger MM et al (2012) ZINC: a free tool to discover chemistry for biology. J Chem Inf Model 52:1757–1768. doi:10.1021/ci3001277
101. Li Y, Li L, Chen J et al (2009) 7-Chloroarctinone-b as a new selective PPARgamma antagonist potently blocks adipocyte differentiation. Acta Pharmacol Sin 30:1351–1358. doi:10.1038/aps.2009.113
102. Hwang BY, Lee J-H, Nam JB et al (2002) Two new furanoditerpenes from *Saururus chinenesis* and their effects on the activation of peroxisome proliferator-activated receptor gamma. J Nat Prod 65:616–617
103. Choi S-S, Cha B-Y, Iida K et al (2011) Artepillin C, as a PPARγ ligand, enhances adipocyte differentiation and glucose uptake in 3T3-L1 cells. Biochem Pharmacol 81:925–933. doi:10.1016/j.bcp.2011.01.002
104. Malapaka RR V, Khoo S, Zhang J et al (2012) Identification and mechanism of 10-carbon fatty acid as modulating ligand of peroxisome proliferator-activated receptors. J Biol Chem 287:183–195. doi:10.1074/jbc.M111.294785
105. Atanasov AG, Blunder M, Fakhrudin N et al (2013) Polyacetylenes from *Notopterygium incisum*—new selective partial agonists of peroxisome proliferator-activated receptor-gamma. PLoS One 8:e61755. doi:10.1371/journal.pone.0061755
106. Pferschy-Wenzig E-M, Atanasov AG, Malainer C et al (2014) Identification of isosilybin a from milk thistle seeds as an agonist of peroxisome proliferator-activated receptor gamma. J Nat Prod 77:842–847. doi:10.1021/np400943b
107. Dang Z-C, Audinot V, Papapoulos SE et al (2003) Peroxisome proliferator-activated receptor gamma (PPARgamma) as a molecular target for the soy phytoestrogen genistein. J Biol Chem 278:962–967. doi:10.1074/jbc.M209483200

Chapter 7
DPP-IV, An Important Target for Antidiabetic Functional Food Design

María José Ojeda, Adrià Cereto-Massagué, Cristina Valls and Gerard Pujadas

7.1 Introduction

7.1.1 Type 2 Diabetes Mellitus

Diabetes is a chronic disease that occurs when the pancreas does not produce sufficient insulin. Diabetes may also arise when the body cannot effectively use the insulin it produces. Hyperglycemia, or increased blood sugar, is a common effect of uncontrolled diabetes. Chronic hyperglycemia leads to serious damage to many body systems, particularly the nerves and blood vessels.

Type 2 diabetes mellitus—formerly referred to as noninsulin-dependent diabetes mellitus (T2DM)—is a chronic metabolic disease that is characterized by hyperglycemia and results from the body's ineffective use of insulin (i.e., a gradual decline in insulin sensitivity and/or insulin secretion). T2DM accounts for 90 % of people with diabetes and has become a worldwide epidemic. Moreover, many countries are now reporting the onset of T2DM at an increasingly young age due to sedentary lifestyles, longer life expectancies, and obesity [1].

G. Pujadas (✉) · M. J. Ojeda · A. Cereto-Massagué · C. Valls
Research Group in Chemoinformatics & Nutrition,
Departament de Bioquímica i Biotecnologia,
Universitat Rovira i Virgili, Campus de Sescelades,
C/ Marcel·lí Domingo s/n, 43007 Tarragona, Catalonia, Spain
e-mail: gerard.pujadas@urv.cat

G. Pujadas
Centre Tecnològic de Nutrició i Salut (CTNS),
TECNIO, CEICS,
Avinguda Universitat 1, 43204 Reus,
Catalonia, Spain

K. Martinez-Mayorga, J. L. Medina-Franco (eds.), *Foodinformatics*,
DOI 10.1007/978-3-319-10226-9_7

The majority of patients with T2DM are obese [2], and many of the current therapeutic options for management of T2DM can cause further weight gain [3, 4]. Concerns about weight gain adversely affect patients' willingness to begin and continue treatment with glucose-lowering medications, such as thiazolidinediones, insulin, and sulfonylureas [5]. In addition to weight gain, a patient's quality of life can be negatively affected by the underlying disease process and its complications, such as polypharmacy, hypoglycemia and micro- and macro-vascular complications [6].

The World Health Organization (WHO) and the International Diabetes Federation (IDF) report that between 347 and 371 million people worldwide currently have diabetes. It is forecasted that the number of diabetes deaths will double between 2005 and 2030, which will make diabetes the seventh leading cause of death in 2030 [7, 8]. According to the WHO and IDF information, this strong correlation between diabetes and death are supported by the following data: (a) between 50 and 80 % of people with diabetes die of cardiovascular disease (primarily heart disease and stroke) [9], (b) diabetes is among the leading causes of kidney failure [10], (c) the overall risk of dying among people with diabetes is at least double the risk of their peers without diabetes [11], and (d) half of all people who die from diabetes are under the age of 60. Moreover, the WHO data also reveal the following: (a) the combination of diabetes with reduced blood flow and neuropathy increases the chance of foot ulcers, infection, and eventual need for limb amputation, and (b) 1 % of global blindness can be attributed to diabetes because it is the result of long-term accumulated damage to the retina's small blood vessels [12].

7.1.2 Current T2DM Incidence in North America and the Caribbean Region

According to the last Diabetes Atlas Update from the IDF [1], approximately 9.6 % of the population between 20 and 79 years old in the North American and Caribbean region (corresponding to 36.8 million people; 24.4 million in the USA) is estimated to be affected by diabetes. By 2035, the number of affected people is expected to increase to 50.4 million. Moreover, 44.2 million people (13.2 % of adults in this region) have impaired glucose tolerance (58.8 million expected by 2035), which increases their risk for developing T2DM. Diabetes-related causes were responsible for 13.5 % (150,000 men and 143,000 women) of all deaths among adults in this region during 2013. In the USA, more than 192,000 people died from diabetes in 2013, which is one of the highest numbers of deaths due to diabetes of any country in the world. The USA is estimated to account for almost half (42 %) of the world's diabetes-related health-care spending.

7.1.3 Pharmacological Treatment of T2DM

There are now ten different drug classes available as adjuncts to diet and exercise for the management of hyperglycemia in T2DM patients in the USA (e.g., sulfonylureas, biguanides, meglitinides, α-glucosidase inhibitors, thiazolidinediones, glucagon-like peptide 1 (GLP-1) agonists, DPP-IV inhibitors, amylin analogs, bile acid sequestrants, and dopamine receptor agonists; Table 7.1) [13]. Despite the many available drugs, there is still a need for new therapies to control glycemia [14]. Many compounds can reduce blood glucose levels. However, clinical use requires an effective antihyperglycemic agent that can meet requirements beyond simply reducing the blood glucose levels [15]. For example, safety profiles (particularly cardiovascular safety) have received significant attention over the past few years.

7.1.4 DPP-IV Inhibition in T2DM Treatment

DPP-IV (also known as adenosine deaminase-binding protein or CD26; EC 3.4.14.5) is a ubiquitous aminodipeptidase that was first described by Hopsu-Havu and Sarimo [16]. It belongs to the α/β hydrolases (family S9B) and is related to the prolyl oligopeptidase [17]. DPP-IV is expressed on the surface of several cell types including lymphocytes and monocytes and in tissues in the pancreas, kidneys, liver, and the gastrointestinal tract [18]. There are different expression levels in different tissue types. Its expression is particularly high in the kidney cortex, the small intestine brush-border membranes, and the epithelial cells of pancreatic ducts [19]. The widespread expression of DPP-IV means that it can easily access and inactivate a wide variety of biological regulatory peptides. The target peptides include glucose-dependent insulinotropic polypeptide (GIP), GLP-1, growth hormone, peptide YY, and neuropeptide Y [20].

The structure of DPP-IV is a homodimeric transmembrane glycoprotein. Each subunit of the protein is anchored to the plasma membrane by a hydrophobic helix consisting of seven N-terminal amino acids. Each subunit has a large globular extracellular region that contains an active site located in the interface between the β-propeller domain (from residues 39 to 508) and the α/β-hydrolase domain (from residues 509–766; Fig. 7.1) [21–24]. The cleavage of the extracellular portion of DPP-IV from the transmembrane section results in a soluble circulating form of approximately 100 kDa. The soluble form is found in plasma and cerebrospinal fluid [18, 25]. DPP-IV is secreted as a mature monomer, but it requires dimerization to undergo normal proteolytic activity [26].

Recent studies indicated that in addition to the regulation of postprandial glycemia, DPP-IV may have pleiotropic effects (e.g., obesity, tumor growth, and HIV infection), which makes it an attractive target for drug discovery research [27–32]. DPP-IV inhibitors block the degradation of GLP-1 and inhibit the inactivation of several other peptides that may have vasoactive and cardioprotective effects

Table 7.1 The ten different drug classes currently available in the USA that serve as adjuncts to diet and exercise in the management of hyperglycemia in T2DM patients

Antidiabetic agents	Examples	Mode of action	Advantages	Adverse effects
Sulfonylureas	Glipide, glyburide, glimepiride	Induction insulin release from β cells by inhibiting potassium flux through ATP-dependent potassium channels (K_{ATP})	Reduced hepatic uptake, inhibition of glucagon and enhanced insulin sensitivity	Hypoglycemia, body weight gain and possible affection to pancreatic function
Biguanides	Metformin	Suppression of hepatic gluconeogenesis by AMPK phosphorylation	Low rates of hypoglycemia, weight stability/loss, better insulin sensitivity	Gastrointestinal side effects and possible affection to renal or hepatic function
Meglitinides	Repaglinide, nateglinide	Interaction with the voltage-dependent K_{ATP} chanels of pancreatic β cells	Induction of an early insulin response to meals decreasing postprandial blood glucose levels, low rates of hypoglycemia	Weight gain and increased on the insulin deficiency
α-glucosidase inhibitors	Acarbose, miglitol	Competitive inhibition of the α-glucosidase in the intestine	No drug-drug interaction, weight loss, no risk of hypoglycemia, cardioprotective effects, stimulated secretion of GLP-1	Gastrointestinal effects: flatulence, diarrhea, abdominal discomfort
Thiazolidinediones or PPAR-γ agonists	Rosiglitazone, pioglitazone	Binding on the PPAR-γ, it activates the transcription of specific genes of lipid metabolism	Sensitivity to insulin, anti-inflammatory effects and amelioration of hypertension, microalbuminuria and hepatic steatosis	Severe liver failure, death and increased cardiac risk
GLP-1 agonists or mimetics	Exenatide, liraglutide	They are modified GLP-1 molecules that are resistant to DPP-IV induced degradation	Stimulate insulin secretion and inhibit glucagon output in a glucose-dependent manner, slow gastric emptying and decrease appetite	Increased risk of pancreatitis, pre-cancerous cellular changes called pancreatic duct metaplasia and of tumor development at the thyroid gland
DPP-IV inhibitors	Sitagliptin, Saxagliptin	Increase circulating GLP-1 and GIP levels prolonging their action (which lead to decreased levels of blood glucose, HbA1c and glucagon)	Better glucose homeostasis with a lower risk of hypoglycemia and without adversely affecting cardiovascular markers	Headache, nausea, vomiting, loss of appetite

Table 7.1 (continued)

Antidiabetic agents	Examples	Mode of action	Advantages	Adverse effects
Amylin analogues	Pramlintide	Amylin binds to calcitonin receptors in the central nervous system that cooperate with receptor activity modifying proteins	Enhanced satiety, diminished glucagon secretion and delayed gastric emptying	Severe hypoglycemia, nausea, vomiting, anorexia and headache
Bile acid sequestrants	Colesevelam	Binding to the nuclear farnesoid X receptor or the membrane receptor TGR5, where it regulates lipids and glucose levels	No toxicity, no dependency of liver and kidney function	Abdominal and muscle pain, nausea, diarrhea and constipating effects. Associated with dysphagia and esophageal obstruction
Dopamine receptor agonists	Bromocriptine	Activation of hypothalamic-pituitary-adrenal axis	No effects on free fatty acids levels or hepatic glucose production	Nausea, vomiting, diarrhea, stomach cramps and depression

[33–42]. Therefore, the growing body of evidence suggests that DPP-IV inhibitors improve several cardiovascular risk factors, including (a) improvement of endothelium-dependent relaxation, (b) reduction of the vascular inflammation and oxidative stress, (c) reduction of total cholesterol levels, (d) prevention of vascular endothelial dysfunction and atherosclerosis, and (e) reduction of myocardial fibrosis and oxidative stress [42]. Major prospective clinical trials involving various DPP-IV inhibitors with predefined cardiovascular outcomes are currently in progress. These studies are examining T2DM patients who have a high-risk cardiovascular profile to confirm this cardiovascular protective effect [40].

7.1.5 Importance of Selectivity in DPP-IV Inhibition

DPP-IV is in a family of ubiquitous atypical serine proteases with numerous functions, including roles in nutrition, metabolism, the endocrine and immune systems, cancer growth, bone marrow mobilization, and cell adhesion [20]. The DPP-IV family includes four enzymes (DPP-IV, fibroblast activation protein (FAP),DPP8, and DPP9) and two nonenzymes (DPP-IV-like protein-6; DPP6, DPL-1, or DPP-X; and DPP10; DPL-2) [20].

The enzyme FAP, also known as seprase, is the most similar family member to DPP-IV. FAP and DPP-IV share 52% amino acid identity (human enzymes) and similar substrate specificity. Despite these similarities, FAP and DPP-IV differ in their expression patterns because FAP expression is confined predominantly to

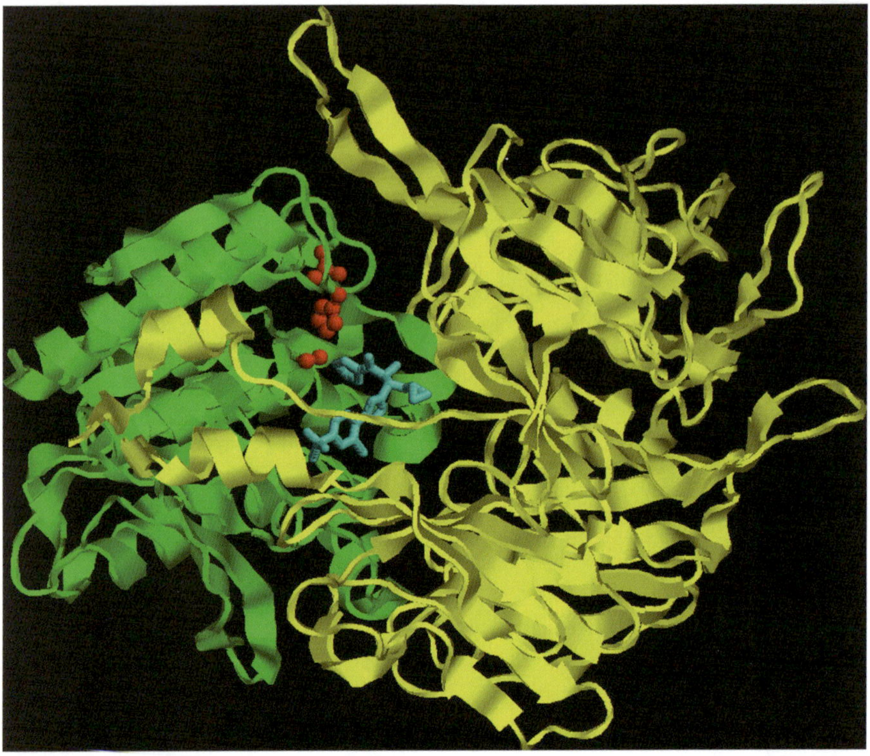

Fig. 7.1 A general overview of the 3D fold of the extracellular region for one of the subunits in the human DPP-IV homodimer. The β-propeller domain is shown in *yellow* whereas the α/β-hydrolase domain is shown in *green*. The location of the active site is indicated by the *red* residues from the catalytic triad (Ser630, Asp708 and His740) and the fluoroolefin inhibitor (*in cyan*). This figure has been built with the PDB structure with 3C45 code [92] and with the molecular visualization software RasMol [208]

activated fibroblasts in diseased tissue (e.g., fibrotic and epithelial tumors, invasive cancers [43], and some fetal mesenchymal tissues), but it is absent in the adult human tissues. The other two catalytically active DPP-IV family members, DPP8 and DPP9, share 26 and 21 % amino acid identity with the protein sequence of DPP-IV and FAP, respectively (human enzymes). DPP8 and DPP9 are soluble monomeric proteins in the cytoplasm and are very similar proteins because they share 61 % amino acid sequence similarity. DPP8 expression is upregulated in activated T cells, and high levels of DPP9 are found in cancer cells, normal skeletal muscle, and the heart and liver [44]. However, their physiological function is not known. Compounds that were previously thought to be specific for DPP-IV could also be inhibitors of other members of the DPP-IV family.

A number of DPP-IV inhibitors have recently been tested for selectivity to DPP-IV, FAP, DPP8, and DPP9 enzymes [45]. In that study, individually selective compounds for DPP-IV, DPP8/9, and FAP were identified, which allowed an evaluation of the potential toxicity and tolerability of each type of inhibition. The DPP8/9 selective inhibitor produced alopecia, thrombocytopenia, reticulocytopenia, multi-organ histopathological changes, enlarged spleen, and mortality in rats. In dogs, the DPP8/9 inhibitor produced gastrointestinal toxicity. However, investigation of the DPP-IV selective inhibitor demonstrated no apparent toxicity [45]. Because inhibition of DPP8 and/or DPP9 has been shown to cause severe toxicity in preclinical species [45], high selectivity is an important criterion in selecting DPP-IV inhibitors for antidiabetic clinical development. Thus, new DPP-IV inhibitors reported on the literature are selective relative to other members of the DPP-IV family [86–105].

7.2 The Incretin System

7.2.1 Overview

Incretin hormones are gut peptides secreted by endocrine cells in the intestinal mucosa in response to nutrient ingestion. These peptides play a key role in the regulation of islet function and blood glucose levels (Fig. 7.2). In humans, the major incretin hormones are GLP-1 and GIP, and, together, they fully account for the incretin effect [46]. The incretin effect is defined as the phenomenon whereby orally ingested glucose elicits a much greater insulin response compared with the response obtained when glucose is infused intravenously to give identical blood glucose levels (the so-called isoglycemic glucose infusion) [47–49]. It has been demonstrated that the incretin effect is responsible for 50–70 % of insulin response in healthy humans [48, 50, 51].

The incretin hormones are released following meal ingestion and are rapidly degraded by DPP-IV [46, 48, 52]. GLP-1 is produced by L cells located in the ileum and in the colon where they are found in high density [49]. In contrast, GIP is secreted by K cells, which are primarily located in the duodenum. Both L cells and K cells are situated in the intestinal mucosa. As a result, these cells can be influenced by direct contact with nutrients from food ingestion [49, 53]. The secretion of GLP-1 and GIP depends not only on the type of macronutrients but also on the rate of gastric emptying and intestinal transit time. Moreover, some evidences show that secretion is modulated by the circadian system, and that higher secretion occurs in the morning than in the afternoon [54, 55]. The incretin metabolites are primarily cleared by the kidneys.

7.2.2 Incretins and Glucose Homeostasis

Both GLP-1 and GIP are able to regulate glucose homeostasis by interacting with G protein-coupled receptors (GPCR) [56, 57]. The GIP receptor is mainly expressed on islet β cells, but it also occurs in adipose tissue and in the central nervous system. Conversely, the GLP-1 receptor is localized on islet α and β cells and in

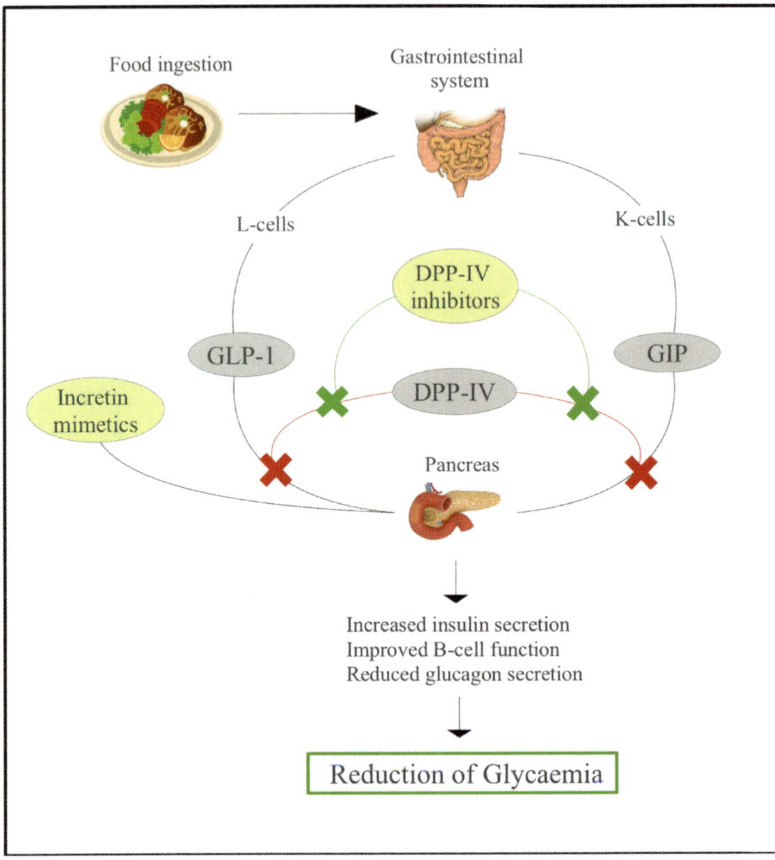

Fig. 7.2 The incretin system. Relationship between the physiological effects of *GLP-1* and *GIP* on insulin secretion and the action of targets implied in T2DM treatment. *GLP-1* and *GIP* are released from enteroendocrine cells after nutrient ingestion to stimulate insulin secretion. However, their activity is reduced because of the cleavage of *DPP-IV* at the second residue of *GLP-1* and *GIP*. Two alternatives to avoid the cleavage are administration of incretin mimetics or *DPP-IV* inhibitors

peripheral tissues, such as the heart, kidneys, lungs, gastrointestinal tract, and peripheral nervous system [57]. As a result of β cell activation, the levels of cAMP and intracellular calcium increase rapidly [57, 58]. This causes insulin secretion in a glucose-dependent manner because of their action after nutrient ingestion [58].

The incretin effect is involved in multiple actions that stimulate all stages of insulin biosynthesis and secretion to reduce the levels of glucose after food ingestion. GLP-1 acts on α cells by suppressing the secretion of glucagon, which has been demonstrated to reduce the risk of hyperglycemia [58]. GLP-1 has a trophic effect on β cells. It not only stimulates their proliferation but also enhances the differentiation of pancreatic cells and reduces apoptosis [49, 59]. Moreover, this gastrointestinal hormone slows gastric emptying and can reduce the postprandial glucose levels. These effects are similar to inhibiting appetite and food intake [49]. In addition, GLP-1 protects against ischemic and reperfused myocardium injury in rats via mechanisms independent of insulin because of the receptors expressed in this tissue. The hormone may also possess neuroprotective effects. GLP-1 has been proposed as a new therapeutic agent for neurodegenerative diseases such as Alzheimer's disease [49, 58, 59].

Similar to GLP-1, GIP increases insulin biosynthesis and secretion and has a protective activity on β cells. In addition, GIP stimulates the release of glucagon, and it is implicated in lipid metabolism and adiposity [60].

7.2.3 Incretins in T2DM Patients

Although patients with T2DM produce normal levels of GIP, the reduced response to the insulinotropic actions may be related to a reduction in receptor expression or reduced β cell sensitivity to GIP. However, GLP-1 maintains full physiological efficacy, despite being produced in lower concentrations [56, 61, 62]. Although GLP-1 and GIP are responsible for 50–70 % of postprandial insulin release in healthy subjects, the incretin effect contributes to only 20–35 % of the insulin response to oral glucose in T2DM patients [48]. A reduced insulinotropic effect is also found in healthy subjects with experimental insulin resistance induced by a combination of a high-fat diet, sedentary lifestyle, and steroid therapy [48, 63].

7.3 DPP-IV Inhibition in Detail

7.3.1 Commercially Available DPP-IV Inhibitors

The inhibition of DPP-IV in humans increases the circulating GLP-1 and GIP levels (and, consequently, prolongs their action), which leads to decreased levels of blood glucose, HbA_{1c}, and glucagon. Therefore, DPP-IV inhibition improves glucose homeostasis with a lower risk of hypoglycemia. As a result, DPP-IV inhibitors are of considerable interest to the pharmaceutical industry [64]. Intensive research activities in this field have resulted in the launch of sitagliptin, saxagliptin, alogliptin, linagliptin, vildagliptin, gemigliptin, and teneligliptin (collectively called as *gliptins*) to the market (Table 7.2) [19, 65].

7.3.2 Side Effects of Commercially Available DPP-IV Inhibitors

A recent post (March 14, 2013) at the sitagliptin [66], saxagliptin [67], and linagliptin [68] pages on MedLinePlus showed that the US Food and Drug Administration (FDA) is evaluating unpublished new findings by a group of academic

Table 7.2 Main features of commercially available DPP-IV inhibitors

Pharmacological name	Commercial name and developer	FDA approval	Advantages	Adverse effects	Selectivity over DPP8/9
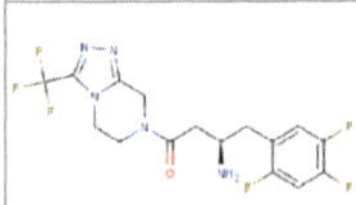 Sitagliptin	**Januvia®** (Merck & Co)	October 17th, 2006	Free from major drug interactions, well-tolerated, moderately efficacious, weight-neutral, low incidence of hypoglycemia, particular role in kidney or liver dysfunction	Abdominal pain, nausea, diarrhea, nasopharyngitis, back pain, osteoarthritis	2600-fold greater affinity
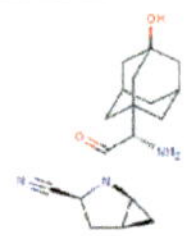 Saxagliptin	**Onglyza®** (BMS & AstraZeneca)	July 31st, 2009	Well tolerated, safe to use in renal failure, not affect blood pressure, lipid levels, body weight or cardiovascular markers	Headache, upper respiratory infections, arthralgia, nausea, cough	390 and 77-fold greater affinity, respectively
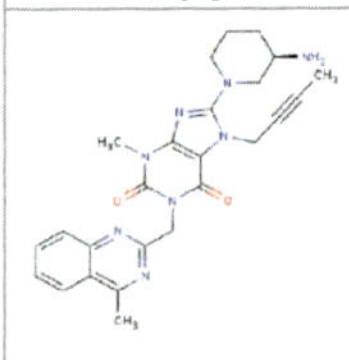 Linagliptin	**Tradjenta®** (Boehringer Ingelheim International GmbH & Co)	May 2nd, 2011	Once-daily oral dosing, high affinity, no dose restriction in patient with nephropathy, no drug-drug interaction, weight neutrality	Muscle pain, headache, nausea, vomiting, loss of appetite	40000 and >10000-fold greater affinity, respectively
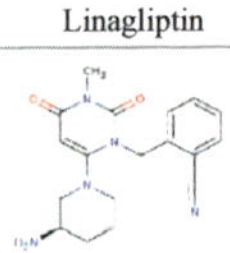 Alogliptin	**Nesina®** (Furiex pharmaceuticals)	January 25th, 2013	No significant interaction with other drugs, absorption is not affected by food ingestion	Headache, dizziness, constipation	>14000-fold greater affinity
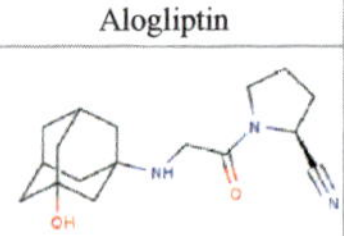 Vildagliptin	Galvus®, Jalra® or Xiliarx® (Novartis Europharm)	(a)	High specificity, durable response	Upper respiratory infection, dizziness, hypoglycemia, headache	270 and 32-fold greater affinity, respectively
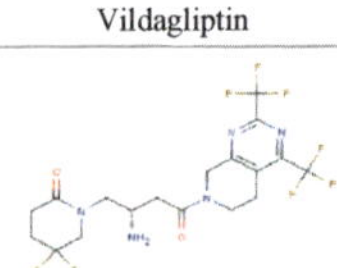 Gemigliptin	Zemiglo® (LG life Sciences)	(b)	Once-daily oral dosing, well tolerated, low rate of hypoglycemia	headache, dizziness, nausea, epistaxis , and possible increased heart rate	3000-fold greater affinity
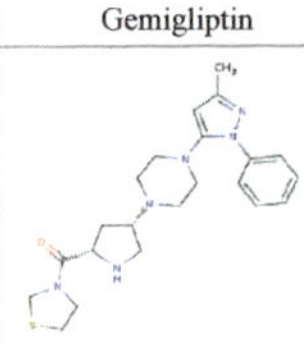 Teneligliptin	TENELIA® Mitsubishi Tanabe Pharma Corporation and Daiichi Sankyo & Co)	(c)	well tolerated, safe, potent and significantly improves glycemic control. inhibited the accumulation of lipids	Risk of hypoglycemia and constipation	700-1500-fold greater affinity

(a) The Europa Union since September 26th, 2007
(b) Korea since June 2012
(c) Japan since September 2012

researchers. The new data suggest an increased risk of pancreatitis and precancerous cellular changes called pancreatic duct metaplasia in patients with T2DM who were treated with these drugs. It is important to note this early communication from the FDA is intended only to inform the public and health-care professionals that the Agency intends to obtain and evaluate the new information before reaching any conclusions about the safety risks of these drugs.

Interestingly, it has been reported that patients with T2DM have a two- to three-fold increased risk of suffering from acute pancreatitis [69]. However, other reported studies suggest no increased risk of pancreatitis or malignancy in clinical trials with these drugs [70–75]. For instance, in a pooled analysis of 19 randomized double-blind clinical trials that included data from 10,246 patients, the incidence of acute pancreatitis was 0.10/100 patient–years in the placebo group and 0.08/100 patient–years in the sitagliptin group [71]. A recent analysis has updated the safety and tolerability of sitagliptin by examining pooled data from 25 double-blind clinical studies that lasted up to 2 years. These studies included data from 14,611 patients and concluded that treatment with sitagliptin is not associated with an increased risk of major adverse cardiovascular events, malignancy, or pancreatitis [72]. Therefore, it is likely that sitagliptin does not play a causal role in the reported instances of pancreatitis [72]. Moreover, clinical trials have not demonstrated an increased risk of renal failure with sitagliptin administration [71], and other studies suggest that sitagliptin, saxagliptin, and linagliptin may be used in patients with advanced kidney disease [76, 77].

7.3.3 DPP-IV-Binding Site Description

The DPP-IV binding site is highly druggable in the sense that tight and specific binding to the enzyme can be achieved using small molecules that have drug-like physicochemical properties [56, 78]. It is accessible in two ways: (1) via an opening in the β-propeller domain or (2) via the large side opening, which is formed at the interface of the β-propeller and α/β-hydrolase domain (Fig. 7.1) [18, 19, 23]. The structural features of DPP-IV suggest that substrates and inhibitors enter or leave the binding site via the side opening. Thus, the ligands can directly reach the active site and are correctly oriented for the subsequent cleavage. However, this possibility has not been fully elucidated [18, 79, 80].

In the active site of a protease, there are subsites labeled according to the peptide residue that they bind [81]. The point of peptide cleavage is between the peptide bond that binds residue P_1 with residue P'_1. As a result, the residues that surround this position are labeled relative to the cleavage site as P_2, P_1, P'_1, P'_2, and so on. Therefore, the protein subsites occupied by residues P_2, P_1, P'_1, and P'_2 are labeled as S_2, S_1, S'_1, and S'_2, respectively.

The analysis of the different DPP-IV/inhibitor complexes available at the protein data bank (PDB) has allowed the following different subsites to be identified for DPP-IV (Fig. 7.3 and Table 7.3) [21, 78, 80, 82–86]: (a) the N-terminal recognition is formed by residues Glu205, Glu206, and Tyr662 where the Glu205 (and, in

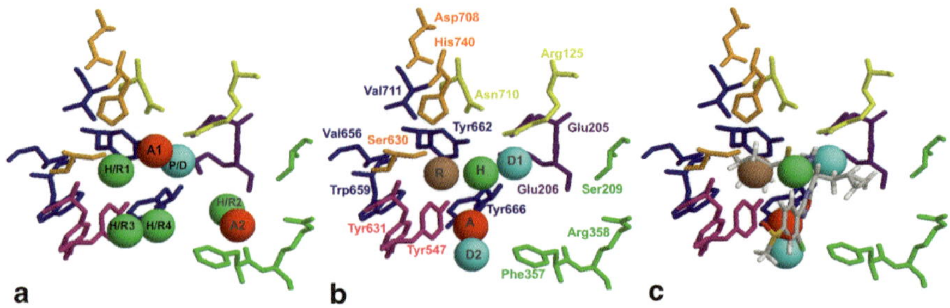

Fig. 7.3 DPP-IV binding site description. Residues belonging to the N-terminal recognition, the S_2 extensive subsite, the S_2 subsite, the S_1 subsite, the catalytic triad, the oxyanion hole and the P_2 amide recognition region are shown in *purple*, *light green*, *green*, *blue*, *orange*, *pink* and *yellow*, respectively. **a** Structure-based energetic pharmacophore built from the PDB structure of 10 complexes of DPP-IV with potent (IC_{50} values ≤ 10 nM) reversible inhibitors of a non-peptide nature [99]; **b** fragment-based energetic pharmacophore built after docking a library of rigid fragments at the DPP-IV binding site and further clustering of the fragments with highest binding energy; **c** the DPP-IV inhibitor from the PDB structure 3C45 in the context of the binding-site and of the fragment-based energetic pharmacophore. Pharmacophore sites are labeled according to their chemical characteristics (*H* hydrophobic, *R* aromatic ring, *P* polar, *D* hydrogen bond donor sites and *A* hydrogen bond acceptor sites; sites labeled as *H/R* and *P/D* accept two different chemical features). All three panels are in the same orientation to facilitate the comparison

some cases, Glu206) forms a salt bridge/hydrogen bond with the peptide's basic amine; (b) the S_2 pocket is formed by the residues Arg125, Ser209, Phe357, Arg358, Tyr547, and Asn710, where Arg125 and Asn710 are essential to coordinate the carbonyl of the P_2 residue and, together with Glu205 and Glu206, align the substrate optimally for the nucleophilic attack by Ser630 [87]; (c) the oxyanion hole is formed by the backbone NH of Tyr631 and the side chain OH of Tyr547 and stabilizes the negatively charged tetrahedral oxyanion intermediate that is generated in the transition state [87]; (d) the S_1 pocket is formed by the residues Tyr631, Val656, Trp659, Tyr662, Tyr666, and Val711; and (e) the catalytic triad is formed by the residues Ser630, Asp708, and His740 (with Ser630 cleaving the peptide bond between P_1 and P'_1 by performing a nucleophilic attack). Although in principle, no subsites are defined further than S_2 in DPP-IV, a recent study has shown that the inhibitors and not the substrates can bind well beyond the S_2 subsite to increase their inhibitory activity [88, 89]. The site beyond S_2 was defined as the S_2 extensive subsite and is formed by Val207, Ser209, Phe357, and Arg358 [23].

Based on the analysis of the DPP-IV crystal structures [90–96] and the interpretation of the structure–activity relationship data, both the lipophilic S_1 pocket and the Glu205/Glu206 dyad can be considered as crucial molecular anchors for DPP-IV inhibition [78]. Moreover, this conclusion is supported by results derived from two different energetic pharmacophores [97, 98] obtained by our group that have quantified the relative contribution of the different pharmacophore sites to the intermolecular interactions with DPP-IV. The first energetic pharmacophore was built from the PDB structure of ten complexes of DPP-IV with potent (IC_{50} values ≤ 10 nM) reversible inhibitors of a nonpeptide nature (Fig. 7.3a) [99]. This study showed that

Table 7.3 Intermolecular interactions between potent (IC_{50} values ≤ 10 nM) and reversible nonpeptide inhibitors in the DPP-IV binding site of available PDB structures

PDB code	Ligand	IC_{50} (nM)	S_2 subsite extensive	N-terminal recognition	S_2 subsite	P_2 amide recognition	Oxyanion hole	S_1 subsite	Enzyme catalitic triad
3C45	317	0.21		SaltB/HBond	Hydroph			Hydroph	
3VJM	W61	0.37	Hydroph	SaltB/HBond	Hydroph	HBond		Hydroph	
3H0C	PS4	0.38		SaltB/HBond	Hydroph	Hydroph		Hydroph	Hydroph
3KWJ	23Q	0.5		SaltB/HBond	Hydroph		Hydroph	Hydroph	
2RGU	356	1		SaltB/HBond			HBond/π-stacking	Hydroph	Hydroph
2QT9	524	2.3	Hydroph	SaltB/HBond	Hydroph	HBond		Hydroph	Hydroph
2IIT	872	2.6		SaltB/HBond	Hydroph	HBond		Hydroph	Hydroph
3HAB	677	4.2		SaltB/HBond	Hydroph			Hydroph	Hydroph
2QTB	474	4.8	Hydroph	SaltB/HBond	HBond/Hydroph	HBond		Hydroph	Hydroph
3G0D	XIH	5		SaltB/HBond		HBond	HBond	Hydroph	
3G0G	KUM	5		SaltB/HBond		HBond	HBond	Hydroph	
3O95	01T	5.3		SaltB/HBond	Hydroph	HBond		Hydroph	HBond
3VJL	W94	5.6		SaltB/HBond	Hydroph	HBond		Hydroph	
2QJR	PZF	6.4	HBond/π-stacking	SaltB/HBond	Hydroph			Hydroph	Hydroph
2IIV	565	6.6		SaltB/HBond	Hydroph			Hydroph	Hydroph
3HAC	361	6.7		SaltB/HBond	Hydroph			Hydroph	
3KWF	B1Q	6.8	Hydroph	SaltB/HBond	Hydroph	Cation-dipole/ HBond		Hydroph	Hydroph
3G0B	T22	7		SaltB/HBond		HBond	HBond	Hydroph	

Rows are sorted according to increasing IC_{50}. The data have been obtained from the literature and from the analysis of the corresponding LigPlot diagrams [207]. *Hydroph*, *SaltB*, and *HBond* refer to hydrophobic contacts, salt bridges, and hydrogen bonds, respectively two of the six sites of the pharmacophore (**P/D** and **H/R1**): (a) were accomplished by all ten inhibitors, (b) accounted for more than 90 % of the inhibitor/DPP-IV binding energy, and (c) were located in the two previously identified crucial molecular anchors for DPP-IV inhibition (**P/D** and **H/R1** are close to the Glu205/Glu206 dyad and the S_1 pocket, respectively). The second energetic pharmacophore (unpublished results) has been obtained after (1) docking a library of rigid fragments at the DPP-IV binding site and (2) further clustering of the fragments with the highest binding energy. This fragment-based energetic pharmacophore is formed by five relevant sites (i.e., two hydrogen-bond donors, one hydrogen-bond acceptor, one hydrophobic site, and one aromatic ring; Fig. 7.3b). According to our results, two of these five sites (**R** and **H**) show a very large contribution to the binding energy (scores of −10.05 and −5.77 kcal/mol, respectively) compared with the remaining three binding energies (scores of −2.71, −2.09 and −1.33 kcal/mol). Interestingly, the comparison of the energetic pharmacophores in Figs. 7.3a and b show that (a) the *P/D* site at Fig. 7.3a matches the *D1* site at Fig. 7.3b; and (b) the *H/R1* site in Fig. 7.3a approximately matches the *R* site at Fig. 7.3b. Therefore, both energetic pharmacophores highlight the importance of the N-terminal recognition performed by the Glu205/Glu206 dyad and the intermolecular interactions at the hydrophobic S_1 site. Moreover, other studies suggest that the binding free energy can be further improved by additional favorable contacts [84] with the following: (a) the catalytic triad, (b) the oxyanion hole, (c) the P_2 amide recognition region (formed by Arg125 and Asn710) where, for instance, Arg125 can stabilize the amide carbonyl moiety of an inhibitor by making a hydrogen bond with it [82], (d) the phenyl rings from Phe357 and Tyr547 (by interacting with different aromatic ligand fragments to give π–π stacking interaction or by making hydrophobic contacts with large aliphatic groups) [80, 84], or (e) Arg358, which uses its positively charged side chain to interact with substituents on the ligand's aromatic rings or to place electronegative groups on the ligands close to its positive-charged side chain [84].

Interestingly, the comparison of Figs. 7.3a and b also shows that there are unexplored ways to inhibit DPP-IV. In the fragment-based pharmacophore sites *A* and *D2* (with scores of −2.71 and −2.09 kcal/mol, respectively) located between the residues Phe357, Tyr547, and Tyr666 (Fig. 7.3b) are not present at the PDB-based energetic pharmacophore (Fig. 7.3a). A similar situation occurs for the *H* site (located at the center of the DPP-IV binding site; Fig. 7.3b) that, as mentioned before, has a very large score (−5.77 kcal/mol). As a result, it is remarkable that only three of the ten experimental poses that were used to derive the structure-based pharmacophore are able to simultaneously fit the *R, H,* and *D1* sites of the fragment-based pharmacophore (unpublished results). Therefore, it can be concluded that the use of the fragment-based pharmacophore in a virtual screening could identify previously undescribed DPP-IV inhibitors in molecular databases by reducing the bias toward the existing covered space of the binding site. Our group is currently using this pharmacophore to identify potent DPP-IV inhibitors in the molecules found in nontoxic mushrooms of the Catalan forests. Our aim is to use extracts rich in these bioactive molecules as food additives for people affected (or potentially affected) by T2DM.

7.3.4 How Differences at the Binding Site Among DPP-IV, DPP8, and DPP9 Explain the Selective Inhibition of DPP-IV

Unlike DPP-IV and FAP, the 3D structures for DPP8 and DPP9 are unknown. However, the structures can be built by homology modeling [100–102]. A comparison of the binding sites in DPP-IV, DPP8, and DPP9 suggests how to look for (or design) potent DPP-IV inhibitors with no (or low) bioactivity on DPP8/9 [103]. This comparison shows the following: (a) the S_1 pocket is significantly smaller in DPP-IV (27.72 $Å^3$) than it is in DPP8 (99.77 $Å^3$) and DPP9 (75.89 $Å^3$) [103, 104–106], which suggests that the excluded volumes obtained for this pocket in DPP-IV can be used to remove DPP8/9 inhibitors during the virtual screening (VS) workflow, (b) the Glu205/Glu206 dyad side chains are oriented towards the ligand site in DPP-IV where they form a salt bridge with ligands whereas in DPP8/9 one of the two equivalent glutamic acids (Glu276 for DPP8 and Glu249 for DPP9) has its side chain oriented away from the active site (consequently, its intermolecular interaction with a ligand hydrophilic group is not as strong as it is in DPP-IV [103, 106], which can result in a lower docking score for the same ligand in DPP8/9 relative to DPP-IV), and (c) whereas the S_2 extensive subsite has not been clearly defined for DPP8/9, it has been shown to be important for the potency and selectivity of DPP-IV inhibitors [23, 27, 78, 88, 100, 105, 107].

7.3.5 How to Predict DPP-IV Selective Inhibition

The relevance of selectivity in the clinical application of DPP-IV inhibitors is an essential step in reducing the toxicity associated with the inhibition of DPP8 and DPP9 [45]. Thus, the importance of computational approaches in designing or looking for selective DPP-IV inhibitors has become indispensable [103]. Various in silico methods have been described, mostly supported by docking studies on DPP8 and DPP9 enzymes [101, 103, 104], which could be subsequently followed either by finding molecules that show a significant higher (i.e., more negative) score for DPP-IV than for DPP8/9 [103], or by a 3D-QSAR study that uses the aligned docked poses to build a predictive model [104]. In contrast, it has been recently used as a conformational-free ligand-based methodology (i.e., holographic QSAR or HQSAR) for predicting DPP-IV selectivity [108] that has the advantage that eliminates the need for generation of the putative binding conformations at the different binding sites and their subsequent 3D-structure alignment. HQSAR involves the investigation of important indications of the molecular fragments that are directly related to biological activity or responsible for the low biological potency of the compounds, and this method is used to propose structural modifications. Therefore, contribution maps indicating the individual contributions to the activity of each atom in a given molecule of the data set can be obtained. Additionally, the most relevant structural fragments can be analyzed.

7.3.6 Natural Products as DPP-IV Inhibitors

Dietary intervention is accepted as a key component in the prevention and management of T2DM [109]. Natural products are useful as bioactive components to develop new functional foods for specific population sectors [110–112]. A functional food has been defined as "any modified food or food ingredient that may provide a health benefit beyond the traditional nutrients it contains" [113]. According to the literature, the capacity to inhibit DPP-IV has been identified in natural nonpeptide (Fig. 7.4) [99, 114–125] and peptide products (Table 7.4) [18, 126–142]. Therefore, they could be used as bioactive ingredients in functional foods for T2DM prevention or treatment [99, 126]. These foods may also serve as lead compounds for deriving more potent DPP-IV inhibitors [99, 117, 143].

7.3.6.1 Natural Products of Nonpeptide Nature

There are presently a limited number of DPP-IV inhibitors that have a nonpeptide nature (see Fig. 7.4 for the most relevant examples). Akiyama et al. [144] isolated sulphostin from the culture broth of *Streptomyces sp.* MK251–43F3. This molecule exhibits an antidiabetic activity that is 100-fold stronger than the well-known DPP-IV peptide inhibitor diprotin A [145]. Berberine [115], trigonelline [116], and eight different DPP-IV inhibitors [119] have been isolated from different plants (e.g., *Coptis chinensis, Trigonella foenum-graecum, Bacopa monnieri,* and *Daphne odora*) and are widely used as antihyperglycemic agents in traditional Chinese medicine (TCM). Moreover, curcumin (isolated from the rhizome of the herb *Curcuma longa*), resveratrol, luteolin, apigenin, flavone, and naringin (commonly found in berry wine blends, citrus, berry, grape, and soybean) are plant phenolic compounds that are also DPP-IV inhibitors [117, 121, 122]. Moreover, different natural extracts inhibit DPP-IV, although the specific nonpeptide molecules that are responsible for this bioactivity have not been fully characterized [114, 123–125].

7.3.6.2 Naturally Derived Peptides

Protein–peptide interactions are vital for life because peptides can take part in nearly 40 % of macromolecular interaction-mediating signals [146]. In recent years, studies on peptides derived from food proteins have shown that their bioactivity can significantly improve human health and prevent chronic diseases [126]. These bioactive peptides are short peptide sequences that are typically less than ten amino acids. They are encrypted within the structure of a food protein and can be released by enzyme hydrolysis, microbial fermentation, or physical and chemical processing [18]. The peptides can interact with specific receptors and regulate a variety of physiological functions. Interestingly, peptides offer certain advantages as drugs due to their high biological activity, high specificity, and low toxicity [147].

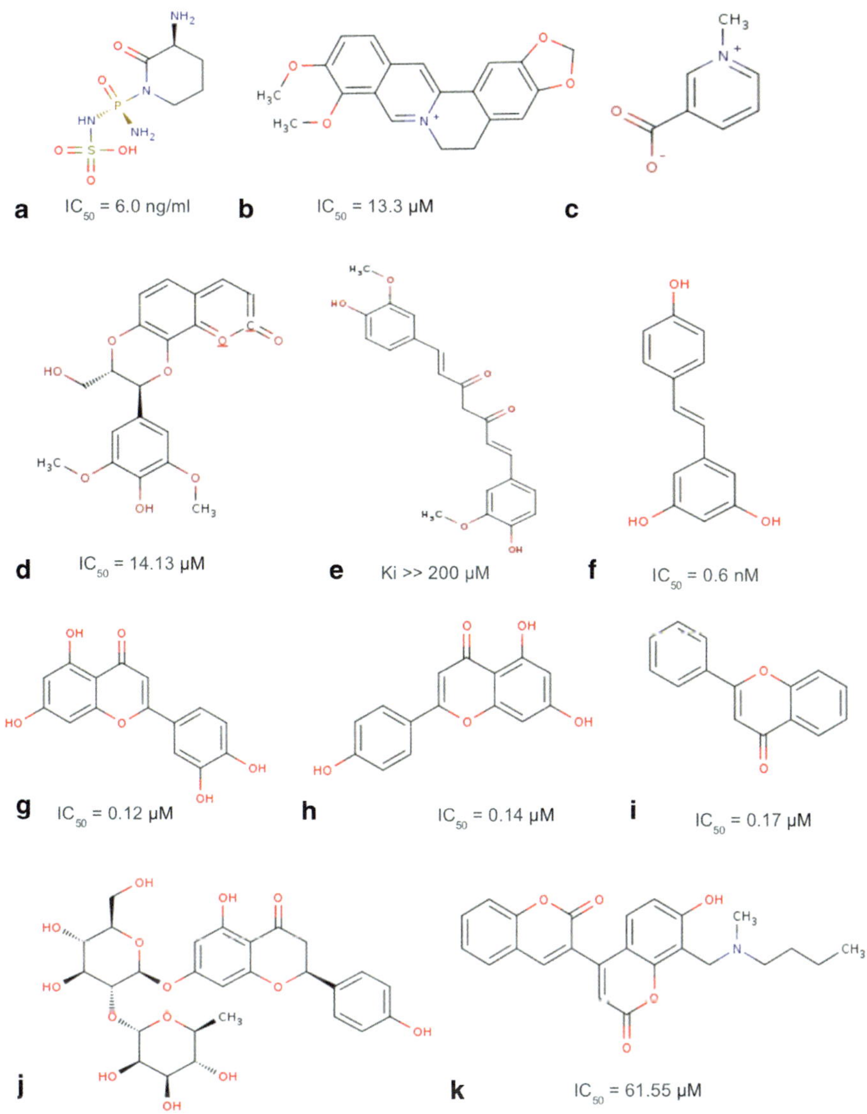

Fig. 7.4 Chemical structures and DPP-IV inhibitory activity for the most relevant natural compounds of non-peptide nature: **a** sulphostin; **b** berberine; **c** trigonelline; **d** compound 4; **e** curcumin; **f** resveratrol; **g** luteolin; **h** apigenin; **i** flavone; **j** naringin; and **k** ZINC02132035

Several recent studies have demonstrated that peptides obtained from proteins from the following sources are able to inhibit DPP-IV: dairy products [126, 127, 129–131, 135, 139–141], defatted rice bran [132], tuna cooking juice [133], dry-cured ham [134], *Amaranthus hypochondriacus* [136], barley [126], canola [126], oat [126], soybean [126], wheat [126], chicken egg [126], bovine meat [126, 142],

Table 7.4 Peptide sequences that inhibit DPP-IV according to the literature

Peptide sequence	IC50 (μM)	Type of inhibition
Ile-*Pro*-Ile (diprotin A)*	3.4–24.7	Competitive
Val-*Pro*-Leu (diprotin B)	5.5	Competitive
Ile-*Pro*-Ile-Gln-Tyr*	35.2	Competitive
Gly-*Pro*-Gly-Ala*	41.9	
Ile-*Pro*-Ala-Val-Phe	44.7	
Leu-Lys-Pro-Thr-Pro-Glu-Gly-Leu-Asp*	45	Un-competitive
Leu-*Pro*-Gln-Asn-Ile-Pro-Pro-Leu	46	
Ile-*Pro*-Ala	49	
Gly-*Pro*-Ala-Glu*	49.6	
Leu-Lys-Pro-Thr-Pro-Glu-Gly-Leu-Asp-Leu-Glu-Ile-Leu*	57	Un-competitive
Trp-Val*	65.69	Non-competitive
Cys-Ala-Tyr-Gln-Trp-Gln-Arg-Pro-Val-Asp-Arg-Ile-Arg*	78	
Leu-*Pro*-Gln	82	
Pro-Ala-Cys-Gly- Gly-Phe-Try-Ile-Ser-Gly-Arg-Pro-Gly*	96.4	
Leu-*Pro*-Tyr-Pro-Tyr *	108.3	Competitive
Val-*Pro*-Ile-Thr-Pro-Thr-Leu	110	
Pro-Gly-Val-Gly-Gly-Pro-Leu-Gly-Pro-Ile-Gly-Pro-Cys-Tyr-Glu*	116.1	
Val-*Pro*-Ile-Thr-Pro-Thr	130	
Trp-Leu-Ala-His-Lys-Ala-Leu-Cys-Ser-Glu-Lys-Leu-Asp-Gln*	141	Un-competitive
Ile-*Pro*-Ala-Val-Phe-Lys	143	
His-Leu*	143.19	Competitive
Ile-*Pro**	149.6	Competitive
Leu-*Pro*-Gln-Asn-Ile-Pro-Pro	160	
Leu-Ala-His-Lys-Ala-Leu-Cys-Ser-Glu-Lys-Leu*	165	Competitive
Thr-Lys-Cys-Glu-Val-Phe-Arg-Glu*	166	Un-competitive
Val-Ala*	168.24	Competitive
Val-Ala-Gly-Thr-Trp-Tyr	174	
Leu-Cys-Ser-Glu-Lys-Leu-Asp-Gln*	186	Non-competitive
Ile-*Pro*-Ala-Val-Phe-Lys-Ile-Asp-Ala*	191	Competitive
Tyr-*Pro*-Tyr-Tyr*	194.4	Competitive
Leu-*Pro*-Leu*	241.4	Competitive
Tyr-*Pro*-Tyr*	243.7	Competitive
Phe-*Pro*-Gly-Pro-Ile-Pro-Asn	260	
Ile-Leu-Asp-Lys-Val-Gly-Ile-Asn-Tyr*	263	Competitive
Trp-Leu-Ala-His-Lys-Ala-Leu*	286	Non-competitive
Thr-*Pro*-Glu-Val-Asp-Asp-Glu-Ala-Leu-Glu-Lys	319.5	
Leu-*Pro*-Leu-Pro-Leu*	325	Competitive
Ile-Val-Gln-Asn-Asn-Asp-Ser-Thr-Glu-Tyr-Gly-Leu-Phe*	337	Non-competitive
Phe-Leu*	399.58	Competitive
Ile-*Pro*	410	Competitive
Val-Leu-Val-Leu-Asp-Thr-Asp-Tyr-Lys	424.4	
Tyr-*Pro**	658.1	Competitive
Tyr-*Pro*-Phe-Pro-Gly-Pro-Ile-Pro-Asn	670	
Leu-*Pro**	712.5	Competitive
Met-*Pro*	870	Competitive
Val-*Pro*	880	Competitive

Table 7.4 (continued)

Peptide sequence	IC50 (μM)	Type of inhibition
Ala-Leu*	882.13	Competitive
Pro-Gly-Pro-Ile-His-Asn-Ser	1000	
Ile-*Pro*-Pro-Leu-Thr-Gln-Thr-Pro-Val	1300	
Pro-Gln-Asn-Ile-Pro-Pro-Leu	1500	
Arg-*Pro*	2240	Competitive
Thr-*Pro*	2370	Competitive
Leu-*Pro*	2370	Competitive
Met*	2381.51	Competitive
Val-*Pro*-Pro-Phe-Ile-Gln-Pro-Glu	2500	
Ser-Leu*	2517.08	Competitive
Lys-*Pro*	2540	Competitive
Gly-Leu*	2615.03	Competitive
His-*Pro*	2820	Competitive
Tyr-*Pro*	3170	Competitive
Glu-Lys*	3216.75	Competitive
Leu*	3419.25	Competitive
Phe-*Pro*	3630	Competitive
Trp*	4280.4	Competitive
Trp-*Pro*	4530	Competitive
Pro-*Pro*	5860	Competitive
Ser-*Pro*	5980	Competitive
Lys-Ala*	6270	
Ala-Ala-Ala-Thr-Pro*	6470	
Ala-*Pro*	7950	Competitive
Ala-Ala-Ala-Ala-Gly*	8130	
Ala-Ala*	9400	
Gly-*Pro**	9690	

Rows are sorted according to increasing IC_{50}. The presence of Pro at the P_1 position of some peptides is highlighted

*The IC_{50} value has been measured with porcine instead of human DPP-IV

and chum and Atlantic salmon (Table 7.4) [126, 138]. They are usually di-, tri-, and oligopeptides that contain proline and/or hydrophobic amino acids within their sequence [126]. Moreover, the sequence of the peptide, not its amino acid composition, influences the DPP-IV inhibitory activity. For instance, the dipeptides Ile–Pro and Trp–Val had DPP-IV inhibitory activity (Table 7.4). However, the reverse peptides Pro–Ile and Val–Trp had no inhibitory activity [130, 132]. Thus, proline is the preferential amino acid residue at the P_1-position. Furthermore, alanine, glycine, and serine are also accepted (Table 7.4). The data in Table 7.4 also show that (a) dipeptides of the general structures Xaa-Pro (except Gly-Pro) are competitive inhibitors of DPP-IV [148], and (b) the residue present at the N-terminus influences inhibitory activity because the dipeptide Leu-Pro has a higher IC_{50} value than Ile-Pro (see Table 7.4) [18].

Longer peptides (larger than 13 residues) have been shown to act as noncompetitive inhibitors by forming interactions at the dimerization interface and blocking the formation of the DPP-IV active dimer [136, 149].

7.4 Using In Silico Tools for Identifying DPP-IV Inhibitors of Natural Origin

The identification of inhibitors with previously undescribed bioactivities in natural extracts exclusively by in vitro or in vivo approaches is a complex and expensive process [114–117, 127, 129–135, 138–140, 144]. The use of in silico approaches can significantly increase this identification of natural extracts. There are successful examples of newly identified DPP-IV inhibitors of natural origin that have been found using either VS workflows [99, 120] or target fishing [119] or sequence similarity tools [126, 141, 142].

7.4.1 Virtual Screening Workflows

7.4.1.1 Defining Virtual Screening Workflows

A VS workflow consists of several sequential *filters* that are used to discern the molecules that share and those that do not share properties that characterize drugs with a specific bioactivity. In a VS workflow, the molecules that survive a filter are then evaluated by the next filter (whereas the rest are rejected). Thus, a VS workflow is described as a funnel shape to indicate the decreasing number of molecules that are evaluated by the successive filters (Fig. 7.5). Some of the most commonly used filters during VS workflows include ADME/Toxicity analysis, protein–ligand docking, pharmacophore matching and similarity/electrostatic comparison (Fig. 7.5) [99, 120].

7.4.1.2 Natural Products Databases

The main goal of using a VS workflow and finding bioactive molecules for functional food design is to find a cheap natural source that can easily provide extracts enriched in the bioactive molecule. Therefore, it is necessary to use databases for naturally occurring molecules that, in addition to showing the molecular structure, include the natural source from which these molecules can be obtained. Examples of such databases are the NuBBE database [150], the TCM database @Taiwan [151], and Reaxys [152].

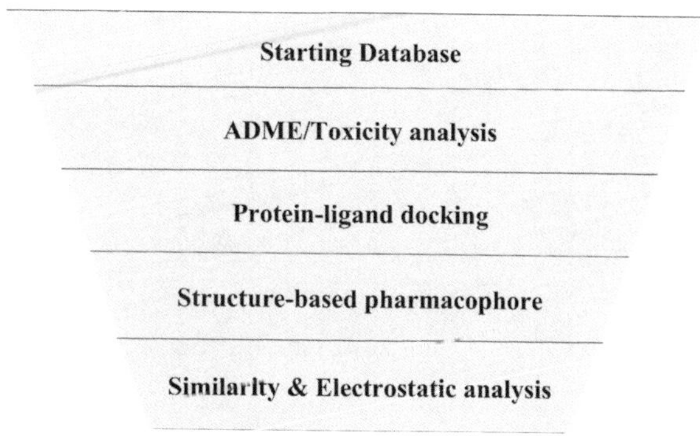

Fig. 7.5 Overview of a typical virtual screening workflow

7.4.1.3 Examples

We have developed a VS workflow to successfully identify molecules that are able to inhibit DPP-IV and molecules that do not inhibit this enzyme [99]. Among other filters, this VS workflow included a structure-based energetic pharmacophore (Fig. 7.3a) that was obtained from the consensus of the different energetic pharmacophores [97] that can be obtained from ten different complexes between human DPP-IV and potent reversible inhibitors (i.e., IC50 values ≤10 nM) of nonpeptide nature available in the PDB [153]. This VS workflow was applied to the *Natural Products* subset of the ZINC database [154]. The results predicted that 446 of the 89,425 molecules present in the database could be potential DPP-IV inhibitors. These 446 molecules were merged with 2,342 known DPP-IV inhibitors, and the resulting set was classified into 50 clusters according to chemical similarity. We found that there were 12 clusters that contained only natural products not previously identified as DPP-IV inhibitors [99]. Nine molecules from 7 of the 12 clusters (from which no antidiabetic activity has been described to date) were selected for in vitro activity testing. The results of the in vitro activity testing showed the following: (a) seven molecules that could be solubilized inhibited DPP-IV, and (b) the most potent compound was ZINC02132035 (with an IC_{50} of 61.55 μM; Fig. 7.4k) [99]. Therefore, we experimentally demonstrated that the VS workflow was able to identify DPP-IV inhibitor molecules that (1) have never been reported to have antidiabetic activity and (2) were not structurally related to any known DPP-IV inhibitor.

We next used a slightly modified version of the VS workflow to evaluate an in-house database of 29,779 natural products annotated with their natural source. We were able to identify 84 molecules (isolated from 95 different natural sources) that were predicted to inhibit DPP-IV [120]. An exhaustive bibliographic search revealed that we predicted 12 potential DPP-IV inhibitors from 12 different plant extracts that are known to have antidiabetic activity (Table 7.5). Six of these 12 molecules are identical or similar to molecules with described antidiabetic activity (although their role as DPP-IV inhibitors has not been suggested as an explanation for their bioactivity; Table 7.5). Therefore, it is plausible that these 12 molecules could be partially responsible for the antidiabetic activity of these extracts through DPP-IV inhibition [120]. In addition, we identified six potential DPP-IV inhibitor molecules from six

Table 7.5 Natural extracts with reported antidiabetic activity that contain molecules predicted to be DPP-IV inhibitors by our VS protocol [120]

Molecule	Name and CAS number (when available)	Extract	Ref. Isolation molecule from extract	Ref. Antidiabetic extract	Ref. Antidiabetic molecule
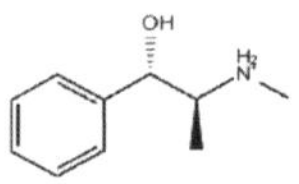	(+)-pseudoephedrine (90-82-4)	*Ephedra alata*	[155]	[156]	[157]
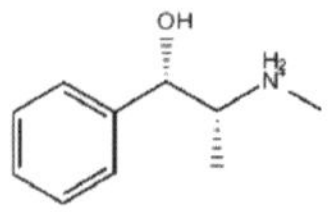	(-)-ephedrine (299-42-3)	*Ephedra distachya*	[158]	[155]	[155]
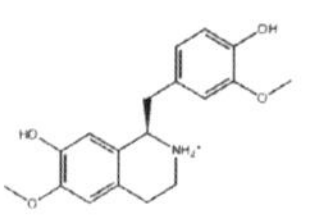	N-nororientalin (29079-44-5)	*Erythrina variegata*	[159]	[160]	[161-163]
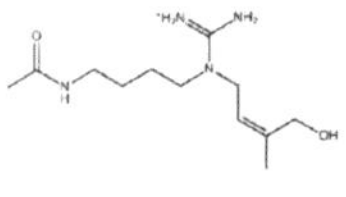	hydroxysmirnovine	*Galega orientalis*	[164]	[165]	
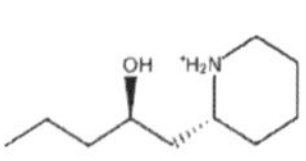	(-)-halosaline (26648-71-5)	*Haloxylon salicornicum*	[166]	[156]	

Table 7.5 (continued)

Molecule	Name and CAS number (when available)	Extract	Ref. Isolation molecule from extract	Ref. Antidiabetic extract	Ref. Antidiabetic molecule
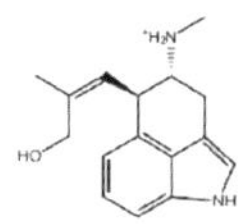	isochanoclavin-(I) (1150-43-2)	*Pennisetum typhoideum*	[167]	[168]	
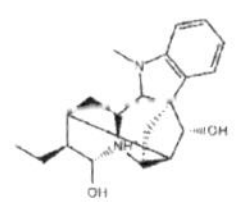	ajmaline (509-37-5)	*Rauwolfia serpentina*	[169]	[170]	
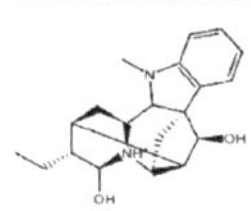	isosandwichine (509-37-5)	*Rauwolfia vomitoria*	[171]	[172]	
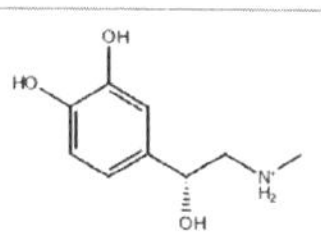	epinephrine (51-43-4)	*Scoparia dulcis*	[173]	[174]	[175]
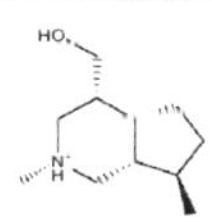	tecostanine	*Tecoma stans*	[176]	[177]	[178]
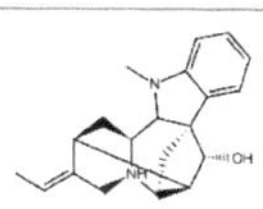	serpinine (509-38-6)	*Vinca major*	[167]	[179]	
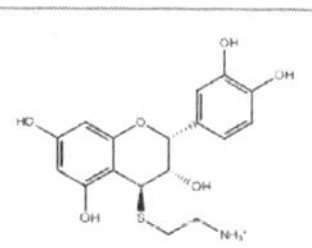	epicatechin derivate	*Vitis vinifera*	[180]	[181]	[182]

The first column shows the 2D structure of each molecule. The second column shows the corresponding common name and the CAS number (when available). The third column shows the scientific name of one of the sources in which the antidiabetic activity has been reported (rows in that table are alphabetically sorted based on this column). Bibliographic references for each molecule are divided into three columns in which (a) the first column presents studies that describe the purification of the molecule from the corresponding extract, (b) the second column lists studies that describe the antidiabetic activity of the corresponding extract; and (c) the third column lists studies, when available, that describe the antidiabetic activity of the corresponding molecule or one that is very similar to it different plants with no described antidiabetic activity. These molecules share the same *genus* as plants with known antidiabetic properties (thus suggesting that they could be new sources for antidiabetic extracts; Table 7.6). Moreover, none of the 18 molecules that we predicted as DPP-IV inhibitors exhibits chemical similarity with any previously known DPP-IV inhibitor [120]. Finally, the same study also predicted 77 other sources with no described antidiabetic activity that contain at least one VS hit. Consequently, this work will permit the discovery of new antidiabetic extracts of natural origin that could be of use in the design of functional foods aimed at preventing/treating T2DM [120].

7.4.2 Target Fishing

7.4.2.1 Defining Target Fishing

Target fishing refers to a computer-assisted methodology used to predict the targets of a specific compound (or a limited set of compounds). Therefore, it can be considered the inverse process of a usual VS workflow. Target fishing has applications in drug repositioning [198] and anticipating potential side effects [199]. Other common synonymous for target fishing are chemogenomics [200], drug repurposing [201], polypharmacology [202], virtual target screening [203], and target profiling [204].

7.4.2.2 Examples

The potential drug target database (PDTD) [205] was searched using the TarFisDock server [206] to identify putative targets for a collection of 19 natural products obtained from *Bacopa monnieri* (L.) Wettst and *Daphne odora* Thunb. var. *marginata* (two plants commonly used by TCM and Ayurvedic medicine in diabetes and inflammation treatment) [119]. This study predicted that from more than 800 drug targets available at PDTD, DPP-IV was one of the most probable for these 19 molecules (consistent with the known therapeutic indications of both plants). Furthermore, an in vitro analysis of the bioactivity of these 19 molecules showed that five have moderate inhibitory activities for DPP-IV (with IC_{50} values ranging from 14.13 to 113.76 μM) [119]. Subsequently, these five molecules were used to identify 27 analogs in the in-house natural products database of the researchers. The in vitro analysis of the bioactivity of these 27 molecules showed that 13 have moderate inhibitory activities for DPP-IV (with IC_{50} values ranging from 22.39 to 87.72 μM) [119].

7.4.3 Sequence Similarity

The aim of these kind of studies consist in performing an in silico evaluation of dietary proteins as potential precursors of biologically active peptides, as well as to

Table 7.6 Natural extracts with no described antidiabetic activity (but from the same *genus* as plants with extracts with described anti-diabetic activity) that contain molecules that are predicted to be DPP-IV inhibitors by our VS protocol [120]

Molecule	CAS number or name	Extract	Ref. Isolation molecule from extract	Extract with antidiabetic activity described	Ref. Antidiabetic extract
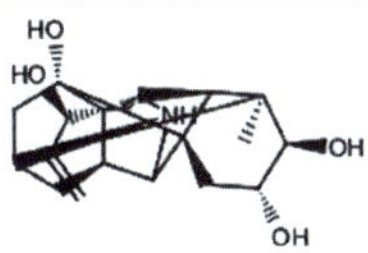	30373-79-6	*Aconitum japonicum*	[183]	*Aconitum carmichaelii*	[184]
				Aconitum moschatum	[185]
				Aconitum violaceum	[185]
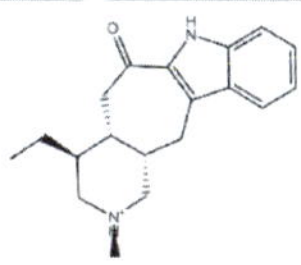	episilicine	*Ervatamia officinalis*	[186]	*Ervatamia microphylla*	[187]
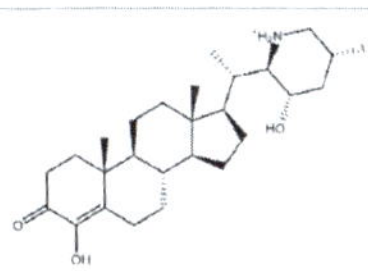	solanudine	*Solanum nudum*	[188]	*Solanum lycocarpum*	[189]
				Solanum nigrum	[190]
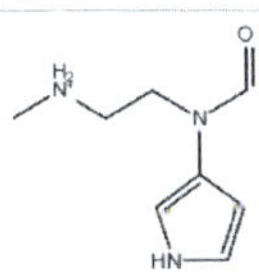		*Solanum sodomaeum*	[191]	*Solanum xanthocarpum*	[192]
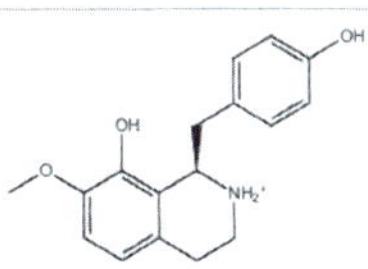	norjuziphine	*Stephania cepharantha*	[193]	*Stephania hernandifolia*	[194]
				Stephania glabra	[195]
				Stephania tetrandra	[196]
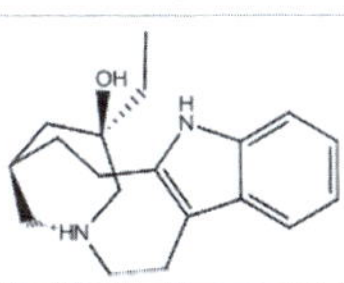	19637-92-4	*Tabernaemontana eglandulosa*	[197]	*Tabernaemontana divaricata*	[187]

The first column shows the 2D structure of each molecule. The second column shows the corresponding common name and/or the CAS number (when available). The third column lists the source from which the VS hits have been purified (rows in that table are alphabetically sorted based on this column). The fourth column lists the studies that describe the purification of the each molecule from the corresponding extract. The fifth column shows the extracts from the same *genus* where the antidiabetic activity has been described. Finally, the last column lists studies that describe the antidiabetic activity of the corresponding extract determine whether such peptides can be released by selected proteolytic enzymes [126, 141, 142]. This approach finds biologically active peptides in the protein sequences that remain inactive in precursor protein sequences. However, when released by proteolytic enzymes, these peptides may interact with selected receptors and regulate physiological functions [141]. Thus, the potential of various dietary proteins to serve as DPP-IV inhibitor precursors is predicted by searching for fragments within the protein chains that match the peptide sequences reported in the literature (Table 7.4) to present an inhibitory activity against DPP-IV. This potential is quantified for each protein by calculating A (the occurrence frequency) as $A = a/N$ (where a is the number of peptides with DPP-IV inhibitory activity within the protein chain and N is the number of amino acid residues in the protein chain) [141]. These studies show that β-casein from cow's milk, collagens from bovine meat, and chum salmon have occurrence frequency values of 0.249, 0.380, and 0.305, respectively, and appeared to be the best potential sources of DPP-IV inhibitory peptides among all of the proteins studied [126, 141]. Moreover, it is also shown that DPP-IV inhibitory peptides can be obtained from milk proteins by using serine endopeptidases (e.g., proteinase K, EC.3.4.21.14; pancreatic elastase, EC 3.4.21.36; prolyl oligopeptidase, EC 3.4.21.26; chymotrypsin C, EC 3.4.21.2; and leukocyte elastase, EC 3.4.21.37) or cysteine endopeptidases (papain, EC 3.4.22.2; ficin, EC 3.4.22.3; and bromelain, EC 3.4.22.4) or thermolysin (EC 3.4.24.27). [141] These proteins also hold special interest for the food industry because proteins from the connective tissue (usually with low commercial value) are rich in proline. Therefore, they can be a very important source for DPP-IV inhibitors (Table 7.4) and may represent a new method of generating profit from food industry byproducts.

7.5 Concluding Remarks and Future Perspectives

DPP-IV inhibition appears to be one of the most effective and secure ways of controlling diabetes and related diseases. Three of the seven gliptins that are currently authorized for human use have been released to the market over the last 2 years (Table 7.2). Moreover, DPP-IV inhibitors are orally administered, which makes them compatible with the food additive concept. Therefore, finding naturally available molecules with bioactivity is an area of high interest for the functional food and nutraceutical industry. VS is an essential (and low-cost) tool for predicting new DPP-IV inhibitors from natural molecule databases and recovering them from food-processing byproducts or biomass with low- or no-economic value. Nevertheless, there are some key points that, in our opinion, could improve the performance of VS on DPP-IV and that need to be addressed in future research: (1) including di- and tripeptides in VS studies; (2) improving VS filters to remove molecules that could inhibit FAP, DPP8, or DPP9; and (3) using the dimerization area as the part of the target where ligand binding is predicted during VS. Our lab is making progress in addressing these challenges and has promising results that will be published elsewhere.

References

1. International Diabetes Federation (2013) IDF Diabetes Atlas, 6th edn. Brussels, Belgium: International Diabetes Federation, http://www.idf.org/diabetesatlas
2. Daousi C, Casson IF, Gill GV, MacFarlane IA, Wilding JPH, Pinkney JH (2006) Prevalence of obesity in type 2 diabetes in secondary care: association with cardiovascular risk factors. Postgrad Med J 82:280–284
3. UK Prospective Diabetes Study (UKPDS) Group (1998) Intensive blood-glucose control with sulphonylureas or insulin compared with conventional treatment and risk of complications in patients with type 2 diabetes (UKPDS 33). Lancet 352:837–853
4. Kahn SE, Haffner SM, Heise MA et al (2006) Glycemic durability of rosiglitazone, metformin, or glyburide monotherapy. N Engl J Med 355:2427–2443
5. Ross SA, Dzida G, Vora J, Khunti K, Kaiser M, Ligthelm RJ (2011) Impact of weight gain on outcomes in type 2 diabetes. Curr Med Res Opin 27:1431–1438
6. Jacobson AM (2004) Impact of improved glycemic control on quality of life in patients with diabetes. Endocr Pract 10:502–508
7. International Diabetes Federation. IDF diabetes atlas. http://www.idf.org/diabetesatlas. Accessed 15 Aug 2013
8. World Health Organization. Diabetes programme. http://www.who.int/diabetes/en/. Accessed 15 Aug 2013
9. Morrish NJ, Wang SL, Stevens LK, Fuller JH, Keen H (2001) Mortality and causes of death in the WHO multinational study of vascular disease in diabetes. Diabetologia 44(Suppl 2):S14–S21
10. World Health Organization (2011). Global status report on noncommunicable diseases 2010. http://www.who.int/nmh/publications/ncd_report2010/en/. Accessed 15 Aug 2013
11. Roglic G, Unwin N, Bennett PH, Mathers C, Tuomilehto J, Nag S, Connolly V, King H (2005) The burden of mortality attributable to diabetes: realistic estimates for the year 2000. Diabetes Care 28:2130–2135
12. World Health Organization (2011). Prevention of blindness and visual impairment. Action plan for the prevention of avoidable blindness. Global data on visual impairment 2010. http://www.who.int/entity/blindness/GLOBALDATAFINALforweb.pdf. Accessed 15 Aug 2013
13. Guthrie RM (2012) Evolving therapeutic options for type 2 diabetes mellitus: an overview. Postgrad Med 124:82–89
14. US Food and Drug Administration (2008). Guidance for industry. Diabetes mellitus—evaluating cardiovascular risk in new anti-diabetic therapies to treat type 2 diabetes. http://www.fda.gov/downloads/Drugs/GuidanceComplianceRegulatoryInformation/Guid-ances/ucm071627.pdf. Accessed 15 Aug 2013
15. Nathan DM, Buse JB, Davidson MB, Ferrannini E, Holman RR, Sherwin R, Zinman B (2009) Medical management of hyperglycemia in type 2 diabetes: a consensus algorithm for the initiation and adjustment of therapy: a consensus statement of the American Diabetes Association and the European Association for the Study of Diabetes. Diabetes Care 32:193–203
16. Hopsu-Havu VK, Sarimo SR (1967) Purification and characterization of an aminopeptidase hydrolyzing glycyl-proline-naphthylamide. Hoppe Seylers Z Physiol Chem 348:1540–1550
17. Rawlings ND, Tolle DP, Barrett AJ (2004) MEROPS: the peptidase database. Nucleic Acids Res 32:D160–D164
18. Power O, Nongonierma AB, Jakeman P, Fitzgerald RJ (2013) Food protein hydrolysates as a source of dipeptidyl peptidase IV inhibitory peptides for the management of type 2 diabetes. Proc Nutr Soc 73:34–46
19. Mendieta L, Tarrago T, Giralt E (2011) Recent patents of dipeptidyl peptidase IV inhibitors. Expert Opin Ther Pat 21:1693–1741
20. Gorrell MD (2005) Dipeptidyl peptidase IV and related enzymes in cell biology and liver disorders. Clin Sci (Lond) 108:277–292
21. Juillerat-Jeanneret L (2014) Dipeptidyl peptidase IV and its inhibitors: therapeutics for type 2 diabetes and what else? J Med Chem 57:2197–2212

22. Mentlein R (1999) Dipeptidyl-peptidase IV (CD26)–role in the inactivation of regulatory peptides. Regul Pept 85:9–24
23. Nabeno M, Akahoshi F, Kishida H, Miyaguchi I, Tanaka Y, Ishii S, Kadowaki T (2013) A comparative study of the binding modes of recently launched dipeptidyl peptidase IV inhibitors in the active site. Biochem Biophys Res Commun 434:191–196
24. Thoma R, Löffler B, Stihle M, Huber W, Ruf A, Hennig M (2003) Structural basis of proline-specific exopeptidase activity as observed in human dipeptidyl peptidase-IV. Structure 11:947–959
25. Doherty AM, Bock MG, Desai MC, Overington J, Plattner JJ, Stamford A, Wustrow D, Young H, Gwaltney SL, Stafford JA (2005) Inhibitors of dipeptidyl peptidase 4. Annu Rep Med Chem 40:149–165
26. Chien C-H, Huang L-H, Chou C-Y, Chen Y-S, Han Y-S, Chang G-G, Liang P-H, Chen X (2004) One site mutation disrupts dimer formation in human DPP-IV proteins. J Biol Chem 279:52338–52345
27. Engel M, Hoffmann T, Wagner L, Wermann M, Heiser U, Kiefersauer R, Huber R, Bode W, Demuth H-U, Brandstetter H (2003) The crystal structure of dipeptidyl peptidase IV (CD26) reveals its functional regulation and enzymatic mechanism. Proc Natl Acad Sci U S A 100:5063–5068
28. Pederson RA, White HA, Schlenzig D, Pauly RP, McIntosh CH, Demuth HU (1998) Improved glucose tolerance in Zucker fatty rats by oral administration of the dipeptidyl peptidase IV inhibitor isoleucine thiazolidide. Diabetes 47:1253–1258
29. Pospisilik JA, Stafford SG, Demuth H-U, McIntosh CHS, Pederson RA (2002) Long-term treatment with dipeptidyl peptidase IV inhibitor improves hepatic and peripheral insulin sensitivity in the VDF Zucker rat: a euglycemic-hyperinsulinemic clamp study. Diabetes 51:2677–2683
30. Cheng JD, Dunbrack RL, Valianou M, Rogatko A, Alpaugh RK, Weiner LM (2002) Promotion of tumor growth by murine fibroblast activation protein, a serine protease, in an animal model. Cancer Res 62:4767–4772
31. Kajiyama H, Kikkawa F, Suzuki T, Shibata K, Ino K, Mizutani S (2002) Prolonged survival and decreased invasive activity attributable to dipeptidyl peptidase IV overexpression in ovarian carcinoma. Cancer Res 62:2753–2757
32. Ho L, Aytac U, Stephens LC et al (2001) In vitro and in vivo antitumor effect of the anti-CD26 monoclonal antibody 1F7 on human CD30+ anaplastic large cell T-cell lymphoma Karpas 299. Clin Cancer Res 7:2031–2040
33. Ussher JR, Sutendra G, Jaswal JS (2012) The impact of current and novel anti-diabetic therapies on cardiovascular risk. Future Cardiol 8:895–912
34. Zhong J, Rao X, Rajagopalan S (2013) An emerging role of dipeptidyl peptidase 4 (DPP4) beyond glucose control: potential implications in cardiovascular disease. Atherosclerosis 226:305–314
35. Patil HR, Al Badarin FJ, Al Shami HA, Bhatti SK, Lavie CJ, Bell DSH, O'Keefe JH (2012) Meta-analysis of effect of dipeptidyl peptidase-4 inhibitors on cardiovascular risk in type 2 diabetes mellitus. Am J Cardiol 110:826–833
36. Frederich R, Alexander JH, Fiedorek FT, Donovan M, Berglind N, Harris S, Chen R, Wolf R, Mahaffey KW (2010) A systematic assessment of cardiovascular outcomes in the saxagliptin drug development program for type 2 diabetes. Postgrad Med 122:16–27
37. Scheen AJ (2013) Cardiovascular effects of gliptins. Nat Rev Cardiol 10:73–84
38. Simsek S, de Galan BE (2012) Cardiovascular protective properties of incretin-based therapies in type 2 diabetes. Curr Opin Lipidol 23:540–547
39. Dai Y, Dai D, Mercanti F, Ding Z, Wang X, Mehta JL (2013) Dipeptidyl peptidase-4 inhibitors in cardioprotection: a promising therapeutic approach. Acta Diabetol 50:827–835
40. Scheen AJ (2013) Cardiovascular effects of dipeptidyl peptidase-4 inhibitors: from risk factors to clinical outcomes. Postgrad Med 125:7–20
41. Yousefzadeh P, Wang X (2013) The effects of dipeptidyl peptidase-4 inhibitors on cardiovascular disease risks in type 2 diabetes mellitus. J Diabetes Res 2013:459821

42. Balakumar P, Dhanaraj SA (2013) Cardiovascular pleiotropic actions of DPP-4 inhibitors: a step at the cutting edge in understanding their additional therapeutic potentials. Cell Signal 25:1799–1803
43. Wang XM, Yao T-W, Nadvi NA, Osborne B, McCaughan GW, Gorrell MD (2008) Fibroblast activation protein and chronic liver disease. Front Biosci 13:3168–3180
44. Kirby M, Yu DMT, O'Connor S, Gorrell MD (2010) Inhibitor selectivity in the clinical application of dipeptidyl peptidase-4 inhibition. Clin Sci (Lond) 118:31–41
45. Lankas GR, Leiting B, Roy RS et al (2005) Dipeptidyl peptidase IV inhibition for the treatment of type 2 diabetes: potential importance of selectivity over dipeptidyl peptidases 8 and 9. Diabetes 54:2988–2994
46. Deacon CF, Ahrén B (2011) Physiology of incretins in health and disease. Rev Diabet Stud 8:293–306
47. Tortosa F, Dotta F (2013) Incretin hormones and beta-cell mass expansion: what we know and what is missing? Arch Physiol Biochem 119:161–169
48. Ahrén B (2013) Incretin dysfunction in type 2 diabetes: clinical impact and future perspectives. Diabetes Metab 39:195–201
49. Opinto G, Natalicchio A, Marchetti P (2013) Physiology of incretins and loss of incretin effect in type 2 diabetes and obesity. Arch Physiol Biochem 119:170–178
50. Brunton S (2013) Integrating incretin-based therapy into type 2 diabetes management. Vital Signs 62:S1–S8
51. Papamargaritis D, Miras AD, le Roux CW (2013) Influence of diabetes surgery on gut hormones and incretins. Nutr Hosp 28(Suppl 2):95–103
52. Meier JJ, Nauck MA, Schmidt WE, Gallwitz B (2002) Gastric inhibitory polypeptide: the neglected incretin revisited. Regul Pept 107:1–13
53. Green BD, Flatt PR, Bailey CJ (2006) Inhibition of dipeptidylpeptidase IV activity as a therapy of type 2 diabetes. Expert Opin Emerg Drugs 11:525–539
54. Lindgren O, Mari A, Deacon CF, Carr RD, Winzell MS, Vikman J, Ahrén B (2009) Differential islet and incretin hormone responses in morning versus afternoon after standardized meal in healthy men. J Clin Endocrinol Metab 94:2887–2892
55. Ahrén B, Carr RD, Deacon CF (2010) Incretin hormone secretion over the day. Vitam Horm 84:203–220
56. Zettl H, Schubert-Zsilavecz M, Steinhilber D (2010) Medicinal chemistry of incretin mimetics and DPP-4 inhibitors. ChemMedChem 5:179–185
57. Drucker DJ, Nauck MA (2006) The incretin system: glucagon-like peptide-1 receptor agonists and dipeptidyl peptidase-4 inhibitors in type 2 diabetes. Lancet 368:1696–1705
58. Holst JJ, Vilsbøll T, Deacon CF (2009) The incretin system and its role in type 2 diabetes mellitus. Mol Cell Endocrinol 297:127–136
59. Holst JJ, Deacon CF (2004) Glucagon-like peptide 1 and inhibitors of dipeptidyl peptidase IV in the treatment of type 2 diabetes mellitus. Curr Opin Pharmacol 4:589–596
60. Baggio LL, Drucker DJ (2007) Biology of incretins: GLP-1 and GIP. Gastroenterology 132:2131–2157
61. Drucker DJ (2003) Therapeutic potential of dipeptidyl peptidase IV inhibitors for the treatment of type 2 diabetes. Expert Opin Investig Drugs 12:87–100
62. Højberg PV, Vilsbøll T, Rabøl R, Knop FK, Bache M, Krarup T, Holst JJ, Madsbad S (2009) Four weeks of near-normalisation of blood glucose improves the insulin response to glucagon-like peptide-1 and glucose-dependent insulinotropic polypeptide in patients with type 2 diabetes. Diabetologia 52:199–207
63. Hansen KB, Vilsbøll T, Bagger JI, Holst JJ, Knop FK (2012) Impaired incretin-induced amplification of insulin secretion after glucose homeostatic dysregulation in healthy subjects. J Clin Endocrinol Metab 97:1363–1370
64. Demuth H-U, McIntosh CHS, Pederson RA (2005) Type 2 diabetes–therapy with dipeptidyl peptidase IV inhibitors. Biochim Biophys Acta 1751:33–44
65. Kim S-H, Lee S-H, Yim H-J (2013) Gemigliptin, a novel dipeptidyl peptidase 4 inhibitor: first new anti-diabetic drug in the history of Korean pharmaceutical industry. Arch Pharm Res 36:1185–1188

66. US National Library of Medicine. National Institutes of Health. MedlinePlus (2014). Sitagliptin. http://www.nlm.nih.gov/medlineplus/druginfo/meds/a606023.html. Accessed 21 Nov 2013
67. US National Library of Medicine. National Institutes of Health. MedlinePlus (2014). Saxagliptin. http://www.nlm.nih.gov/medlineplus/druginfo/meds/a610003.html. Accessed 21 Nov 2013
68. US National Library of Medicine. National Institutes of Health. MedlinePlus (2014). Linagliptin. http://www.nlm.nih.gov/medlineplus/druginfo/meds/a611036.html. Accessed 21 Nov 2013
69. Noel RA, Braun DK, Patterson RE, Bloomgren GL (2009) Increased risk of acute pancreatitis and biliary disease observed in patients with type 2 diabetes: a retrospective cohort study. Diabetes Care 32:834–838
70. Engel SS, Williams-Herman DE, Golm GT, Clay RJ, Machotka S V, Kaufman KD, Goldstein BJ (2010) Sitagliptin: review of preclinical and clinical data regarding incidence of pancreatitis. Int J Clin Pract 64:984–990
71. Williams-Herman D, Engel SS, Round E, Johnson J, Golm GT, Guo H, Musser BJ, Davies MJ, Kaufman KD, Goldstein BJ (2010) Safety and tolerability of sitagliptin in clinical studies: a pooled analysis of data from 10,246 patients with type 2 diabetes. BMC Endocr Disord 10:7
72. Engel SS, Round E, Golm GT, Kaufman KD, Goldstein BJ (2013) Safety and tolerability of sitagliptin in type 2 diabetes: pooled analysis of 25 clinical studies. Diabetes Ther 4:119–145
73. Monami M, Dicembrini I, Mannucci E (2014) Dipeptidyl peptidase-4 inhibitors and pancreatitis risk: a meta-analysis of randomized clinical trials. Diabetes Obes Metab 16:48–56
74. Scheen A (2013) Gliptins (dipeptidyl peptidase-4 inhibitors) and risk of acute pancreatitis. Expert Opin Drug Saf 12:545–557
75. Deacon CF, Holst JJ (2013) Dipeptidyl peptidase-4 inhibitors for the treatment of type 2 diabetes: comparison, efficacy and safety. Expert Opin Pharmacother 14:2047–2058
76. Zanchi A, Lehmann R, Philippe J (2012) Anti-diabetic drugs and kidney disease–recommendations of the Swiss Society for Endocrinology and Diabetology. Swiss Med Wkly 142:w13629
77. Ramirez G, Morrison AD, Bittle PA (2013) Clinical practice considerations and review of the literature for the use of DPP-4 inhibitors in patients with type 2 diabetes and chronic kidney disease. Endocr Pract 19:1025–1034
78. Kuhn B, Hennig M, Mattei P (2007) Molecular recognition of ligands in dipeptidyl peptidase IV. Curr Top Med Chem 7:609–619
79. Engel M, Hoffmann T, Manhart S, Heiser U, Chambre S, Huber R, Demuth H-U, Bode W (2006) Rigidity and flexibility of dipeptidyl peptidase IV: crystal structures of and docking experiments with DPIV. J Mol Biol 355:768–783
80. Li C, Shen J, Li W, Lu C (2011) Possible ligand release pathway of dipeptidyl peptidase IV investigated by molecular dynamics simulations. Proteins Struct Funct Bioinforma 79:1800–1809
81. Schechter I, Berger A (2012) On the size of the active site in proteases. I. Papain. 1967. Biochem Biophys Res Commun 425:497–502
82. Weber AE (2004) Dipeptidyl peptidase IV inhibitors for the treatment of diabetes. J Med Chem 47:4135–4141
83. Wallace MB, Feng J, Zhang Z, Skene RJ, Shi L, Caster CL, Kassel DB, Xu R, Gwaltney SL (2008) Structure-based design and synthesis of benzimidazole derivatives as dipeptidyl peptidase IV inhibitors. Bioorg Med Chem Lett 18:2362–2367
84. Patel B, Ghate M (2013) Computational studies on structurally diverse dipeptidyl peptidase IV inhibitors: an approach for new anti-diabetic drug development. Med Chem Res 22:4505–4521
85. Al-Masri IM, Mohammad MK, Taha MO (2008) Discovery of DPP IV inhibitors by pharmacophore modeling and QSAR analysis followed by in silico screening. ChemMedChem 3:1763–1779
86. Aertgeerts K, Ye S, Tennant MG, Kraus ML, Rogers J, Sang B-C, Skene RJ, Webb DR, Prasad GS (2004) Crystal structure of human dipeptidyl peptidase IV in complex with a decapeptide reveals details on substrate specificity and tetrahedral intermediate formation. Protein Sci 13:412–421
87. Bjelke JR, Christensen J, Branner S, Wagtmann N, Olsen C, Kanstrup AB, Rasmussen HB (2004) Tyrosine 547 constitutes an essential part of the catalytic mechanism of dipeptidyl peptidase IV. J Biol Chem 279:34691–34697

88. Yoshida T, Akahoshi F, Sakashita H, et al (2012) Discovery and preclinical profile of teneligliptin (3-[(2S,4S)-4-[4-(3-methyl-1-phenyl-1H-pyrazol-5-yl)piperazin-1-yl]pyrrolidin-2-ylcarbonyl]thiazolidine): a highly potent, selective, long-lasting and orally active dipeptidyl peptidase IV inhibitor for t. Bioorg Med Chem 20:5705–5719
89. Yoshida T, Akahoshi F, Sakashita H, Sonda S, Takeuchi M, Tanaka Y, Nabeno M, Kishida H, Miyaguchi I, Hayashi Y (2012) Fused bicyclic heteroarylpiperazine-substituted L-prolylthiazolidines as highly potent DPP-4 inhibitors lacking the electrophilic nitrile group. Bioorg Med Chem 20:5033–5041
90. Edmondson SD, Mastracchio A, Cox JM et al (2009) Aminopiperidine-fused imidazoles as dipeptidyl peptidase-IV inhibitors. Bioorg Med Chem Lett 19:4097–4101
91. Edmondson SD, Mastracchio A, Mathvink RJ et al (2006) (2S,3S)-3-Amino-4-(3,3-difluoropyrrolidin-1-yl)-N, N-dimethyl-4-oxo-2-(4-[1,2,4]triazolo[1,5-a]-pyridin-6-ylphenyl)butanamide: a selective alpha-amino amide dipeptidyl peptidase IV inhibitor for the treatment of type 2 diabetes. J Med Chem 49:3614–3627
92. Edmondson SD, Wei L, Xu J et al (2008) Fluoroolefins as amide bond mimics in dipeptidyl peptidase IV inhibitors. Bioorg Med Chem Lett 18:2409–2413
93. Biftu T, Scapin G, Singh S et al (2007) Rational design of a novel, potent, and orally bioavailable cyclohexylamine DPP-4 inhibitor by application of molecular modeling and X-ray crystallography of sitagliptin. Bioorg Med Chem Lett 17:3384–3387
94. Eckhardt M, Langkopf E, Mark M et al (2007) 8-(3-®-aminopiperidin-1-yl)-7-but-2-ynyl-3-methyl-1-(4-methyl-quinazolin-2-ylmethyl)-3,7-dihydropurine-2,6-dione (BI 1356), a highly potent, selective, long-acting, and orally bioavailable DPP-4 inhibitor for the treatment of type 2 diabetes. J Med Chem 50:6450–6453
95. Kaelin DE, Smenton AL, Eiermann GJ et al (2007) 4-arylcyclohexylalanine analogs as potent, selective, and orally active inhibitors of dipeptidyl peptidase IV. Bioorg Med Chem Lett 17:5806–5811
96. Nordhoff S, Cerezo-Gálvez S, Deppe H, Hill O, López-Canet M, Rummey C, Thiemann M, Matassa VG, Edwards PJ, Feurer A (2009) Discovery of beta-homophenylalanine based pyrrolidin-2-ylmethyl amides and sulfonamides as highly potent and selective inhibitors of dipeptidyl peptidase IV. Bioorg Med Chem Lett 19:4201–4203
97. Salam NK, Nuti R, Sherman W (2009) Novel method for generating structure-based pharmacophores using energetic analysis. J Chem Inf Model 49:2356–2368
98. Loving K, Salam NK, Sherman W (2009) Energetic analysis of fragment docking and application to structure-based pharmacophore hypothesis generation. J Comput Aided Mol Des 23:541–554
99. Guasch L, Ojeda MJ, González-Abuín N et al (2012) Identification of novel human dipeptidyl peptidase-IV inhibitors of natural origin (part I): virtual screening and activity assays. PLoS One 7:e44971
100. Rummey C, Metz G (2007) Homology models of dipeptidyl peptidases 8 and 9 with a focus on loop predictions near the active site. Proteins 66:160–171
101. Janardhan S, Reddy YP (2011) Homology modeling and molecular docking studies of human DPP8 and DPP9. Int J Pharma Res Dev 2:131–146
102. Pitman MR, Menz RI, Abbott CA (2006) Prediction of dipeptidyl peptidase (DP) 8 structure by homology modelling. Adv Exp Med Biol 575:33–42
103. Tanwar O, Deora GS, Tanwar L, Kumar G, Janardhan S, Alam MM, Shaquiquzzaman M, Akhter M (2014) Novel hydrazine derivatives as selective DPP-IV inhibitors: findings from virtual screening and validation through molecular dynamics simulations. J Mol Model 20:2118
104. Kang NS, Ahn JH, Kim SS, Chae CH, Yoo S-E (2007) Docking-based 3D-QSAR study for selectivity of DPP4, DPP8, and DPP9 inhibitors. Bioorg Med Chem Lett 17:3716–3721
105. Patel BD, Ghate MD (2014) Recent approaches to medicinal chemistry and therapeutic potential of dipeptidyl peptidase-4 (DPP-4) inhibitors. Eur J Med Chem 74:574–605
106. Ghate M, Jain SV (2013) Structure based lead optimization approach in discovery of selective DPP4 inhibitors. Mini Rev Med Chem 13:888–914

107. Fukuda-Tsuru S, Anabuki J, Abe Y, Yoshida K, Ishii S (2012) A novel, potent, and long-lasting dipeptidyl peptidase-4 inhibitor, teneligliptin, improves postprandial hyperglycemia and dyslipidemia after single and repeated administrations. Eur J Pharmacol 696:194–202
108. Ghate M, Jain S (2014) Fragment based HQSAR modeling and docking analysis of conformationally rigid 3-azabicyclo hexane derivatives to design selective DPP-4 inhibitors. Lett Drug Des Discov 11:184–198
109. American Diabetes Association (2014) Standards of medical care in diabetes–2014. Diabetes Care 37(Suppl 1):S14–S80
110. Rollinger JM, Stuppner H, Langer T (2008) Virtual screening for the discovery of bioactive natural products. Prog drug Res 65:211, 213–249
111. Schuster D, Wolber G (2010) Identification of bioactive natural products by pharmacophore-based virtual screening. Curr Pharm Des 16:1666–1681
112. Martinez-Mayorga K, Medina-Franco JL (2009) Chemoinformatics-applications in food chemistry. Adv Food Nutr Res 58:33–56
113. Ferguson LLR (2009) Nutrigenomics approaches to functional foods. J Am Diet Assoc 109:452–458
114. Pascual I, Lopéz A, Gómez H, Chappé M, Saroyán A, González Y, Cisneros M, Charli JL, Chávez MDLA (2007) Screening of inhibitors of porcine dipeptidyl peptidase IV activity in aqueous extracts from marine organisms. Enzyme Microb Technol 40:414–419
115. Al-masri IM, Mohammad MK, Tahaa MO (2009) Inhibition of dipeptidyl peptidase IV (DPP IV) is one of the mechanisms explaining the hypoglycemic effect of berberine. J Enzyme Inhib Med Chem 24:1061–1066
116. Hamden K, Bengara A, Amri Z, Elfeki A (2013) Experimental diabetes treated with trigonelline: effect on key enzymes related to diabetes and hypertension, β-cell and liver function. Mol Cell Biochem 381:85–94
117. Antonyan A, De A, Vitali L, Pettinari R, Marchetti F, Gigliobianco MR, Pettinari C, Camaioni E, Lupidi G (2014) Evaluation of (arene)Ru(II) complexes of curcumin as inhibitors of dipeptidyl peptidase IV. Biochimie 99:146–152
118. González-Abuín N, Martínez-Micaelo N, Blay M, Pujadas G, Garcia-Vallvé S, Pinent M, Ardévol A (2012) Grape seed-derived procyanidins decrease dipeptidyl-peptidase 4 activity and expression. J Agric Food Chem 60:9055–9061
119. Zhang S, Lu W, Liu X, Diao Y, Bai F, Wang L, Shan L, Huang J, Li H, Zhang W (2011) Fast and effective identification of the bioactive compounds and their targets from medicinal plants via computational chemical biology approach. MedChemComm 2:471
120. Guasch L, Sala E, Ojeda MJ, Valls C, Bladé C, Mulero M, Blay M, Ardévol A, Garcia-Vallvé S, Pujadas G (2012) Identification of novel human dipeptidyl peptidase-IV inhibitors of natural origin (part II): in silico prediction in anti-diabetic extracts. PLoS One 7:e44972
121. Fan J, Johnson MH, Lila MA, Yousef G, de Mejia EG (2013) Berry and citrus phenolic compounds inhibit dipeptidyl peptidase IV: implications in diabetes management. Evid Based Complement Alternat Med 2013:479505
122. Parmar HS, Jain P, Chauhan DS et al (2012) DPP-IV inhibitory potential of naringin: an in silico, in vitro and in vivo study. Diabetes Res Clin Pract 97:105–111
123. Geng Y, Lu Z-M, Huang W, Xu H-Y, Shi J-S, Xu Z-H (2013) Bioassay-guided isolation of DPP-4 inhibitory fractions from extracts of submerged cultured of Inonotus obliquus. Molecules 18:1150–1161
124. Bharti SK, Krishnan S, Kumar A, Rajak KK, Murari K, Bharti BK, Gupta AK (2012) Antihyperglycemic activity with DPP-IV inhibition of alkaloids from seed extract of *Castanospermum australe*: investigation by experimental validation and molecular docking Phytomedicine 20:24–31
125. Bellé LP, Bitencourt PER, Abdalla FH, Bona KS de, Peres A, Maders LDK, Moretto MB (2013) Aqueous seed extract of *Syzygium cumini* inhibits the dipeptidyl peptidase IV and adenosine deaminase activities, but it does not change the CD26 expression in lymphocytes in vitro. J Physiol Biochem 69:119–124

126. Lacroix IME, Li-Chan ECY (2012) Evaluation of the potential of dietary proteins as precursors of dipeptidyl peptidase (DPP)-IV inhibitors by an in silico approach. J Funct Foods 4:403–422
127. Nongonierma AB, Fitzgerald RJ (2014) Susceptibility of milk protein-derived peptides to dipeptidyl peptidase IV (DPP-IV) hydrolysis. Food Chem 145:845–852
128. Rahfeld J, Schierhorn M, Hartrodt B, Neubert K, Heins J (1991) Are diprotin A (Ile-Pro-Ile) and diprotin B (Val-Pro-Leu) inhibitors or substrates of dipeptidyl peptidase IV? Biochim Biophys Acta 1076:314–316
129. Tulipano G, Sibilia V, Caroli AM, Cocchi D (2011) Whey proteins as source of dipeptidyl dipeptidase IV (dipeptidyl peptidase-4) inhibitors. Peptides 32:835–838
130. Nongonierma AB, FitzGerald RJ (2013) Dipeptidyl peptidase IV inhibitory and antioxidative properties of milk protein-derived dipeptides and hydrolysates. Peptides 39:157–163
131. Silveira ST, Martínez-Maqueda D, Recio I, Hernández-Ledesma B (2013) Dipeptidyl peptidase-IV inhibitory peptides generated by tryptic hydrolysis of a whey protein concentrate rich in β-lactoglobulin. Food Chem 141:1072–1077
132. Hatanaka T, Inoue Y, Arima J, Kumagai Y, Usuki H, Kawakami K, Kimura M, Mukaihara T (2012) Production of dipeptidyl peptidase IV inhibitory peptides from defatted rice bran. Food Chem 134:797–802
133. Huang S-L, Jao C-L, Ho K-P, Hsu K-C (2012) Dipeptidyl-peptidase IV inhibitory activity of peptides derived from tuna cooking juice hydrolysates. Peptides 35:114–121
134. Gallego M, Aristoy M-C, Toldrá F (2013) Dipeptidyl peptidase IV inhibitory peptides generated in Spanish dry-cured ham. Meat Sci 96:757–761
135. Lacroix IME, Li-Chan ECY (2012) Dipeptidyl peptidase-IV inhibitory activity of dairy protein hydrolysates. Int Dairy J 25:97–102
136. Velarde-Salcedo AJ, Barrera-Pacheco A, Lara-González S, Montero-Morán GM, Díaz-Gois A, González de Mejia E, Barba de la Rosa AP (2013) In vitro inhibition of dipeptidyl peptidase IV by peptides derived from the hydrolysis of amaranth (*Amaranthus hypochondriacus* L.) proteins. Food Chem 136:758–764
137. Nongonierma AB, Mooney C, Shields DC, Fitzgerald RJ (2013) Inhibition of dipeptidyl peptidase IV and xanthine oxidase by amino acids and dipeptides. Food Chem 141:644–653
138. Li-Chan ECY, Hunag S-L, Jao C-L, Ho K-P, Hsu K-C (2012) Peptides derived from atlantic salmon skin gelatin as dipeptidyl-peptidase IV inhibitors. J Agric Food Chem 60:973–978
139. Uenishi H, Kabuki T, Seto Y, Serizawa A, Nakajima H (2012) Isolation and identification of casein-derived dipeptidyl-peptidase 4 (DPP-4)-inhibitory peptide LPQNIPPL from gouda-type cheese and its effect on plasma glucose in rats. Int Dairy J 22:24–30
140. Uchida M, Ohshiba Y, Mogami O (2011) Novel dipeptidyl peptidase-4-inhibiting peptide derived from β-lactoglobulin. J Pharmacol Sci 117:63–66
141. Dziuba M, Dziuba B, Iwaniak A (2009) Milk proteins as precursors of bioactive peptides. Acta Sci Pol Technol Aliment 8(1):71–90 (http://www.food.actapol.net/volume8/issue1/abstract-7.html)
142. Minkiewicz P, Dziuba J, Michalska J (2011) Bovine meat proteins as potential precursors of biologically active peptides—a computational study based on the BIOPEP database. Food Sci Technol Int 17:39–45
143. Abe M, Akiyama T, Umezawa Y, Yamamoto K, Nagai M, Yamazaki H, Ichikawa Y-I, Muraoka Y (2005) Synthesis and biological activity of sulphostin analogues, novel dipeptidyl peptidase IV inhibitors. Bioorg Med Chem 13:785–797
144. Akiyama T, Abe M, Harada S et al (2001) Sulphostin, a potent inhibitor for dipeptidyl peptidase IV from Streptomyces sp. MK251–43F3. J Antibiot (Tokyo) 54:744–746
145. Umezawa H, Aoyagi T, Ogawa K, Naganawa H, Hamada M, Takeuchi T (1984) Diprotins A and B, inhibitors of dipeptidyl aminopeptidase IV, produced by bacteria. J Antibiot (Tokyo) 37:422–425
146. Trellet M, Melquiond A, Bonvin A (2013) A unified conformational selection and induced fit approach to protein peptide docking. PLoS One 8:e58769
147. Albericio F, Kruger HG (2012) Therapeutic peptides. Future Med Chem 4:1527–1531

148. Yan TR, Ho SC, Hou CL (1992) Catalytic properties of X-prolyl dipeptidyl aminopeptidase from *Lactococcus lactis* subsp. cremoris nTR. Biosci Biotechnol Biochem 56:704–707
149. Lorey S, Stöckel-Maschek A, Faust J et al (2003) Different modes of dipeptidyl peptidase IV (CD26) inhibition by oligopeptides derived from the N-terminus of HIV-1 Tat indicate at least two inhibitor binding sites. Eur J Biochem 270:2147–2156
150. Valli M, dos Santos RN, Figueira LD, Nakajima CH, Castro-Gamboa I, Andricopulo AD, Bolzani VS (2013) Development of a natural products database from the biodiversity of Brazil. J Nat Prod 76:439–444
151. Chen CY-C (2011) TCM database@Taiwan: the world's largest traditional Chinese medicine database for drug screening in silico. PLoS One 6:e15939
152. Elsevier Reaxys chemistry workflow solution. http://www.reaxys.com. Accessed 20 Jan 2014
153. Parasuraman S (2012) Protein data bank. J Pharmacol Pharmacother 3:351–352
154. Irwin JJ, Sterling T, Mysinger MM, Bolstad ES, Coleman RG (2012) ZINC: a free tool to discover chemistry for biology. J Chem Inf Model 52:1757–1768
155. Black OF, Kelly JW (1927) Pseudo ephedrine from *Ephedra alata*. Am J Pharm 99:748–751
156. Shabana MM, Mirhom YW, Genenah AA, Aboutabl EA, Amer HA (1990) Study into wild Egyptian plants of potential medicinal activity. Ninth communication: hypoglycaemic activity of some selected plants in normal fasting and alloxanised rats. Arch Exp Veterinarmed 44:389–394
157. Konno C, Mizuno T, Hikino H (1985) Isolation and hypoglycemic activity of ephedrans A, B, C, D and E, glycans of *Ephedra distachya* herbs. Planta Med 51:162–163
158. Grue-Sorensen G, Spenser ID (1989) The biosynthesis of ephedrine. Can J Chem 67:998–1009
159. Ito K, Haruna M, Furukawa H (1975) Studies on the erythrina alkaloids. X. Alkaloids of several Erythrina plants from Singapore (author's transl). Yakugaku Zasshi 95:358–362
160. Kumar A, Lingadurai S, Shrivastava TP, Bhattacharya S, Haldar PK (2011) Hypoglycemic activity of *Erythrina variegata* leaf in streptozotocin-induced diabetic rats. Pharm Biol 49:577–582
161. Oh WK, Lee C-H, Seo JH, Chung MY, Cui L, Fomum ZT, Kang JS, Lee HS (2009) Diacylglycerol acyltransferase-inhibitory compounds from *Erythrina senegalensis*. Arch Pharm Res 32:43–47
162. Na M, Jang J, Njamen D, Mbafor JT, Fomum ZT, Kim BY, Oh WK, Ahn JS (2006) Protein tyrosine phosphatase-1B inhibitory activity of isoprenylated flavonoids isolated from *Erythrina mildbraedii*. J Nat Prod 69:1572–1576
163. Bae EY, Na M, Njamen D, Mbafor JT, Fomum ZT, Cui L, Choung DH, Kim BY, Oh WK, Ahn JS (2006) Inhibition of protein tyrosine phosphatase 1B by prenylated isoflavonoids isolated from the stem bark of *Erythrina addisoniae*. Planta Med 72:945–948
164. Benn MH, Shustov G, Shustova L, Majak W, Bai Y, Fairey NA (1996) Isolation and characterization of two guanidines from *Galega orientalis* Lam. Cv. Gale (fodder galega). J Agric Food Chem 44:2779–2781
165. Vuksan V, Sievenpiper JL (2005) Herbal remedies in the management of diabetes: lessons learned from the study of ginseng. Nutr Metab Cardiovasc Dis 15:149–160
166. Michel KH, Sandberg F, Haglid F, Norin T (1967) Alkaloids of *Haloxylon salicornicum* (Moq.-Tand.) Boiss. Acta Pharm Suec 4:97–116
167. Brack A (1962) Verlauf der Alkaloidbildung durch den Clavicepsstamm von Pennisetum typhoideum Rich. in saprophytischer Kultur. 54. Mitteilung über Mutterkornalkaloide. Arch Pharm (Weinheim) 295:510–515
168. Shukla K, Narain JP, Puri P, Gupta A, Bijlani RL, Mahapatra SC, Karmarkar MG (1991) Glycaemic response to maize, bajra and barley. Indian J Physiol Pharmacol 35:249–254
169. Sheludko Y, Gerasimenko I, Kolshorn H, Stöckigt J (2002) New alkaloids of the sarpagine group from *Rauvolfia serpentina* hairy root culture. J Nat Prod 65:1006–1010

170. Benzi G, Villa RF, Dossena M, Vercesi L, Gorini A, Pastoris O (1984) Cerebral and cerebellar metabolic changes induced by drugs during the recovery period after profound hypoglycemia. Farmaco Sci 39:44–56
171. Ronchetti F, Russo G, Bombardelli E, Bonati A (1971) A new alkaloid from *Rauwolfia vomitoria*. Phytochemistry 10:1385–1388
172. Campbell JIA, Mortensen A, Mølgaard P (2006) Tissue lipid lowering-effect of a traditional Nigerian anti-diabetic infusion of *Rauwolfia vomitoria* foilage and *Citrus aurantium* fruit. J Ethnopharmacol 104:379–386
173. Phan MG, Phan TS, Matsunami K, Otsuka H (2006) Chemical and biological evaluation on scopadulane-type diterpenoids from *Scoparia dulcis* of Vietnamese origin. Chem Pharm Bull (Tokyo) 54:546–549
174. Latha M, Pari L, Sitasawad S, Bhonde R (2004) *Scoparia dulcis*, a traditional anti-diabetic plant, protects against streptozotocin induced oxidative stress and apoptosis in vitro and in vivo. J Biochem Mol Toxicol 18:261–272
175. Ly TT, Hewitt J, Davey RJ, Lim EM, Davis EA, Jones TW (2011) Improving epinephrine responses in hypoglycemia unawareness with real-time continuous glucose monitoring in adolescents with type 1 diabetes. Diabetes Care 34:50–52
176. Andrews KM, Beebe D a, Benbow JW et al (2011) 1-((3S,4S)-4-amino-1-(4-substituted-1,3,5-triazin-2-yl) pyrrolidin-3-yl)-5,5-difluoropiperidin-2-one inhibitors of DPP-4 for the treatment of type 2 diabetes. Bioorg Med Chem Lett 21:1810–1814
177. Aguilar-Santamaría L, Ramírez G, Nicasio P, Alegría-Reyes C, Herrera-Arellano A (2009) Anti-diabetic activities of *Tecoma stans* (L.) Juss. ex Kunth. J Ethnopharmacol 124:284–288
178. Hammouda Y, Rashid A-K, Amer MS (1964) Hypoglycaemic properties of tecomine and tecostanine. J Pharm Pharmacol 16:833–834
179. Van de Venter M, Roux S, Bungu LC et al (2008) Anti-diabetic screening and scoring of 11 plants traditionally used in South Africa. J Ethnopharmacol 119:81–86
180. Torres JL, Bobet R (2001) New flavanol derivatives from grape (*Vitis vinifera*) byproducts. Antioxidant aminoethylthio–flavan-3-ol conjugates from a polymeric waste fraction used as a source of flavanols. J Agric Food Chem 49:4627–4634
181. Pinent M, Blay M, Bladé MC, Salvadó MJ, Arola L, Ardévol A (2004) Grape seed-derived procyanidins have an antihyperglycemic effect in streptozotocin-induced diabetic rats and insulinomimetic activity in insulin-sensitive cell lines. Endocrinology 145:4985–4990
182. Song E-K, Hur H, Han M-K (2003) Epigallocatechin gallate prevents autoimmune diabetes induced by multiple low doses of streptozotocin in mice. Arch Pharm Res 26:559–563
183. Takayama H, Okazaki T, Yamaguchi K, Aimi N, Haginiwa J et al (1988) Structure of two new diterpene alkaloids, 3-epi-ignavinol and 2,3-dehydrodelcosine. Chem Pharm Bull (Tokyo) 36(8):3210–3212
184. Konno C, Murayama M, Sugiyama K, Arai M, Murakami M, Takahashi M, Hikino H (1985) Isolation and hypoglycemic activity of aconitans A, B, C and D, glycans of *Aconitum carmichaeli* roots. Planta Med 51:160–161
185. Howes M, Simmonds M (2005) Plants used in the treatment of diabetes. In: Soumyanath A (ed) Traditional medicines for modern times. CRC, Boca Raton.
186. Zhang H, Wang X-N, Lin L-P, Ding J, Yue J-M (2007) Indole alkaloids from three species of the Ervatamia genus: *E. officinalis*, *E. divaricata*, and *E. divaricata* Gouyahua. J Nat Prod 70:54–59
187. Fujii M, Takei I, Umezawa K (2009) Anti-diabetic effect of orally administered conophylline-containing plant extract on streptozotocin-treated and Goto-Kakizaki rats. Biomed Pharmacother 63:710–716
188. Usubillaga A (1988) Solanudine, a steroidal alkaloid from *Solanum nudum*. Phytochemistry 27:3031–3032
189. Yoshikawa M, Nakamura S, Ozaki K, Kumahara A, Morikawa T, Matsuda H (2007) Structures of steroidal alkaloid oligoglycosides, robeneosides A and B, and antidiabetogenic constituents from the Brazilian medicinal plant *Solanum lycocarpum*. J Nat Prod 70:210–214

190. Villaseñor IM, Lamadrid MRA (2006) Comparative anti-hyperglycemic potentials of medicinal plants. J Ethnopharmacol 104:129–131
191. El Sayed KA, Hamann MT, Abd El-Rahman HA, Zaghloul AM (1998) New pyrrole alkaloids from *Solanum sodomaeum*. J Nat Prod 61:848–850
192. Kar DM, Maharana L, Pattnaik S, Dash GK (2006) Studies on hypoglycaemic activity of *Solanum xanthocarpum* Schrad. & Wendl. fruit extract in rats. J Ethnopharmacol 108:251–256
193. Kashiwaba N, Morooka S, Ono M, Toda J, Suzuki H et al (1997) Alkaloidal constituents of the leaves of *Stephania cepharantha* cultivated in Japan: structure of cephasugine, a new morphinane alkaloid. Chem Pharm Bull (Tokyo) 45(3):545–548
194. Mosihuzzaman M, Nahar N, Ali L, Rokeya B, Khan AK et al (1994) Hypoglycemic effects of three plants from eastern himalayan belt. Diabetes Res 26(3):127–138
195. Semwal DK, Rawat U, Semwal R, Singh R, Singh GJP (2010) Anti-hyperglycemic effect of 11-hydroxypalmatine, a palmatine derivative from *Stephania glabra* tubers. J Asian Nat Prod Res 12:99–105
196. Tsutsumi T, Kobayashi S, Liu YY, Kontani H (2003) Anti-hyperglycemic effect of fangchinoline Isolated from *Stephania tetrandra* radix in streptozotocin-diabetic mice. Biol Pharm Bull 26:313–317
197. Beek TAV, Verpoorte R, Svendsen AB (1984) Alkaloids of *Tabernaemontana eglandulosa*. Tetrahedron 40(4):737
198. Ma D-L, Chan DS-H, Leung C-H (2013) Drug repositioning by structure-based virtual screening. Chem Soc Rev 42:2130–2141
199. Meslamani J, Bhajun R, Martz F, Rognan D (2013) Computational profiling of bioactive compounds using a target-dependent composite workflow. J Chem Inf Model 53:2322–2333 doi:10.1021/ci400303n
200. Peng S, Lin X, Guo Z, Huang N (2012) Identifying multiple-target ligands via computational chemogenomics approaches. Curr Top Med Chem 12:1363–1375
201. Swamidass SJ, Lu Z, Agarwal P, Butte AJ (2014) Computational approaches to drug repurposing and pharmacolog- session introduction. Pac Symp Biocomput 19:110–113
202. Peters J-U (2013) Polypharmacology—foe or friend? J Med Chem 56:8955–8971
203. Santiago DN, Pevzner Y, Durand AA, Tran M, Scheerer RR, Daniel K, Sung S-S, Woodcock HL, Guida WC, Brooks WH (2012) Virtual target screening: validation using kinase inhibitors. J Chem Inf Model 52:2192–2203
204. Yue R, Shan L, Yang X, Zhang W (2012) Approaches to target profiling of natural products. Curr Med Chem 19:3841–3855
205. Gao Z, Li H, Zhang H, Liu X, Kang L, Luo X, Zhu W, Chen K, Wang X, Jiang H (2008) PDTD: a web-accessible protein database for drug target identification. BMC Bioinformatics 9:104
206. Li H, Gao Z, Kang L et al (2006) TarFisDock: a web server for identifying drug targets with docking approach. Nucleic Acids Res 34:W219–W224
207. Laskowski RA, Swindells MB (2011) LigPlot+: multiple ligand-protein interaction diagrams for drug discovery. J Chem Inf Model 51:2778–2786
208. Sayle RA, Milner-White EJ (1995) RASMOL: biomolecular graphics for all. Trends Biochem Sci 20:374

Chapter 8
Comparison of Different Data Analysis Tools to Study the Effect of Storage Conditions on Wine Sensory Attributes and Trace Metal Composition

Helene Hopfer, Susan E. Ebeler and Hildegarde Heymann

8.1 Introduction

Multivariate data analysis, i.e., the simultaneous analysis of more than one measured variable, includes different statistical methods that study both the impact of a measured variable on the samples and the interaction and correlation among the measured variables. In that sense, multivariate data analysis, or more generally, multivariate statistics, may be more effective in evaluating naturally occurring events than univariate statistics, since in nature, things are connected and impact each other. Especially in recent years, study designs which generate large data sets which involve complex connections among experimental variables frequently require multivariate analysis methods in order to fully evaluate the complex research questions involved [1].

Multivariate data statistics is applied in many sciences, including natural and life sciences as well as social sciences and each area uses slightly different techniques to study similar problems. In an applied and interdisciplinary field, such as food science, the challenge is to make use of all these different fields for useful and applicable methods for one's own research.

Generally, two types of questions are asked when applying multivariate data analysis techniques; one question aims to explore the gathered data without any preconceived assumptions or notions, while the second question relates to sample classification and finding valid and powerful models for prediction purposes.

H. Hopfer (✉) · S. E. Ebeler · H. Heymann
Department of Viticulture and Enology, University of California, One Shields Ave., Davis, CA 95616, USA
e-mail: hhopfer@ucdavis.edu

K. Martinez-Mayorga, J. L. Medina-Franco (eds.), *Foodinformatics*,
DOI 10.1007/978-3-319-10226-9_8

The essence of exploratory data analysis methods is to filter relevant information from the gathered data, and present these important features, often in a visual way.

One of the most commonly used exploratory methods is principal component analysis (PCA), which is also an unsupervised technique. PCA is a lower-dimensional representation of the multidimensional data space, using linear combinations (so-called principal components PCs) of the existing variables that explain most of the variance in the data set [1, 2]. These PCs are also orthogonal to each other, which means that they are uncorrelated and perpendicular to each other [2]. Creating the PCs is independent of any assumptions, and simply based on the gathered data, thus called *unsupervised.* Using just a few PCs, one typically is able to explain most of the variance in the data. Relationships among samples as well as between samples and variables are then displayed in so-called *score plots* (the positions of the samples in the lower-dimensional space) and *loadings plots* (the positions of the measured variables). In the *score plot* samples that are similar to each other show similar scores and are positioned close to each other, while dissimilar samples are positioned further apart from each other. Similarly with the loadings, measured variables that are positively correlated to each other are close to each other in the *loadings plot,* while negatively correlated variables are positioned opposite of each other.

PCA is a widely used technique in food science, and is used in nearly every subfield within food science such as sensory and consumer science [3] and food component profiling [4, 5], and is now part of a typical workflow, in general to gain a deeper understanding of the differences among a set of samples, how these differences relate to each of the measured variables, and which variables explain more of the observed differences. Specific examples are, e.g., the use of PCA to analyze and correlate instrumental and sensory measurements of cooked wheat noodles with varying degrees of gluten and glyceryl monostearate [6], where the authors used PCA besides PLSR and general procrustes' analysis (GPA) to study how the changes in the physical properties affected the appearance and texture. PCA was also used to study the impact of stabilization on the changes in volatile patterns of food packaging materials over time [7].

In contrast to exploratory techniques, classification methods are used to test if samples group together based on prior assumptions, and to model data for future prediction. In that sense, classification techniques are *supervised* and the researcher has a testable hypothesis about relationships within the gathered data prior to running the analysis. One example of a supervised classification method is canonical variate analysis (CVA), sometimes also called Fisher's linear discriminant analysis (LDA). A CVA tries to find linear combinations of the measured variables that maximize the variance ratio by minimizing the variance within the group and maximizing the variance between the groups [2]. In contrast to a PCA, a CVA highlights the differences between the groupings, e.g., different wine regions [8]. The linear combinations of the measured variables, the so-called canonical variates (CVs), are not necessarily orthogonal as in a PCA, and the angle between the CVs can be calculated [2], but are in most cases close to 90°. The importance of the various axes of a CVA solution can be tested statistically using the Bartlett's test, thus, helping to select the appropriate number of dimensions for interpretation; this is not possible for a PCA. Additionally, confidence intervals around the group means can be easily calculated and incorporated into the CVA product plot, providing a visual statistical significance test (confidence interval circles that do not overlap are statistically different at the chosen significance level, e.g., 5 %).

Classification problems are numerous in food science, for example, classification methods are used to determine a food's origin based on chemical fingerprints [9], but can also discriminate among different fig cultivars with sensory attributes, independent of the source and harvest date of the different cultivars [10]. CVA is just one of many classification techniques, and the reader is referred to specific articles, e.g., partial least squares discriminant analysis (PLS-DA) [11], artificial neural networks (ANNs) [12], and support vector machines (SVMs) [13].

Besides using classification techniques to identify separate groups in a given sample set, classifying methods can also be used for creating prediction models. Typically, one uses a set of given samples with known properties to create the model, which is then tested with a second set of new samples. In this chapter, we use one data set to predict the second data set, as a way to study the correlation between the variables of the two data sets. This is done using partial least squares regression (PLSR) [1]. PLSR combines PCA and regression, and can be used to predict a group of so called dependent (i.e., predicted) variables by a second set of independent, or predicting variables. In contrast to multiple regression, PLSR is trying to select so-called latent vectors (LVs) that explain most of the covariance between the predicting and the predicted data sets [14]. PLSR attempts to find LVs that maximize the covariance between the two data sets and that capture most of the variance in both data sets at the same time [1].

PLSR is commonly used to correlate different data sets to each other (e.g., sensory to chemical measurements), as well as for prediction purposes (i.e., substitution of various wet chemistry methods by a near-infrared spectroscopy (NIR)-based model). One example for the former case is the correlation of the sensory and instrumental flavor perception in ice creams with different flavor compounds and additionally also varying in fat levels [15]. A quantitative and validated prediction model for fatty acid profile, fat and water content, retrogradation, and viscosity was developed by [16] for the characterization of potato, maize, wheat, rice, and tapioca starches for industrial purposes.

Using two defined data sets, we will apply these three data analysis techniques to show the differences and similarities between different multivariate methods. The data consists of trace elemental and sensory measurements of wines that have been stored at different conditions, varying in temperature and packaging type. Comparing the outcome of different data analysis methods is something not very often done. For example, Heymann and Noble compared PCA and CVA outcomes of sensory data [2], while Zhao and Maclean compared the same two techniques for spectral transformations in satellite image preprocessing [17].

8.2 Methods and Materials

8.2.1 Samples

Twelve sample treatments were realized, storing one Cabernet Sauvignon wine (vintage 2009, from Northern California) in four different packaging configurations at three constant storage temperatures (10, 20 and 40 °C) for a period of 6 months. The four packaging configurations were (1) a 3-L bag-in-box container (BIB; Durashield 34ES, Scholle Packaging, Northlake, IL, USA), (2) a 0.75-L dark-green glass bottle closed with natural cork (AC-1 grade, 29 mm × 49 mm, ACI Cork, Fairfield, CA, USA), a 0.75-L dark-green glass bottle capped with an aluminum screw cap (Federfin Tech S.R.L., Tromello, Italy) with a tin-PVDC liner (Oenoseal, Chazay D'Azergues, France) with either (3) a normal filling height (headspace was 15 mm) or (4) filled to the very top of the bottle. Further details about the samples and how they were prepared can be found in [18].

8.2.2 Sensory Analysis

Ten unpaid volunteers were recruited based on their availability and agreement to serve on the sensory panel (mean age 33.8 years, nine females), and included students, staff, and retirees of the UC Davis campus. The UC Davis institutional review board approved the study. All panelists completed six training sessions of 1 h each, spread over a period of 2 weeks. During these training sessions, the panelists created, chose, and agreed upon the descriptors and descriptor references to describe differences among the samples, using different subsets of the samples for each training session. The panel chose 16 aroma descriptors (*red fruit, cherry, jammy, grapefruit, fresh veggie, canned veggie, earthy, wood, black pepper, spice, molasses/soy sauce, brown flavor, dried fruit, oxidized, chemical, floral*), three taste descriptors (*sour, sweet, bitter*), and three mouthfeel descriptors (*astringent, hot mouthfeel, viscous*), all with corresponding reference standards (see [18] for details). Panelists also completed scaling exercises to ensure that the panel perceived differences among the samples in a similar way, both in quality and magnitude. Following training, all samples were tasted in triplicate over a period of 3 weeks in separate tasting booths. Each panelist tasted six samples during each of the evaluation sessions. Samples were presented in a randomized William Latin Square design to control for carry-over effects. Panelists rated each descriptor for each sample on an unstructured line scale, anchored on the left with "low" and on the right with "high," using a dedicated sensory computer software (FIZZ, Biosystemes, Couteron, France).

8.2.3 Trace Element Analysis

All samples were profiled for their elemental composition using inductively coupled plasma mass spectrometry (ICP-MS). An Agilent 7700x ICP-MS (Santa Clara, CA, USA) was equipped with a MicroMist nebulizer, a double-quartz spray chamber, and a peristaltic pump (0.1 rps). Argon was used as carrier gas (1.03 L/min), while Helium was used in the octapole reaction cell at a flow rate of 4.3 or 10 mL/min. All monitored isotopes (^{51}V, ^{52}Cr, ^{55}Mn, ^{56}Fe, ^{57}Fe, ^{58}Ni, ^{59}Co, ^{60}Ni, ^{66}Zn, ^{75}As, ^{78}Se, ^{111}Cd, ^{117}Sn, ^{118}Sn, ^{119}Sn, ^{120}Sn, ^{133}Cs, ^{205}Tl, ^{208}Pb) were measured in helium mode, with ^{75}As and ^{78}Se in high-energy helium mode (flow of 10 mL/min). Samples were prepared in triplicate by diluting them 1:3 in 1 % nitric acid (HNO_3; Optima, Fisher Scientific, Pittsburgh, PA, USA). Quality control samples were prepared by spiking wine samples with 0.5, 1, or 10 µg/L tin (Inorganic Ventures, Christiansburg, VA, USA), and measured together with the samples. An internal standard (IS) mix consisting of six elements (SPEX CertiPrep, Metuchen, NJ, USA) covered the whole mass range between 6 and 238 amu, and was constantly fed into the sample stream using a mixing tee. All monitored elements were quantified between 0 and 500 µg/L in a matrix-matched solution (1 % HNO_3 and 4 % ethanol). Limits of detection (LOD) and quantification (LOQ) were determined via the standard deviation of seven calibration blank runs. Further details with regard to the ICP-MS method can be found in [19].

8.2.4 Data Analysis

The sensory data (10 judges × 12 samples × 3 replicates = 360 observations of 22 descriptors) as well as the ICP-MS data (12 samples × 3 replicates = 36 observations of 19 isotopes) were statistically evaluated with a fixed effect analysis of variance (ANOVA), after a multivariate analysis of variance (MANOVA) for a sample effect showed significant differences ($P \leq 0.05$). For the sensory data all three main effects and all two-way interactions were added to the model, while for the ICP-MS data only the two main effects were included.

All significant sensory descriptors that showed a significant sample effect together with a significant *sample × judge* interaction were treated with a pseudo-mixed model, with the interaction as the new error term, as suggested by Gay in [20]. All significant descriptors and elements were retained for further analysis.

Exploratory data analysis was conducted using PCA on the correlation matrix (i.e., scaled to unit variance) of the averaged data sets (over judges and replicates) to account for scaling and concentration differences in the two data sets. A classification technique, CVA, based on the MANOVA model with a sample effect was also conducted. The main difference between PCA and CVA lies in the interpretation of sample differences; while a PCA algorithm attempts to maximize the sample differences, the CVA algorithm maximizes the ratio of the between-group to the within-group sums of squares (the groups in our case are the samples). Additionally, confidence intervals (e.g., at the 95 % level) can easily be constructed as circles around the sample means, providing visual significance testing. Circles that overlap are not significantly different from each other, and were calculated using the algorithm described by Owen and Chmielewski [21]. Due to the nature of CVA using a MANOVA model, a Bartlett's test for the number of significant dimensions can be included.

In a last analysis step, the two data sets were compared to each other with PLSR to find correlations between the descriptors and the elements.

All analyses were conducted in RStudio [22], running in the R language environment [23], with several add-on packages, including FactoMineR [24, 25], candisc [26], plotrix [27], and pls [28].

8.3 Results and Discussion

8.3.1 Descriptive Analysis Panel

Significant differences among the samples were revealed by MANOVA, and in the subsequent individual ANOVAs, 11 aroma descriptors were found to differ significantly among the treatments ($P \leq 0.05$). These significant descriptors were subsequently used in all analyses (for further details see [18]).

8.3.2 PCA of the Sensory Data Set

A PCA was conducted using the significant 11 sensory descriptors, and the resulting biplot is shown in Fig. 8.1. In the scree plot (dimensions over eigenvalues) a large drop and a knee was observed after two dimensions (data not shown). Additionally, over 80 % of the total variance was explained within the first two PCs, thus, the first two dimensions were kept for the interpretation of the PCA. Samples were separated in the PCA to a large degree due to their storage temperature, and to a smaller degree by their packaging configuration. Along the first principal component (PC 1), explaining 67 % of the total variance, samples stored at 40 °C were well separated from the 10 and 20 °C samples. Samples on the right-hand side of the PCA plot, which were stored at 40 °C, were described by the sensory panel with the descriptors *dried fruit, brown flavor, spice, oxidized, molasses/soysauce, canned veggie,* and *earthy*. All these sensory descriptors were previously reported as ageing and/or oxidation attributes in red wine [29–31]. On the left-hand side of the PCA plot, samples that were stored at 10 and 20 °C are positioned. These treatments were scored higher in *red fruit, cherry, grapefruit,* and *black pepper*. Fresh fruit attributes as well as citrus aromas were previously described in young Cabernet Sauvignon wine [30].

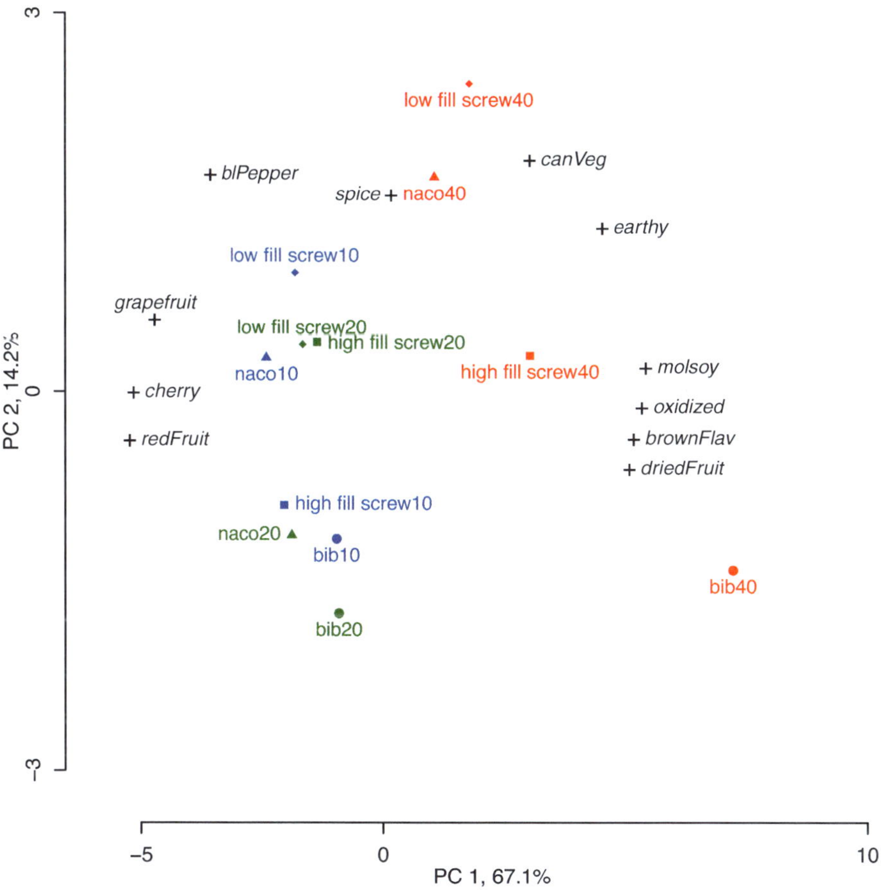

Fig. 8.1 PCA biplot of the DA data, showing the significant descriptors (in *black*) projected into the score plot of the samples. Samples are color-coded according to their storage temperature (*blue* 10 °C, *green* 20 °C, *red* 40 °C), and different symbols represent different packaging configurations (*filled circle* 3 L bag-in-box (BIB), *filled triangle* 0.75 L green glass bottle with natural cork (naco), *filled square* 0.75 L *green* glass bottle with a screw cap and filled to the top of the bottle (high fill screw), and *filled diamond* 0.75 L *green* glass bottle with a screw cap and filled to a normal fill height (low fill screw))

Along the second PC, an additional 14 % of the total variance was explained, and PC 2 captures mostly the differences due to the different packaging configurations. All BIB samples (*bib10, bib20, bib40*) are positioned at the bottom of the plot, while all screw-capped samples with a low-fill height (*low-fill screw10, low-fill screw20, low-fill screw40*) are positioned towards the top of the PCA plot. In between those treatments the remaining two packaging configurations (natural cork closure and high-fill screw-capped bottles) are located.

With increasing storage temperature, the differences between the four packaging configurations become larger. Samples stored at 40 °C form three subgroups, with the natural cork sealed bottles and the low-fill screw-capped bottles forming one group and scoring higher in *canned veggie* and *earthy,* while high-fill screw cap and BIB samples formed two separate groups. The latter two samples were more described by *oxidized, brown flavor, dried fruit,* and *molasses/soysauce* characters, with the high-fill screw cap sample stored at 40 °C being positioned in between the BIB sample and the other two samples stored at the same temperature.

The PCA on the DA data shows a clear separation of the samples due to their storage conditions; storage temperature had the largest impact on the sensory properties of the stored wines, while the packaging configuration altered the sensory profile to a lesser extent, especially at lower storage temperatures. The most oxidized wine in the sample set was the combination of a highly oxygen-permeable wine packaging, such as BIB, with high storage temperature.

8.4 CVA of the Sensory Data Set

Similar to the PCA, only the significantly different sensory descriptors were used in the CVA. As CVA is a classification technique, an a priori grouping is needed. We chose the most basic model, and used a MANOVA model with only the *sample* effect. Bartlett's test for the determination of significant canonical dimensions revealed that only the first CV was significantly different ($P \leq 0.05$). However, a knee in the scree plot was observed after the first two CVs, thus, the first two dimensions were kept for interpretation (data not shown).

Nearly 90 % of the total variance ratio is explained within the first two CVs shown in Fig. 8.2. Along the first dimension (CV 1), explaining 75 % of the variance ratio, treatments are somewhat separated due to their storage temperature, with samples stored at 40 °C more on the left-hand side of the plot, and all 10–20 °C samples clustering together on the right side. The BIB sample stored at 40 °C is the main driver for the observed separation among the samples, while the other three 40 °C treatments are not significantly different from each other (their confidence interval circles overlap). The descriptors *oxidized, molasses/soysauce, brown flavors, dried fruit, earthy,* and *grapefruit* are close to the 40 °C treatments, while samples stored at lower temperatures were described by the attributes *spice, cherry, red fruit, black pepper,* and *canned veggie,* with the latter two being expressed in the bottle treatments stored at 40 °C as well.

The second CV, accounting for an additional 14 % of the variance ratio, is mainly expressing the differences between the 40 °C bottle treatments and the 10–20 °C samples, with the latter group being higher in *cherry* and *red fruit* and *spice* characters, while the 40 °C bottle samples showed increasing ratings in *canned veggie*.

In contrast to the PCA, the CVA is mostly driven by the extreme changes observed in the BIB stored at 40 °C, which is responsible for the separation along the first (and only significant) CV. Additionally, the addition of the confidence intervals around the sample means provides a visual significance test, and reveals that the natural cork samples stored at 40 °C showed a larger variability than the

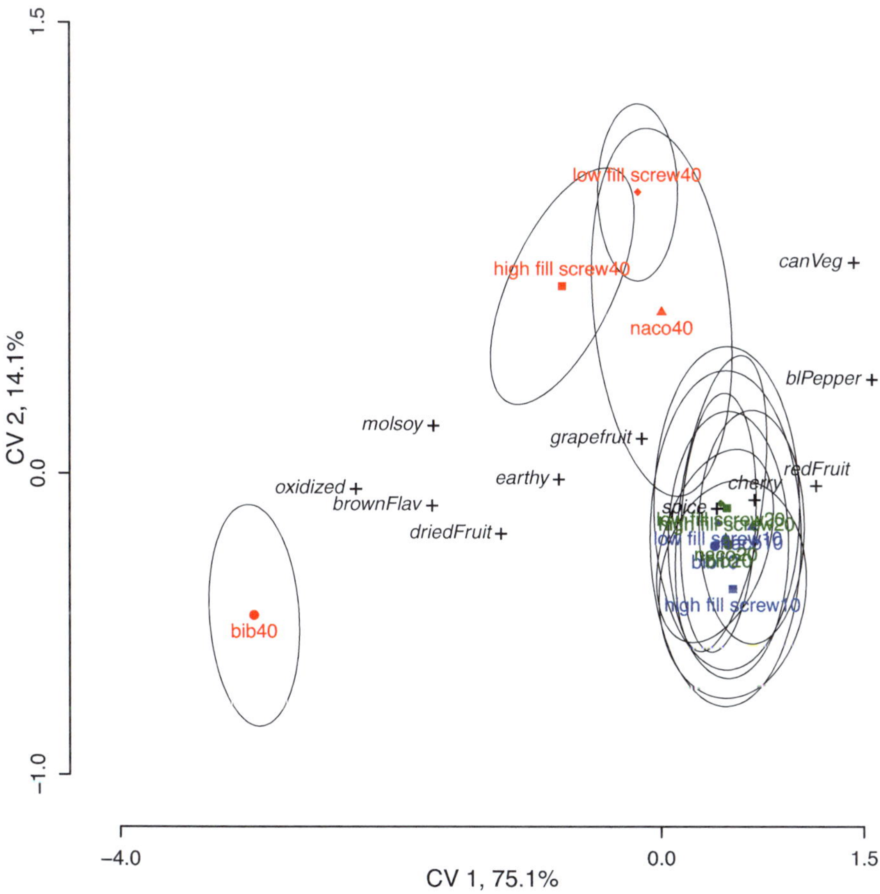

Fig. 8.2 CVA biplot of the DA panel, showing the significant descriptors (in *black*) projected into the score plot of the samples. Samples are color-coded according to their storage temperature (*blue* 10 °C, *green* 20 °C, *red* 40 °C), and different symbols represent different packaging configurations (*filled circle* 3 L bag-in-box (BIB), *filled triangle* 0.75 L *green* glass bottle with natural cork (naco), *filled square* 0.75 L *green* glass bottle with a screw cap and filled to the top of the bottle (high fill screw), and *filled diamond* 0.75 L *green* glass bottle with a screw cap and filled to a normal fill height—low fill screw). 95 % confidence intervals around the sample means are shown as *gray* circles

screw cap and the BIB samples stored at the same temperature, which could be the result of cork being a natural product with an inherently higher product variability. Comparing the results from the PCA to the CVA results, one might also conclude that the differences among the bottle treatments at 40 °C were more significant in the PCA than they are statistically–e.g., the low fill screw cap and the natural cork samples stored at 40 °C seem different from the high fill screw cap sample which seems different from the BIB sample in the PCA, while in the CVA the confidence intervals for the three bottle treatments at 40 °C overlap, and only the BIB treatment at 40 °C is statistically different from all the other 40 °C samples. Similar were the differences in the packaging at lower temperatures; in the PCA the samples seem more different than in the CVA where the confidence intervals overlap for all samples stored at 10–20 °C.

8.4.1 *Elemental Profiling*

Significant differences in the elemental composition among the samples were revealed by MANOVA, and in the subsequent individual ANOVAs, five elements differed significantly among the treatments ($P \leq 0.05$), and were subsequently used in all analyses (for further details see [19]).

8.5 PCA of the Elemental Profile Data Set

The resulting biplot from the PCA on the five elements that differed significantly among the samples is shown in Fig. 8.3. Similar to the DA data set, a very high proportion (over 90 %) of the total variance is explained within the first two PCs. In contrast to the DA data, sample separation in the elemental data set is driven by the packaging configuration, explaining 69 % of the total variance in PC 1. All BIB samples are positioned close to each other on the left side of the PCA biplot, followed by the natural cork samples, the low fill height screw cap samples and the high fill screw samples when moving to the right-hand side of the plot. An additional 21 % of the total variance is explained by PC 2, which separates the treatments due to their storage temperature; the higher the storage temperature the more the samples are positioned at the top of the PCA biplot. Sample separation is driven by higher levels of all five elements in the bottle treatments compared to the BIB samples, which showed the lowest concentrations in all elements. Lead (Pb), copper (Cu), and vanadium (V) showed higher correlations to the high fill screw cap samples stored at 10–20 °C, while chromium (Cr) and Pb were more correlated to the high fill screw cap samples stored at 40 °C. Previously, V and Cr were measured in wine, and their presence was explained due to the use of stainless steel equipment in the winery, for which these two elements are known alloy elements [32, 33]. Cu present in wine can be the result of both viticultural and enological practices, as copper sulfate is a known fungicide used in the vineyard, and Cu itself is a fining agent used in winemaking [32, 34]. Pb, which is still present in the ambient environment due to its former use in gasoline, could also end up in wine due to its use in winery equipment [35]. The presence of tin in wines was just recently described [19], most likely the result of using a tin liner in the screw caps. None of the metal concentrations were above the allowable levels defined by the International Organization of Vine and Wine (OIV) [36]

Another interesting fact is the degree of changes observed in each packaging configuration with increasing storage temperature; while the BIB samples barely

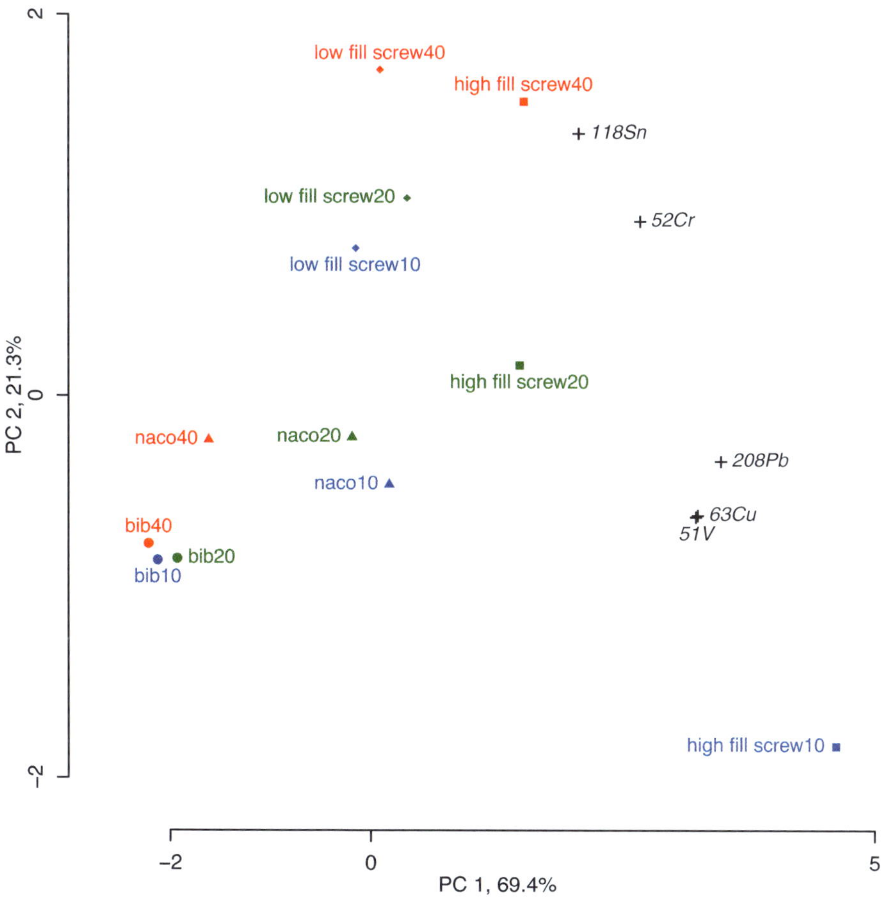

Fig. 8.3 PCA biplot of the elemental data, showing the significantly different elements (in *black*) projected into the score plot of the samples. Samples are color-coded according to their storage temperature (*blue* 10 °C, *green* 20 °C, *red* 40 °C), and different symbols represent different packaging configurations (*filled circle* 3 L bag-in-box (BIB), *filled triangle* 0.75 L *green* glass bottle with natural cork (naco), *filled square* 0.75 L green glass bottle with a screw cap and filled to the top of the bottle (high fill screw), and *filled diamond* 0.75 L *green* glass bottle with a screw cap and filled to a normal fill height—low fill screw)

change in their elemental composition as a function of temperature, the high fill screw cap samples showed large changes in their elemental composition. Changes in the elemental composition in the wines can be explained in two ways: At lower temperatures, metals present in the wine form complexes with other wine components, such as polyphenols or proteins, and these complexes precipitate at higher storage temperatures [32, 34], which could be the explanation for the observed differences in Cr, V, and Pb. In contrast to that, the tin levels increased with increasing storage temperature, which could be the result of increased leaching of tin from the liner when the wine expanded at higher storage temperatures, or, in case of the high fill screw cap samples, even touched the liner [19].

8.6 CVA of the Elemental Profile Data

Using the significantly different elements, a CVA biplot was created and is shown in Fig. 8.4. The Bartlett's test revealed that the first four CVs were significantly different from each other, but in the scree plot a knee was observed after the second CV, thus, only the first two CVs are used for further interpretation (data not shown).

Within the first two dimensions, over 88 % of the total variance ratio is explained, and along CV 1, samples are separated in a different way than in the PCA.

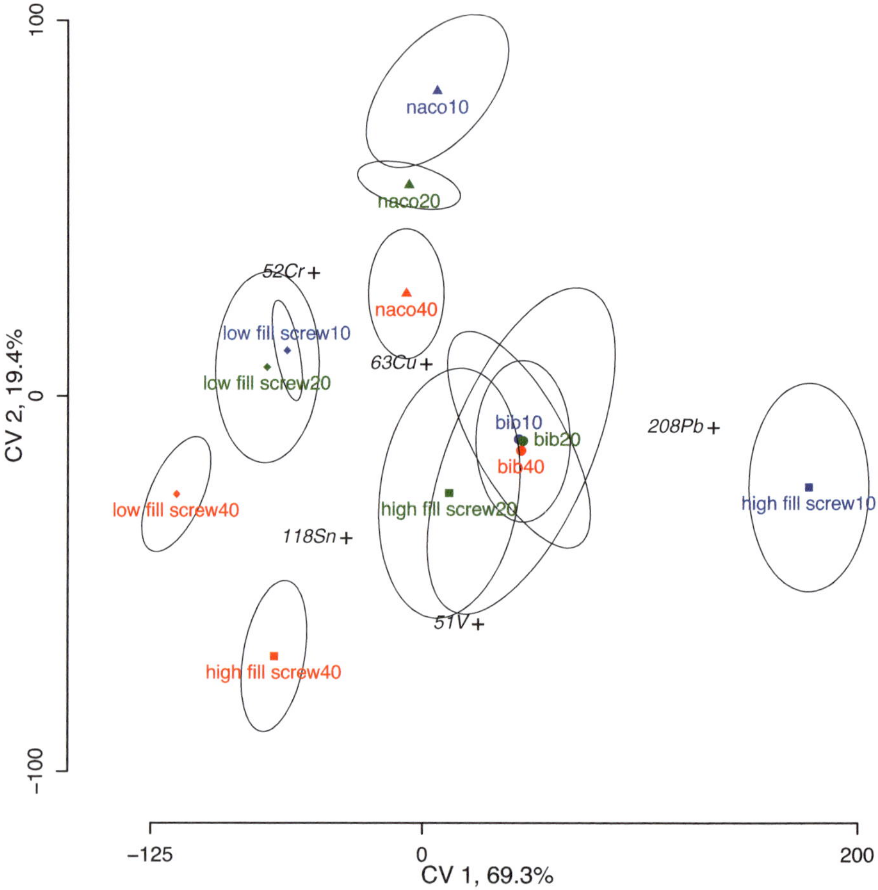

Fig. 8.4 CVA biplot of the elemental data, showing the significant descriptors (in *black*) projected into the score plot of the samples. Samples are color-coded according to their storage temperature (*blue* 10 °C, *green* 20 °C, *red* 40 °C), and different symbols represent different packaging configurations (*filled circle* 3 L bag-in-box (BIB), *filled triangle* 0.75 L *green* glass bottle with natural cork (naco), *filled square* 0.75 L *green* glass bottle with a screw cap and filled to the top of the bottle (high fill screw), and *filled diamond* 0.75 L *green* glass bottle with a screw cap and filled to a normal fill height—low fill screw). 95 % of confidence intervals around the sample means are shown as *gray* circles

While in the PCA all the BIB samples were positioned together at the left-hand side of the plot, all the BIB samples are clustered in the middle of the CVA plot, and similar to the PCA, they show a low correlation to all the elements. All other packaging types are close to each other with the exception of the high fill level screw caps, which show again large differences between the three storage temperatures. Along CV 2, explaining nearly 20 % of the total variance ratio, samples are separated due to storage temperature, with samples stored at lower storage temperatures positioned at the top of each packaging type.

Elements responsible for the sample separation are similarly correlated to the individual treatments, with tin being positioned close to the two screw cap samples stored at 40 °C, Cr being positively correlated to the low-fill screw-capped wines and the natural cork samples, while V and Pb show a high positive correlation to the high fill screw cap treatments at 20–10 °C.

The main differences between the PCA and the CVA for the elemental data lies in the slightly different interpretation, while the temperature effect for the low fill crew cap and the natural cork samples are statistically significant in the CVA (the confidence intervals of these treatments do not overlap),this effect is not so apparent in the PCA. Also, tin is very clearly associated with all of the screw cap samples stored at 10–20 °C in the CVA—this is somewhat harder to tell in the PCA. One might come to slightly different conclusions on the changes in metal composition with the different packaging types based on the two methods—e.g., the loadings for Cu, V, Cr loadings are somewhat different between the two methods.

8.6.1 Comparison of the Two Data Sets

We hypothesized that changes in the metal content could relate to sensory differences since metals act as catalysts for many chemical reactions (e.g., oxidation) [37]. Therefore, in order to compare the two different data sets to each other and identify correlations between the variables, which could then be tested for causality, a PLSR was conducted. All sensory descriptors were used as predicted, and all elements as predicting variables in the PLSR model. The PLSR model was evaluated with a leave-one-out bootstrapping algorithm. Using the first three model components (LVs), over 99 % of the total variance of the predictor matrix (i.e., the elements) was explained. On average, 43 % of the predicted matrix (Y) was explained by the first three components of the PLS regression, with each sensory descriptor being at least 26 % explained (Table 8.1). The model did not improve by adding more components, and additionally, the validation plots (Fig. 8.5) show for each sensory descriptor minimum root mean squared error or prediction (RMSEP) with three LVs, except for *canned veggie* and *earthy*, which have their minimum RMSEP with two LVs. It was decided to keep the first three LVs of the PLS model as most of the sensory variables had their minimum RMSEP there, indicating the best fit, and overfitting by including more model dimensions.

Table 8.1 Percentages of explained variance for the predictor matrix (*X*), the average of the predicted matrix (*Y*) and each of the predicted sensory variables for the first five components (comps) of the PLS regression model

(Percent variance explained)	1 comps	2 comps	3 comps	4 comps	5 comps
X	66.0	85.8	99.4	99.6	100.0
Y	15.4	29.2	42.8	59.0	62.4
red fruit	28.9	48.9	52.1	70.0	71.3
cherry	31.0	45.6	58.7	64.8	72.2
grapefruit	32.0	35.4	55.9	72.7	74.4
canned veggie	5.0	52.3	55.0	56.2	58.4
earthy	12.2	44.9	46.7	51.5	51.6
black pepper	8.1	8.7	50.4	60.5	76.5
spice	1.1	2.3	26.3	28.4	31.1
molasses/soysauce	14.8	26.5	29.7	60.3	61.8
brown flavor	10.8	14.9	28.0	56.0	58.1
dried fruit	10.0	19.3	38.7	68.2	68.4
oxidized	15.8	22.5	29.6	60.9	62.3

However, the sensory descriptors are not well predicted by the elements, and this lack of correlation is also represented in the correlation plots shown in Fig. 8.6.

Despite the good modeling of the variability of the elemental data (i.e., predicting data set), only four (*red fruit, cherry, canned veggie, earthy*) of the 11 sensory descriptors (i.e., predicted data set) are sufficiently explained by the model (i.e., falling within the dotted lines as shown in Fig. 8.6a, b, with over 50 % variance accounted for in the first two model dimensions (see also Table 8.1). Adding another dimension to the model only slightly improves the number of descriptors explained; in Fig. 8.6b, only *grapefruit* and *black pepper* were additionally explained with at least 50 % of the variance explained by adding a third model component.

Some correlation was found between the elements and the sensory descriptors, such as a negative correlation of V, Pb and Cr to *molasses/soysauce, dried fruit, oxidized,* and *brown flavors,* and a somewhat positive correlation between copper and *grapefruit, cherry,* and *red fruit,* while tin shows a negative correlation to these sensory descriptors.

However, due to the poor model quality, the observed correlations are more likely coincidental than causal. Despite a clear hypothesis that metals could play a major role in the formation of oxidative sensory characters [37], the observed correlations were poor, and the observed sensory changes are more likely due to oxygen ingress through the packaging. Generally, one should always be careful in interpretation of statistical models and inferring causality. A robust hypothesis and

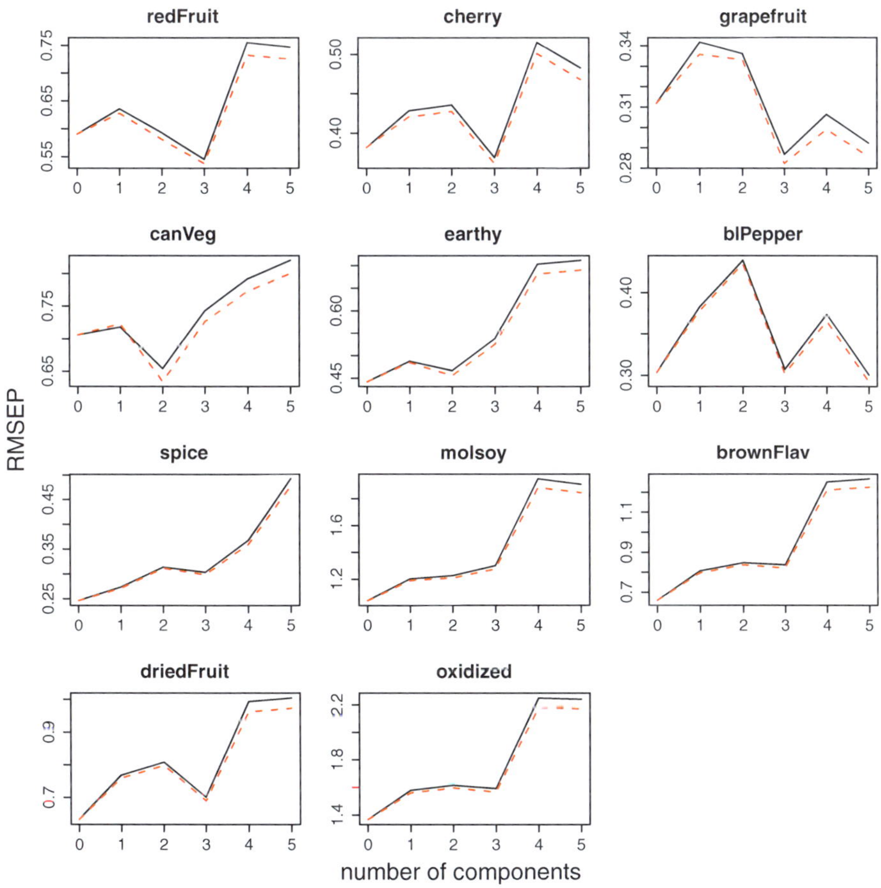

Fig. 8.5 PLS validation plots showing for each predicted variable (i.e., sensory descriptor) the root mean squared error of prediction (RMSEP) over the first five model dimensions. RMSEP values were obtained from a leave-one-out bootstrapping algorithm, and both the cross-validated estimate (*black solid line*) and the bias-adjusted cross-validation estimate (*red dotted line*) are shown [38]

a real understanding of the chosen variables is crucial for later interpretation and it creates a PLS model that is also useful in exploring relationships between variables and samples.

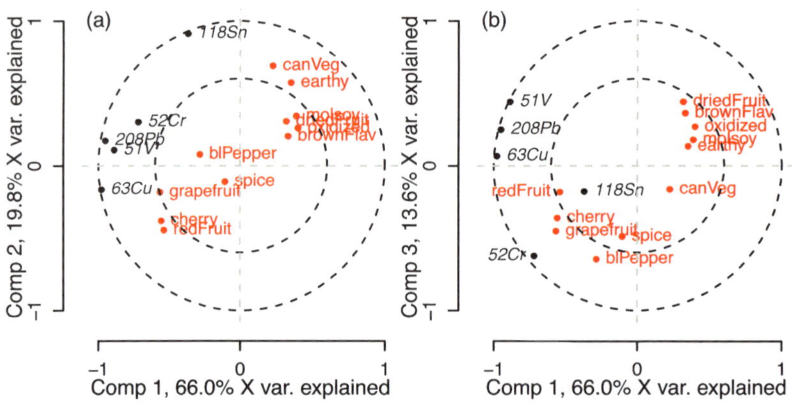

Fig. 8.6 PLS correlation plots for **a** the first and second model component, and **b** the first and third model component. Predicting variables (i.e., the elements) are shown in italicized *black* font, and the predicted variables (i.e., the sensory descriptors) are shown in *red* font

8.7 Conclusion

Choosing one data analysis technique over another can be a challenging task, with often no clear or "correct" answer. To help with this decision, analyzing defined and well-studied data sets with different techniques can enhance the understanding of the strengths and weaknesses of each method. In this chapter, we analyzed two related data sets individually and together, using unsupervised exploratory and supervised classification techniques, including PCA, CVA, and PLSR.

Depending on the goal of the data analysis, each method provides useful insight into the underlying pattern of the data, but highlighted different aspects of the studied data.

Using a rather simple data set, we discussed the different outcomes of various multivariate data analysis techniques from an applied standpoint. We have shown that each method has its justification, but a critical evaluation of the obtained results is necessary for high quality and reliable research, and a basic understanding of how these techniques work will help with this evaluation.

In the end, which method is applied to a certain data set is governed by the research question one seeks to answer, as well as the data itself. Ideally, the data analysis methods used after the data collection step would be decided upon before any data is collected, during the experimental design stage. Only then is one able to correct the data collection plan to being able to use certain data analysis methods. Especially with more and more variables measured in less time than ever before, the importance of a solid experimental design in combination with a thought-out data analysis plan at the beginning of an experiment (i.e., prior to any data collection) is increased, and additionally decreases the risk of data that cannot be analyzed properly. The actual analysis of data is in most cases trivial, but choosing the proper analysis method is the part where sufficient understanding of the different methods is crucial. The man who created the word "chemometrics," Svante Wold, summarized the problem every scientist faces today below, as one needs to (1) extract information from measured data by (2) creating a mathematical analogy for the problem one seeks to solve, followed by (3) selecting appropriate mathematical models [39]:

> The art of extracting chemically relevant information from data produced in chemical experiments is given the name of "chemometrics" in analogy with biometrics, econometrics, etc. Chemometrics, like other "met-rics," is heavily dependent on the use of different kinds of mathematical models (high information models, ad hoc models, and analogy models). This task demands knowledge of statistics, numerical analysis, operation analysis, etc., and in all, applied mathematics. However, as in all applied branches of science, the difficult and interesting problems are defined by the applications; in chemometrics the main issue is to structure the chemical problem to a form that can be expressed as a mathematical relation. The connected mathematical problems are rather simple. (Today, 1994, I would like to add: "as the statistical problems usually are.") Therefore, chemometrics must not be separated from chemistry, or even be allowed to become a separate branch of chemistry; it must remain an integral part of all areas of chemistry.

Acknowledgments We thank everyone who helped with the data collection, especially Jenny Nelson for help with the elemental analysis, as well as all sensory panelists.

References

1. Wehrens R (2011) Chemometrics with R: multivariate data analysis in the natural sciences and life sciences. Springer, Berlin
2. Heymann H, Noble AC (1989) Comparison of canonical variate and principal analyses of wine descriptive analysis component data. J Food Sci 54:1355–1358
3. Naes T, Brockhoff PB, Tomic O (2010) Statistics for sensory and consumer science. Wiley, West Sussex
4. Cevallos-Cevallos JM, Reyes-De-Corcuera JI, Etxeberria E et al (2009) Metabolomic analysis in food science: a review. Trends Food Sci Technol 20:557–566
5. Skov T, Engelsen SB (2013) Chemometrics, mass spectrometry, and foodomics. In: Cifuentes A (ed) Foodomics advanced mass spectrometry modern food science and nutrition. Wiley, Hoboken, pp 507–538
6. Tang C, Hsieh F, Heymann H, Huff HE (1999) Analyzing and correlating instrumental and sensory data: A multivariate study of physical properties of cooked wheat noodles. J Food Qual 22:193–211
7. Hopfer H, Haar N, Stockreiter W, Sauer C, Leitner E (2012) Combining different analytical approaches to identify odor formation mechanisms in polyethylene and polypropylene. Anal Bioanal Chem 402:903–919
8. Tomasino E, Harrison R, Sedcole R, Frost A (2013) Regional differentiation of New Zealand pinot noir wine by wine professionals using canonical variate analysis. Am J Enol Vitic 3:357–363
9. Kelly S, Heaton K, Hoogewerff J (2005) Tracing the geographical origin of food: the application of multi-element and multi-isotope analysis. Trends Food Sci Technol 16:555–567
10. King ES, Hopfer H, Haug MT, Orsi JD, Heymann H, Crisosto GM, Crisosto CH (2012) Describing the appearance and flavor profiles of fresh fig (Ficus carica L.) cultivars. J Food Sci 77:S419–S429

11. Brereton RG, Lloyd GR (2014) Partial least squares discriminant analysis: taking the magic away. J Chemom. doi:10.1002/cem.2609
12. Ferrier JG, Block DE (2001) Neural-network-assisted optimization of wine blending based on sensory analysis. Am J Enol Vitic 52:386–395
13. Zomer S, Brereton RG, Carter JF, Eckers C (2004) Support vector machines for the discrimination of analytical chemical data: application to the determination of tablet production by pyrolysis-gas chromatography-mass spectrometry. Analyst 129:175
14. Abdi H (2003) Partial Least Squares (PLS) Regression. In: Lewis-Beck M, Bryman A (eds) Encyclopedia social sciences research methods. Sage, Thousand Oaks, pp 1–7
15. Chung S-J, Heymann H, Grün IU (2003) Application of GPA and PLSR in correlating sensory and chemical data sets. Food Qual Prefer 14:485–495
16. Schrampf E, Leitner E (2010) Prediction of rheological and chemical properties of different starches used in the paper industry by near infrared spectroscopy (NIRS). Macromol Symp 296:154–160
17. Zhao G, Maclean AL (2000) A comparison of canonical discriminant analysis and principal component analysis for spectral transformation. Photogramm Eng Remote Sens 66:841–847
18. Hopfer H, Buffon PA, Ebeler SE, Heymann H (2013) The combined effects of storage temperature and packaging on the sensory, chemical, and physical properties of a cabernet sauvignon wine. J Agric Food Chem 61:3320–3334
19. Hopfer H, Nelson J, Mitchell AE et al (2013) Profiling the trace metal composition of wine as a function of storage temperature and packaging type. J Anal At Spectrom 28:1288–1291
20. Gay C (1998) Invitation to comment. Food Qual Prefer 9:166
21. Owen JG, Chmielewski MA (1985) On canonical variates analysis and the construction of confidence ellipses in systematic studies. Syst Zool 34:366–374
22. RStudio (2012) RStudio: integrated development environment for R
23. R Core Team (2013) R: A language and environment for statistical computing. R Foundation for Statistical Computing, Vienna
24. Lê S, Josse J, Husson FF (2008) FactoMineR: an R package for multivariate analysis. J Stat Softw 25:1–18
25. Husson F, Josse J, Lê S, Mazet J (2012) FactoMineR: multivariate exploratory data analysis and data mining with R
26. Friendly M, Fox J (2010) candisc: generalized canonical discriminant analysis. R package
27. Lemon J (2006) Plotrix: a package in the red light district of R. R-News 6:8–12
28. Mevik B-H, Wehrens R (2007) The PLS package: principal component and partial least squares regression. R. J Stat Softw 18:1–24
29. Balboa-Lagunero T, Arroyo T, Cabellos JM, Aznar M (2011) Sensory and olfactometric profiles of red wines after natural and forced oxidation processes. Am J Enol Vitic 62:527–535
30. Lee D-H, Kang B-S, Park H-J (2011) Effect of oxygen on volatile and sensory characteristics of cabernet sauvignon during secondary shelf life. J Agric Food Chem 59:11657–11666
31. Robinson AL, Mueller M, Heymann H et al (2010) Effect of simulated shipping conditions on sensory attributes and volatile composition of commercial white and red wines. Am J Enol Vitic 61:337–347
32. Almeida CMR, Vasconcelos MTSD (2003) Multielement composition of wines and their precursors including provenance soil and their potentialities as fingerprints of wine origin. J Agric Food Chem 51:4788–4798
33. Kristl J, Veber M, Slekovec M (2002) The application of ETAAS to the determination of Cr, Pb and Cd in samples taken during different stages of the winemaking process. Anal Bioanal Chem 373:200–204
34. Ugliano M, Kwiatkowski M, Vidal S et al (2011) Evolution of 3-mercaptohexanol, hydrogen sulfide, and methyl mercaptan during bottle storage of Sauvignon blanc wines. Effect of glutathione, copper, oxygen exposure, and closure-derived oxygen. J Agric Food Chem 59:2564–2572
35. Almeida CMR, Vasconcelos MTSD (2003) Lead contamination in Portuguese red wines from the Douro region: from the vineyard to the final product. J Agric Food Chem 51:3012–3023

36. International Organization of Vine and Wine (OIV) (2011) OIV-MA-C1-01: maximum acceptable limits of various substances contained in wine
37. Pohl P (2007) What do metals tell us about wine? TrAC Trends Anal Chem 26:941–949
38. Mevik B-H, Cederkvist HR (2004) Mean squared error of prediction (MSEP) estimates for principal component regression (PCR) and partial least squares regression (PLSR). J Chemom 18:422–429
39. Wold S (1995) Chemometrics; what do we mean with it, and what do we want from it? Chemom Intell Lab Syst 30:109–115

Chapter 9
Software and Online Resources: Perspectives and Potential Applications

Karina Martinez-Mayorga, Terry L. Peppard and José L. Medina-Franco

9.1 Databases

In the food chemistry field, a number of databases have been compiled; some, though not all, contain chemical structures. In certain cases, food components in databases are not single chemicals, but rather mixtures [1]. Nonetheless, conducting useful analyses without necessarily reporting all chemical structures is still feasible, as has been reported by us [2] and by others [3]. When chemical structures are available, however, additional analyses and comparisons can be performed [2, 4]. In other cases, food databases do not contain chemical information, but instead other food-related information, for example, databases containing specific diets to be followed in hospitals or information about food items shelf life, etc.

Typically, each database aims to be unique and serve specific purposes, although in practice there is a fair amount of redundancy and duplication among them. In many cases, chemical databases of commercially available compounds are built and freely distributed. The purpose of such databases is to provide readily useful information (chemical, physicochemical, organoleptic, toxicological, etc.) to the user.

K. Martinez-Mayorga (✉) · J. L. Medina-Franco
Departamento de Fisicoquímica, Instituto de Química, Universidad Nacional Autónoma de México, Av. Universidad 3000, 04510 Mexico City, Mexico

Torrey Pines Institute for Molecular Studies, 11350 SW Village Parkway, Port St. Lucie, FL 34987, USA
e-mail: kmtzm@unam.mx

T. L. Peppard
Robertet Flavors Inc., 10 Colonial Dr., Piscataway, NJ 08854, USA

K. Martinez-Mayorga, J. L. Medina-Franco (eds.), *Foodinformatics*,
DOI 10.1007/978-3-319-10226-9_9

9.1.1 General Food/Flavor-related Databases

Three of the most comprehensive flavor-related databases are described in this section. These databases are accessed worldwide by member companies or via annual subscription.

The Flavor and Extract Manufacturers Association (FEMA) assesses and maintains the generally recognized as safe (GRAS) database [5] of flavoring substances. It comprises a compilation of flavoring materials, whose safety has been reviewed by an expert panel of toxicologists and other specialists [6, 7]. As such, the materials are considered GRAS for human consumption within specified product categories and at/or below listed maximum usage levels. Materials on the GRAS list, together with certain Food and Drug Administration (FDA)-approved food additives, are those that are legally permitted for use as flavorings (and for related purposes, such as taste modification) in the USA. Certain other countries have also adopted the GRAS list in their flavor legislation.

New additions to the GRAS list (originally published about 50 years ago) appear in Food Technology every year or two. For example, GRAS 26 was published in August 2013 and included approximately 50 botanicals and discrete chemical entities. For each material, a FEMA #, principal name, and synonyms are listed, along with permitted food and beverage applications, including anticipated average usual and average maximum use levels (in ppm). To date, of the approximate 2800 GRAS materials, ca. 83 % are discrete chemical entities. In some cases, however, stereochemistry and even geometrical configuration has not been fully specified. In other cases, materials are actually mixtures of isomers.

The FEMA GRAS database is available on FEMA's web site, though exclusively to member companies, and while it is searchable online, entries comprise only a very few fields, as shown in the example below:

Principal name or synonym	[2-(1-Propoxyethoxy)ethyl]benzene
CAS number	7493-57-4
FEMA number	2004
GRAS publication	GRAS 3
Most recent NUL/FC published (normal use level/food category)	GRAS 25

The Research Institute for Fragrance Materials (RIFM)/FEMA Fragrance and Flavor Database [8] is maintained by the RIFM and is available online by annual subscription. The database is an extremely comprehensive, worldwide source of toxicology data, literature, and general information on fragrance and flavor ingredients, classifying more than 5100 materials. RIFM claims to review more than 50 journals every month, conducts literature searches, and regularly collects member company data. According to the RIFM web site, the database has more than 54,000 references and houses more than 112,000 human health and environmental studies. Basic material information includes: Chemical Abstracts Service (CAS) registry numbers, synonyms, chemical structures, simplified molecular-input line-entry system (SMILES) notation, molecular formulas, molecular weights, and physical properties (both measured and estimated). The database also contains material relationships such as isomers and metabolites, as well as commercial usage data. Finally, a vast amount of regulatory and compliance information, both domestic and international, is also contained within the database. RIFM recently released an enhanced version of their database, which features an improved interface, additional content, etc.

The International Organization of the Flavor Industry [9] (IOFI) maintains an online database of chemically defined substances (as well as natural complex substances, i.e., botanicals, extracts, etc.) used by the flavor industry worldwide. Access to this database is restricted to IOFI member associations and their member companies. According to the IOFI web site, the database comprises up-to-date regulatory and analytical information on almost 2800 flavoring substances used in global commerce. The regulatory information includes legal status in the USA, the EU, Japan, China, Russia, and other major markets. Also included are synonyms, CAS registry numbers (and other unique numeric identifiers), chemical structure, etc.

9.1.2 Databases of Flavorings Permitted for use in Individual Countries or Economic Regions

In some cases, lists of flavoring materials approved for use in individual countries or economic regions have been placed in the public domain and are readily accessible online and/or are available for download. An example is the EU's so-called EC Flavor Register, a list of more than 2500 flavoring substances which can be used in food [10]. The EU flavoring database includes name, CAS registry number, and various other numeric identifiers, plus purity criteria. It is available online [11] though it can also be downloaded as a searchable portable document format PDF file [12] which may optionally be extracted into a spreadsheet or a database program.

The FDA "Everything Added to Food in the United States" (EAFUS) Database [13] is freely available online, being generated from a database maintained by the US FDA Center for Food Safety and Applied Nutrition (CFSAN). The database comprises administrative, chemical, and toxicological information on more than 2000 substances directly added to food, including substances regulated by the US FDA as direct, "secondary" direct, and color additives, as well as GRAS and prior-sanctioned substances. The database contains additionally, less than 1000 substances, for which only administrative and chemical information is available. EAFUS contains only a partial list of all legally permitted food ingredients, because under federal law some ingredients may be added to food under a GRAS determination made independently of the FDA; the list does contain many, but not all such substances

The Food Chemicals Codex [14] (FCC) is a compendium of internationally recognized standards for the purity and identity of food ingredients. Originally published in 1966 and now available for purchase through the United States

Pharmacopeia (USP) it comprises more than 1200 monographs of food-grade chemicals, processing aids, certain foods (e.g., fructose, vegetable oils, etc.), flavoring agents, vitamins, and functional food ingredients (e.g., lycopene, olestra, etc.). For each monograph, FCC provides ingredient name, chemical structure, chemical formula, molecular weight, and CAS registry number, plus information on each ingredient's function, packaging, storage, and labeling requirements, as well as information concerning identification and assay (e.g., by ultraviolet (UV) and/or infrared (IR) spectrum). Most recently, information on USP's food fraud database has been added. FCC is published every two years in print and online formats, and is offered as a subscription that includes a main edition and intervening supplements.

Flavor-Base Database of Flavoring Materials and Food Additives [15], written and marketed by John Leffingwell & Associates, provides one of the most comprehensive and wide-ranging collections of flavor, regulatory, toxicological, and related data relevant to the flavor, food, beverage, and tobacco industries. Flavor-Base (version 9) includes all flavor chemicals (and natural flavor materials, e.g., botanicals and derivatives) on the FDA and FEMA GRAS lists through mid-2012, plus all flavor chemicals on the EU's EC Flavor Register. Selected other national jurisdictions and international regulatory bodies are also referenced. Additionally included are direct food additives approved by the FDA, as well as those approved by the European Commission.

The types of information included in the database are illustrated in Fig. 9.1. Molecular structures and other properties for all flavor chemicals are documented. In addition, a wealth of sensory descriptors and flavor thresholds (including as Odor

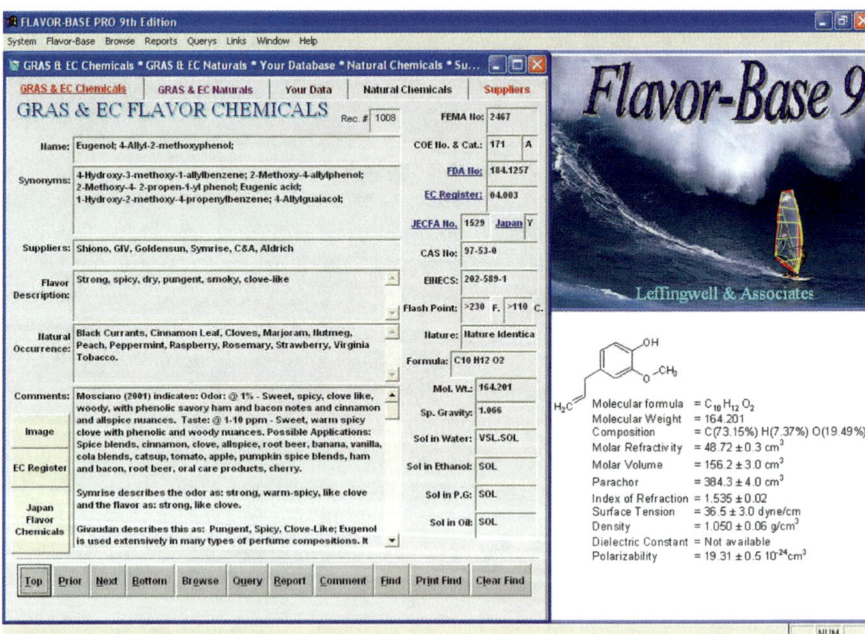

Fig. 9.1 Screenshot of eugenol entry from Flavor-Base 9 software illustrating some of the information available

Activity Values or Flavor Units) are given. Also available are the flavor chemicals' occurrence in foodstuffs and/or natural products (including some data on the levels at which they occur). When available commercially, suppliers of listed flavor chemicals are provided. Finally, the program includes a bibliographic database file with 5000+ references to pertinent flavor literature published through mid-2012.

One very nice feature of Flavor-Base is the ability to export data (selected materials, or indeed all of them) into spreadsheet format using the *find* and then the *report* functions. But while the database, as mentioned above, does provide molecular structures, given in the form of on-screen graphic images, it does not currently include this information in SMILES (or similar) notation, or provide any other means of importing chemical structure information by structure editors for conversion back into 2D or 3D molecular models.

In addition to Flavor-Base, the Leffingwell website [16] provides a wealth of highly useful, pertinent, and up-to-date flavor and fragrance-related information, of both a scientific and technical nature, as well as legislative and business related. Leffingwell also publishes some original articles, e.g., updates of the sensory properties of flavor molecules recently added to the GRAS list [16]. Finally, aside from Flavor-Base itself, Leffingwell offers a number of other useful flavor and fragrance software/database programs, some of which are also written by his group, while others are products of outside organizations. Some of these are briefly described below.

VCF: Volatile Compounds in Food Database [17]. TNO (The Netherlands Organization for Applied Scientific Research) long ago established a database designed for the collection of literature-based information on the natural occurrence of volatile compounds in food products. The VCF database, published for many years in book form, is nowadays available by online subscription.

The VCF database comprises 13 product groups (e.g., vegetables) representing 102 product categories (e.g., Allium spp.) and containing altogether about 500 products (e.g., chive, garlic, scallion, etc.); additionally 175 single products are tabulated. Volatile compounds are enumerated for each product, with more than 8000 volatile compounds grouped in 18 chemical classes, such as hydrocarbons, aldehydes, esters, etc. To be included, specific compounds must have been identified by at least two analytical methods, e.g., gas chromatographic retention time and mass spectrum. Quantitative data are provided if available. In all, the database lists more than 5500 literature references.

For individual named compounds, additional information comprises synonyms, unique identifiers (CAS registry number, FEMA GRAS number, etc.), molecular weight and molecular formula; molecular structures are also shown when available. More than 18,500 Kovats' Retention Indices are given, on four types of gas chromatographic columns (differing in polarity). Finally, approximately 2800 odor values are cataloged.

ESO: The (Complete) Database of Essential Oils [18]. This database, originally published by the Boelens Aroma Chemical Information Service (BACIS) appears to be most readily available through the Leffingwell web site, as indicated above. (The database was apparently updated in 2006, though we have no direct experience with this version.) ESO comprises more than 4100 quantitative analyses of essential oils, including in some cases multiple samples of the same oil from different sources, e.g., from different parts of the same plant (leaves, roots, etc.) or having different countries of growing origin. Each oil entry includes name and/or botanical name, CAS registry number (where applicable), and literature references.

The essential oils' quantitative analyses list a total of more than 4200 naturally occurring chemicals. For each analysis, components are listed in a decreasing order of total gas chromatography (GC) peak area %. Chemicals are specified by name, synonym(s), and CAS registry number. In addition, for approximately 2500 compounds, retention indices on various GC columns are listed (up to six stationary phases, each of differing polarity). One very nice feature of ESO is the ability to reverse search all of the oils containing one or more particular chemicals, based on a user-specified threshold amount. For example, just four oils were listed, when searching for a combination of linalool and linalyl acetate, and using a composition threshold concentration of 35 % for each compound.

FFM: Allured's Flavor and Fragrance Materials [19]. Access to this online database is through Allured, the publisher of Perfumer & Flavorist magazine. It should be noted that we have direct experience only with FFM 2008, a PC-based version of the product. The database contains information collected from a variety of sources, including flavor and fragrance suppliers, industry and government organizations, as well as related texts. Aside from access to materials' names, synonyms, identifiers (e.g., FEMA number, CAS registry number, FDA number, etc.), and empirical formula (or botanical name, as appropriate) functionality in our opinion is somewhat limited. For example, no structural information is provided. However, FFM is an excellent resource for finding suppliers of desired flavor materials (suppliers' names and contact details are provided). Also, the database usefully includes the status of listed materials in terms of whether natural, nature-identical, or synthetic.

Flavornet database [20, 21]. Flavornet is a compilation of aroma compounds found in human odor space, meaning at suprathreshold concentrations where they are likely to stimulate human olfactory receptor neurons [22]. Access to the online database (sponsored by DATU, Inc.) is freely available in the public domain.

Flavornet is based on articles published since 1984 (though data has apparently not been added since 2004) concerning the use of gas chromatography–olfactometry (GC–O) to detect odorants in natural products. Therefore, to be included in Flavornet, an odorant must have been detected in a natural product or real environment by some form of quantitative GC–O, e.g., dilution analysis (Aroma Extraction Dilution Analysis or CharmAnalysis™), or perceived intensity analysis (e.g., Osme), or detection frequency analysis (e.g., SNIFF). The database comprises more than 730 flavor molecules (identified by CAS registry number) for which both Kovats' and ethyl ester-based GC retention indices are provided (four stationary phases, varying in polarity) as well as characteristic odor note descriptions.

The SuperScent Database [23]. Developed and maintained by Preissner et al., SuperScent makes available a database containing 2D and 3D structures of approximately 2100 volatiles. An important feature is the standardization of odor description; accordingly, SuperScent includes around 9200 synonyms. Originally designed as an information source for users/customers looking for odor components, this database is a good reference for comparative studies, as has been reported by

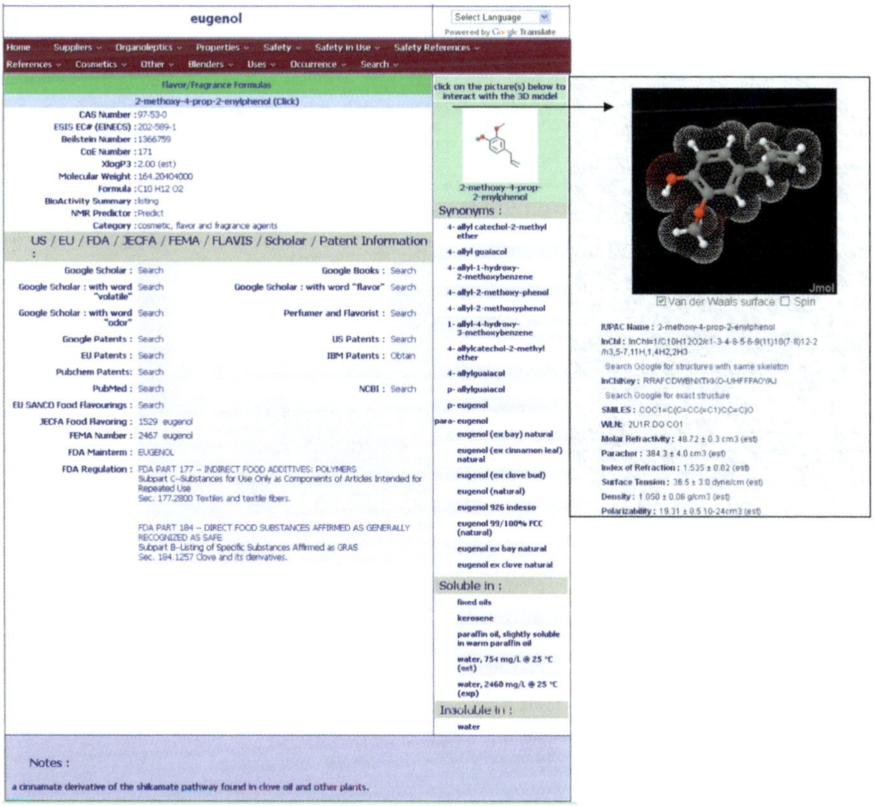

Fig. 9.2 Screenshot of eugenol entry from Good Scents Company web site illustrating some of the information available

us [4, 24]. For easy analysis, it includes physicochemical properties, commercial availability, and references [25].

The Good Scents Company Information System [26]. Originally setup years ago as one perfumer's card-index system for information archiving and retrieval, and progressing through dBase, the current public domain online database is truly a cornucopia of valuable flavor and fragrance data, with handy features absent in many commercial products. The website contains links to scientific and industry associations, and even useful flavor-related books. Information available for individual flavoring materials is searchable by multiple parameters, including: name, various identifiers, odor descriptors, etc. Figure 9.2 illustrates just a fraction of the information available for, for example, eugenol (note that the list of synonyms has been truncated for the sake of brevity). As indicated, visible directly on a chemical's main web page, or easily accessed via links, are supplier information, safety data, physicochemical properties, chemical structures (both 2D and 3D) and application data. The menu shown towards the center of Fig. 9.2 directs users to search engines, and contains links to the literature, including patents, scientific articles, related books, and regulations.

Phenol-Explorer [27]. Collected from more than 1300 scientific publications, Phenol-Explorer contains more than 500 different polyphenols in over 400 foods. In addition to online searching, the database is available for download. The current version includes data on polyphenol metabolism, as well as the effects on food processing and cooking.

9.1.3 Other Online Databases

Rather beyond the scope of what was originally intended to be included in this review, though useful nonetheless, are several databases which link taste or odor receptors to their cognate ligands, at least in the case of those receptors which have been deorphaned to date. For example, in the taste domain, BitterDB comprises a free searchable online database of currently more than 600 bitter compounds obtained from the literature (individual structures can be downloaded, e.g., in SMILES or SDF format) as well as their associated 25 human bitter taste receptors (hT2Rs) for which sequence data is also available [28, 29]. One can search for specific bitter compounds, or by selected ligand properties, or (using substructure searching) by structural similarity to a query compound. Alternatively, one can search by specific bitter receptors or combinations of receptors. So caffeine, for example, is a known cognate ligand of T2R7, 10, 14, 43, and 46, whereas individually these receptors are associated with as few as six to more than 40 listed bitter molecules.

In the case of odor, the SenseLab Project, part of the Human Brain Project, involves novel informatics approaches to constructing databases and database tools for collecting and analyzing neuroscience information, using the olfactory system as a model [30]. SenseLab relates odor molecules in the OdorDB database to ORDB, a database of olfactory receptors (which also contains data on the genes and sequences for olfactory receptor proteins). So 2-hexanone, for example, is a known cognate ligand of both ORL2156 and ORL2157, whereas both of these receptors are associated with 20 or more listed odor molecules.

In addition to some of the databases containing sensory attributes of flavor molecules, already discussed earlier in this section, there are a number of additional useful sources of such information existing in the public domain, represented by vendors' websites. For example, both Sigma-Aldrich [31] and FrutArom [32] feature online lists (catalogs) of flavor molecules, searchable by their principal taste and/or odor qualities.

Even though chemical structures are reported in many of the public domain and commercially available databases described above, they are not readily available for download as structure files, for instance in .MOL2 or Structure data format (.SDF). Nonetheless, there is software available that can convert structure names, SMILES, SMARTS, or InChI notation into molecular or structural files. This is discussed in the next section of this chapter.

9.2 Software and Online Resources

Software for chemoinformatic studies. Software designed specifically to perform chemoinformatic studies has been developed; one of the main applications has been drug discovery, though it is not restricted to that. There are different options to access the software, ranging from perpetual or annual renewal-based commercial licenses (often available at no or low cost to academic institutions) to freely available. Some of the underlying principles and capabilities of the software are common among companies' offerings. However, each usually provides features that make it unique. The different types of software required to develop a chemoinformatic study can be broadly arranged into two classes. Examples of each class are summarized in Table 9.1.

The first class consists of data generators and analysis. Programs to produce fingerprint representations or descriptors belong to this category. The program Dragon is well recognized as generating one of the largest numbers of descriptors. ChemAxon, MOE, and Schrödinger are also able to produce a large number of

Table 9.1 Representative software used in chemoinformatic studies

Name	Description	Reference
Data generators and analysis		
Dragon	Application for the calculation of molecular descriptors. Used to evaluate SAR or SPR, as well as for similarity analysis and HTS of molecule databases	[50]
mMaya Tools		[51]
ChemAxon	Cheminformatics and life science research	[52]
MOE	Drug discovery software package	[53]
Schrödinger	Computational chemistry for life sciences and materials research	[54]
Data analysis, processing, statistical modeling, and visualization		
Statistica	Merging, aggregating, stacking, and unstacking of data, transformations, and smoothing of data, for cleaning/recoding/imputing of missing data, for identifying duplicate records, finding and recoding outliers, etc. Comprehensive selection of advanced data mining algorithms in a single package, options for text mining, comprehensive options for quality control charting, multivariate control methods, model-based quality control methods (including PLS-based methods for monitoring of batch processes in real time), and simple and advanced process monitoring algorithms. Even advanced simulation and general optimization algorithms are provided, to solve complex risk modeling problems and/or perform multi-goal optimization of data mining or STATISTICA models.	[55]
Spotfire	Data discovery and visualization, predictive analytics	[56]
Miner3D	Provides interactive 3D and 2D visual data analysis, data mining, navigation, cherry picking, sonification, chart, and report creation	[57]

SAR structure–activity relationship, *SPR* structure–property relationships, *HTS* high-throughput screening descriptors and fingerprint representations. These last three are multipurpose platforms, capable of running a number of applications, ranging from bioinformatics to molecular modeling and chemoinformatics. These multipurpose programs allow one to transition from one application to another, in a seamless manner, without conflicts of formatting and without requiring additional editing of input files. Due to the frequent necessity of complementing one program with another, it is both possible and worthwhile keeping files in generic formats that may be recognized by other software. This can be done by saving the files in, e.g., .MOL, .PDB, .SDF, or .TXT format, or directly in a format to be used within other software.

The second class is devoted to data analysis and visualization. Robust software is available to perform these tasks. Statistica by StatSoft Inc. allows executing from data preparation to statistical models, with a number of options at each step. Spotfire and Miner3D are mainly devoted to data visualization as a means of analysis. Each of these programs can handle huge databases.

It is worth mentioning that there are overlaps among the tasks that each program can perform. Although software companies stand apart on many aspects, interconnection among software platforms is fortunately not uncommon. For example, Schrödinger allows for data analysis through Spotfire, though, of course, licenses for both programs are required.

Table 9.1 does not purport to be comprehensive, but rather representative of software commonly used in chemoinformatic studies. Additionally, software developed and maintained by research groups abound. There are justified reasons for the proliferation of such software. Since the chemoinformatic field is relatively new, the implementation of novel analyses and concepts requires developing scripts to automate the handling of data and its analysis, which justify generating in-house programs. These programs can be accessed from the researchers' websites or by request. Another reason for in-house software development can be related to cost. This can be a viable route when getting a license is an issue and the research group is able to produce its own scripts. However, the benefits of experience, troubleshooting, and testing provided by the software companies must not be overlooked. On this point, it cannot be stressed enough that it is necessary to have in-depth knowledge on the theory and algorithms employed in each program to be used. This provides the required knowledge to properly employ, complement, and analyze the data.

In the area of molecular modeling, there are websites that perform calculations online. For example, DockBlaster [33] performs automated docking of compounds with minimal intervention; it was developed as a tool for medicinal chemists with an interest in docking. In the area of bioinformatics, servers to perform different steps in the modeling of biomacromolecules are plentiful; some of them have gained strong reputations and are widely used. Examples of these servers are UniProt [34] and PredictProtein [35], used for different stages in modeling studies of biomacromolecules. Chemoinformatic methodologies and concepts are also increasingly employed. A relevant example is the use of similarity principles to search for and select compounds or proteins in databases, such as in the Protein Data Bank. In addition, direct implementation of chemoinformatics on the web is the use of search engines.

Table 9.2 Representative online servers to perform chemoinformatic studies

	Search engines	
Chemicalize	Find chemical structures on web pages and provide data for each structure (by ChemAxon)	[58]
ChemSpider	Free chemical structure database providing fast text and structure searches to over 29 million structures	[59]
Reaxys	Online chemistry workflow, provides access to information including chemical compounds, chemical reactions, and synthesizing compounds	[60]
Online applications and services		
NCI/CADD group	Provides structures, data, tools, programs, and other useful information to the public	[61]
Biopep	Sequence databases of proteins and bioactive peptides	[62]
MOLPRINT 2D	A molecular fingerprint method for similarity searching	[63]
SEA	Similarity Ensemble Approach	[64]
VCCLab	Virtual Chemistry Lab. Online calculation of physicochemical properties	[65]
PASS	Prediction of Activity Spectra for Substances	[66]
FAF-Drugs	Free ADME/tox Filtering	[67]

Online resources: Online programs and services have become increasingly used as part of the various steps employed in investigations. The advantages of such methods are: updates can be performed by the developers at any time; for services, there is no need to download software or databases; for users, there is no need for large hardware requirements to perform calculations. There are, however, disadvantages, for example, the user has limited or no access to the predefined settings. Unfortunately, it is not uncommon that the user has restricted knowledge of how the calculations are performed; this is, of course, the user's responsibility.

The online services vary widely; they can be classified as search engines for chemical information and online services.

Search engines are typically part of other software, such as Chemicalize by ChemAxon, or are managed by editorial groups, like ChemSpider and Reaxys, which belong to the Royal Society of Chemistry and to Elsevier, respectively. Table 9.2 provides the corresponding websites.

Online services are dedicated to data generators and data mining from different sources. The computer-aided drug discovery group at the National Cancer Institute (NCI/CADD), managed by the US federal government through the National Institutes of Health (NIH), provides chemoinformatics tools and user services to handle chemical structures and associated biological activity. For example, it is possible to calculate properties, convert graphical representations of chemical structures in journal articles, and perform chemical searches, among other tasks.

Directly related to food chemistry, an interesting web server called biopep, performs proteolysis simulation of endogenous enzymes, based on the recognition and cut sequence. This simulation allows the prediction of bioactive products by the in silico hydrolysis of proteins by selection of endopeptidases launched on the server.

As detailed in Chap. 1 of this book, similarity searching is at the core of chemoinformatics, and multiple articles are published frequently on this topic. As expected, commercial software as well as programs developed by various research groups are available. For instance, ChemAxon, mentioned above, is a chemoinformatics platform and has a robust implementation of similarity searching methods. MOE and Schrödinger have also implemented structural similarity methods. Online resources are also available. For instance, MOLPRINT 2D and SEA (listed in Table 9.2) provide for similarity searching, the former for ligands and the latter for proteins.

Finally, an important commonly pursued goal is the prediction of bioactivity as well as ADME/tox properties. Physicochemical properties and bioactivities, and ADME/tox properties can be calculated through online services such as VCCLab, PASS, FAF-Drugs. Table 9.2 provides the corresponding websites, for more information.

Being a major concern, a number of initiatives are dedicated to food safety issues. Some programs are maintained by, or in cooperation with, universities while others are consortia involving, in many cases, governments. Some of them are: the *Centers for Disease Control* [36], the *Food Safety Research Consortium* [37], *ComBase* [38], and *Bits* [39]. Developed and/or maintained by universities are: *FareMicrobial* [40], and the *Center for Food Safety* [41].

In addition, other online services are specifically focused on food information (food informatics). In these cases, the information contained is not necessarily directly related to chemical structures, though it does illustrate the versatility of and need for using information technology to excel on tasks having a direct impact on health and well-being through food and nutrition.

Vision Software assists the organization, storage and use of information, data and knowledge for food and nutrition-related problem solving and decision making. One direct application is in the area of diets for hospitals, hotels, etc. [42]. A related novel piece of work (albeit somewhat controversial) concerns so-called food pairing theory. The hypothesis is that a pair of ingredients which share many flavor compounds accompany each other better than those that do not, e.g., bacon and cheese, asparagus and butter, and chocolate and blue cheese. This food pairing concept is useful to understand and further develop culinary practice [43].

The Food & Biobased Research Company (Wageningen UR) has five major projects. One of them, called Food Informatics, focuses directly on food research. This project is conducted in cooperation with the Top Institute-Food & Nutrition (TIFN), Unilever Research, TNO Quality of Life research center, and Friesland Foods. Their focus is on the modeling of knowledge-intensive processes and the development of corresponding applications. Based on ontologies, Top et al. have focused on methods and tools for extracting knowledge in the food industry domain [44]. For example, using this approach, they have developed an on-line system for searching the properties and practical applications of five natural antimicrobial preservatives and their relationships to a large number of microbes and food types [44]. Another example consists of how these methods can help as decision support in the fruit and vegetable supply chain [45, 46].

Another online service is called *Nutrition Informatics*. Nutrition informatics is defined as "the effective retrieval, organization, storage, and optimum use of information, data, and knowledge for food and nutrition-related problem solving and decision making. Informatics is supported by the use of information standards, information processes, and information technology." As part of the Academy of Nutrition and Dietetics, the Nutrition Informatics group provides a service to registered members. It is the intersection of information, nutrition, and technology. The hugely data-rich food-related information includes food/nutrient analysis tables. This provides registered dietitians web-based tools, allowing them to use their knowledge and skills more efficiently in making dietary recommendations [47].

Another service that can be classified in this category is that provided by the O'Neill Institute for National & Global Health Law. It is a free online database of law, from around the world, relating to health and human rights. The database offers an interactive, searchable, and fully indexed website of case law, national constitutions, and international instruments [48].

9.3 Perspectives and Potential Applications

While the exploitation of chemical information in the food chemistry field is still emerging, this has already proven to constitute a useful approach, as illustrated through several examples described in the second section of this book and as also reported in the literature elsewhere.

The use of similarity to compare and explore food-related databases, described in Chap. 1 and exemplified in Chap. 3 clearly demonstrate the applicability of these methods and alludes to exploring other applications; for example, expanding the studies to diseases, methodologies, and databases beyond those explored in Chap. 3. In addition, methods such as artificial neural networks proved useful when exploring the effects of foods on cancer cell growth, suppression activity, antiviral activity and antioxidant stress activity; this, of course, suggests exploring other diseases.

The use of information technology in the food and beverage field is not limited to chemical structures. In fact, data mining is widely used to collect, organize, analyze, and archive diets in hospitals and restaurants, just as it is applied to chemical structures in the area of chemical information. The theory behind the methods devised to perform these tasks is general and can be applied to datasets regardless of origin. Therefore, the software employed in drug discovery can readily be used or adapted to food chemical applications. In the same way that new concepts and methodologies are developed on an almost daily basis and reported in chemical information journals, the food chemical information field can be expected to grow significantly during the coming decade. Taking into account the knowledge and applicability of chemical information devoted to drug discovery, and considering the inherent complexity of the food chemistry field, it is also expected that concepts now developed in the food chemical information area will feed back into the drug discovery arena. This can be exemplified by the well-known complexity of odor perception, where multiple odor receptors are activated by multiple ligands, ultimately to produce specific percepts. The multiple receptor/multiple ligand notion is central to the polypharmacology concept that is gaining attention nowadays. The idea of multiple target responses is neither new nor unexpected, however, only relatively recently is the paradigm change from single target to multitarget being recognized [49]. This interplay between drug discovery and food chemical information is not only promising but has also proven to be useful and may yet further expand our knowledge and boost our creativity in developing new methods to deal with complex multivariable systems.

In the light of the discussion above, there is clearly a need for professionals with skills in information technology and a strong background in food chemistry. To fill this need, it will be necessary to explore the suitability of various chemoinformatic methods, selecting and developing the most useful candidates, and to design appropriate programs and courses in universities. This has been the path of the chemical information field, but with emphasis in drug discovery. For example, chemoinformatics courses leading to a Masters in Science have been implemented at the University of Sheffield, the University of Manchester and Indiana University. How fast we respond to this need will have an impact not only on the development of the field but also on how we take advantage of the emerging field of food informatics.

Acknowledgments K.M-M. thanks the Institute of Chemistry-UNAM and DGAPA-UNAM for funding (PAPIIT IA200513-2). The authors also wish to thank Robertet Flavors for permission to publish this chapter.

References

1. Martinez-Mayorga K, Medina-Franco JL (2009) Chemoinformatics—applications in food chemistry, vol. 58. Elsevier, Burlington
2. Martínez-Mayorga K, Peppard TL, Yongye AB, Santos R, Giulianotti M, Medina-Franco JL (2011) Characterization of a comprehensive flavor database. J Chemometr 25:550–560
3. Sprous DG, Salemme FR (2007) A comparison of the chemical properties of drugs and FEMA/FDA notified GRAS chemical compounds used in the food industry. Food Chem Toxicol 45:1419–1427
4. Medina-Franco JL, Martínez-Mayorga K, Peppard TL, Del Rio A (2012) Chemoinformatic analysis of GRAS (Generally Recognized as Safe) flavor chemicals and natural products. PLoS ONE 7:e50798
5. Femaflavor. http://www.femaflavor.org
6. Hallagan JB, Hall RL (1995) FEMA GRAS—a GRAS assessment program for flavor ingredients. Regul Toxicol Pharm 21:422
7. Hallagan JB, Hall RL (2009) Under the conditions of intended use—new developments in the FEMA GRAS program and the safety assesment of flavor ingredients. Food Chem Toxicol 47:267–278
8. http://www.rifm.org/index.php
9. http://www.iofi.org
10. http://eur-lex.europa.eu/JOHtml.do?uri=OJ:L:2012:267:SOM:EN:HTML
11. https://webgate.ec.europa.eu/sanco_foods/main/?event=display

12. http://eur-lex.europa.eu/LexUriServ/LexUriServ.do?uri=OJ:L:2012:267:FULL:EN:PDF
13. http://www.accessdata.fda.gov/scripts/fcn/fcnNavigation.cfm?filter=&sortColumn=&rpt=eafusListing&displayAll=false#1
14. http://www.usp.org/food-ingredients/food-chemicals-codex
15. http://www.leffingwell.com/flavbase.htm
16. Leffingwell J, Leffingwell D (2014) Perfumer Flavorist 39:26–37
17. http://www.vcf-online.nl/VcfHome.cfm
18. http://www.leffingwell.com/baciseso.htm
19. http://dir.perfumerflavorist.com/main/login.html;jsessionid=9EC896163AA3A88037DD0BC0E2CE6F65
20. http://www.flavornet.org
21. http://acree.foodscience.cornell.edu/flavornet.html
22. Arnam H, Acreeb TE (1998) Flavornet: a database of aroma compounds based on odor potency in natural products. Dev Food Sci 40:27
23. http://bioinf-applied.charite.de/superscent
24. López-Vallejo F, Peppard TL, Medina-Franco JL, Martínez-Mayorga K (2011) Computational methods for the discovery of mood disorder therapies. Expert Opin Drug Discov 6:1227–1245
25. Dunkel M, Schmidt U, Struck S, Berger L, Gruening B, Hossbach J, Jaeger IS, Effmert U, Piechulla B, Eriksson R, Knudsen J, Preissner R (2008) SuperScent–a database of flavors and scents. Nucleic Acids Res 37:D291–D294
26. http://www.thegoodscentscompany.com/index.html
27. http://www.phenol-explorer.eu/
28. Wiener A, Shudler M, Levit A, Niv MY (2011) BitterDB: a database of bitter compounds. Nucleic Acids Res 40:D413–D419
29. http://bitterdb.agri.huji.ac.il/bitterdb
30. http://senselab.med.yale.edu
31. http://www.sigmaaldrich.com/chemistry/flavors-and-fragrances.html
32. http://frutaromfandfingredients.com/ingredients4u/Templates/showpage.asp?DBID=1&LNGID=1&TMID=84&FID=517
33. Irwin JJ, Shoichet BK, Mysinger MM, Huang N, Colizzi F, Wassam P, Cao Y (2009) Automated docking screens: a feasibility study. J Med Chem 52:5712–5720
34. Consortium TU (2014) Activities at the universal protein resource (UniProt). Nucleic Acids Res 32:W321–W326
35. Rost B, Yachdav G, Liu J (2004) The PredictProtein server. Nucleic Acids Res 32:W321–W326
36. http://www.cdc.gov/
37. http://www.rff.org/news/features/pages/food-safety-research-consortium.aspx
38. http://www.combase.cc/index.php/en/
39. http://bites.ksu.edu/
40. http://foodrisk.org/exclusives/faremicrobial/
41. http://www.ugacfs.org/
42. http://www.vstech.com/healthcare-initiatives/food-nutrition-informatics.php
43. Ahn Y-Y, Ahnert SE, Bagrow JP, Barabási A-L (2011) Flavor network and the principles of food pairing. Sci Rep 1:196
44. Koenderink NJJP, Hulzebos JL, Roller S, Egan B, Top JL Antimicrobials on-line: concept and application for multidisciplinary knowledge exchange in the food domain. http://www.koenderink.info/nicole/pdf/Koenderink2003AFO1.pdf
45. Top JL, Rijgersberg H (2003) Modelling for decision support in the vegetable and fruit supply chain. Acta Hort 604:189–197
46. http://www.afsg.nl/InformationManagement/index.php?Itemid=0&id=21&option=com_content&task=view
47. http://www.eatright.org/HealthProfessionals/content.aspx?id=6442471521
48. http://www.law.georgetown.edu/oneillinstitute/

49. Medina-Franco JL, Giulianotti MA, Welmaker GS, Houghten RA (2013) Shifting the single-target to the multi-target paradigm in drug discovery. Drug Discov Today 18:495–501
50. http://www.vcclab.org/
51. http://www.molecularmovies.com/toolkit/
52. http://www.chemaxon.com/
53. http://www.chemcomp.com/
54. http://schrodinger.com/
55. http://www.statsoft.com/
56. http://spotfire.tibco.com/
57. www.miner3d.com/
58. www.chemicalize.org/
59. www.chemspider.com/
60. www.elsevier.com/online-tools/reaxys
61. http://cactus.nci.nih.gov/index.html
62. http://www.uwm.edu.pl/biochemia/index.php/en/biopep
63. http://www.molprint.com/
64. http://sea.bkslab.org/
65. http://www.vcclab.org/lab/
66. http://www.pharmaexpert.ru/passonline/index.php
67. http://bioserv.rpbs.univ-paris-diderot.fr/Help/FAFDrugs.html

Index

K. Martinez-Mayorga, J. L. Medina-Franco (eds.), *Foodinformatics*,
DOI 10.1007/978-3-319-10226-9

Printed by Printforce, the Netherlands